T0224364

Parasitology

Volume 115   *Supplement 1997*

Survival of parasites, microbes and tumours: strategies for evasion, manipulation and exploitation of the immune response

EDITED BY

M. J. DOENHOFF *and* L. H. CHAPPELL

CAMBRIDGE
UNIVERSITY PRESS

CAMBRIDGE UNIVERSITY PRESS
Cambridge, New York, Melbourne, Madrid, Cape Town, Singapore, São Paulo

Cambridge University Press
The Edinburgh Building, Cambridge CB2 8RU, UK

Published in the United States of America by Cambridge University Press, New York

www.cambridge.org
Information on this title: www.cambridge.org/9780521645829

First published 1997

A catalogue record for this publication is available from the British Library

ISBN 978-0-521-64582-9 paperback

Transferred to digital printing 2007

For the purposes of this digital printing the colour figures within this volume
have been reproduced as black and white images

Front Cover Illustration: Section of large intestine of a baboon infected with *Schistosoma mansoni*. The section shows a schistosome egg encapsulated in a granuloma located in the muscularis mucosae en route to the gut lumen. Picture courtesy of Professor Ray Damian.

Contents

Contents

Contents   vi

List of contributions

Survival of parasites, microbes and tumours: strategies for evasion, manipulation and exploitation of the immune response

EDITED BY M. J. DOENHOFF AND L. H. CHAPPELL

Preface

The papers in this volume draw attention to both new and recent information on the mechanisms employed by infectious pathogens to underpin their survival in the immunocompetent host and to facilitate their transmission between hosts. Classical survival strategies include induction of immuno-suppression, antigenic variety and variation, host antigen sequestration, molecular mimicry, antibody destruction and invasion of cells or privileged sites. To these we can now add novel and diverse mechanisms with which the invader may manipulate the host for its own ends. They range from making use of a single molecular component of the immune system, through more sophisticated mechanisms of evasion to modulation of the immune response in the pathogen's favour, particularly by manipulation of T cell subsets and cytokine fluxes. There are then examples of pure exploitation of the adaptive immune system by the generation of specific humoral and cell-mediated responses that prolong the invader's survival and aid transmission. The order of the chapters in this volume is intended to reflect this increasing level of complexity.

It is our hope that the studies described in this multi-disciplinary assemblage of papers will stimulate further research in this important area. We would like to thank all our contributing authors and the anonymous referees for their commitment and help in seeing this project through to completion.

M. J. DOENHOFF
L. H. CHAPPELL
November 1997

Exploitation of immune and other defence mechanisms by parasites: an overview

R. M. LOCKSLEY

Departments of Medicine and Microbiology/Immunology, Howard Hughes Medical Institute, University of California San Francisco, San Francisco, CA 94143-0654, USA

SUMMARY

Parasites have established numerous strategies for evading host immunity. Such tactics promote persistence and chronicity – factors that favour completion of the life cycle and transmission to vectors or the environment. This volume explores many facets of the interaction between host and parasite that will be critical to the development of meaningful vaccines capable of interrupting what has been accomplished over centuries of co-evolution.

Key words: Parasite immunity, T cells, cytokines, superantigens.

The papers in this special issue of *Parasitology* address the very essence of parasitism – evasion of host immunity. The co-evolution of parasite and host has resulted in such profound effects on our own genomes – witness the intense selection for such otherwise deleterious genes as haemoglobin S and the cystic fibrosis transporter mutations – that it is certain to have mutually affected the genomes of these pathogenic organisms. Studies with viruses have yielded a wealth of information regarding critical cellular components essential for the survival of these relatively simple parasites (Barry & McFadden, this volume). In general, viruses inhibit cellular apoptosis (in order to complete their own life cycles before the host cell extinguishes the factory), lymphocyte recognition (by impairing host MHC expression and expressing decoy MHC molecules to impede NK cell killing) and inflammation (by impairing the production or activities of inflammatory cytokines). In most cases, these viral gene products represent capture of host immune genes that favour survival, resulting in a reciprocating race between virus and host to express altered products seeking to bypass the advantage of the other (Murphy, 1993).

Sterile immunity occurs rarely, if ever, for the parasites discussed in this volume. Most inhabit their human hosts for decades, despite their ready recognition by the immune system. After years of studying the parasites themselves, recent investigations have turned increasingly to the host. As more features of a productive immune response become defined, it will be easier to identify the critical measures taken by parasites for evasion and prolonged infestation. Gene capture seems an unlikely mechanism for evasion by parasites, but convergent evolution of parasite and host genomes might be favoured by the constant pressure of cohabitation. Although older paradigms suggested a more passive role for parasites in immune evasion – e.g. sequestration, antigenic masking by host proteins, global immunosuppression, etc. – more recent studies suggest the active deployment of strategies that manipulate the host response (Beverley, 1996). Collectively, these papers give a first look at evidence supporting such concepts.

Should we be surprised that an intestinal helminth can express an IFN-γ-like molecule given that IFN-γ prolongs residence in the intestine (Grencis & Entwistle, this volume)? Or that female schistosomes utilize host TNF-α to signal egg-laying, given that TNF-α mediates the exuberant granulomatous response required for the successful transit of eggs through the intestinal lumen to the bowel for expulsion (McKerrow, this volume)? The role of granuloma formation in host defence remains woefully understudied but of certain importance (Doenhoff, this volume). It seems more than coincidental that TNF-α is required both for granuloma formation and for the establishment of germinal centres in lymphoid organs (Marino *et al.* 1997). Are granulomas the evolutionary forerunner of germinal centres, the innate immune system's method for trapping and presenting antigen and stimulating recognition by host cells, analogous to the role of germinal centres in the adaptive system?

The adaptation of parasites to host makes these organisms ideal probes for investigating the very nature of immunity. Additionally, the availability of inbred strains of mice should allow the location of susceptibility genes that affect the course of disease in certain individuals. Nowhere has this been quite so exploited as in the murine *Leishmania major* model, a system in which otherwise self-limited infection leads to overwhelming disease in mice on a BALB background (Reiner & Locksley, 1995). As reviewed in this volume (Launois, Louis & Milon), studies with this system have led to insights regarding the genetics of CD4+ helper T cell development, disease resistance loci, and mechanisms for counter-regulation among effector T cell populations. The capacity of these organisms to evade induction of IL-12 during invasion of macrophages (Reiner *et al.* 1994) had yielded new insights whereby this inflammatory cytokine can be down-modulated when pathogens are ingested by complement- or antibody-dependent mechanisms, providing some explanation by which cellular and humoral immune responses might be cross-regulated (Mosser & Brittingham, this volume; Sutterwala *et al.* 1997).

Whereas *Leishmania* can redirect host T cell effector function, malaria may abrogate T cell effector function through the generation of polymorphic peptides that serve as altered ligands, essentially blocking the accumulation of T cell receptor-mediated signals sufficiently strong to generate the complete arsenal available to the lymphocyte (Kersh & Allen, 1996). If confirmed, such findings bode gravely for the capacity to construct effective vaccines, as pointed out by Hill and colleagues (Plebanski, Lee & Hill, this volume). And invasive bacteria, such as *Shigella*, simply kill the messenger, the macrophage, that might serve as antigen presenting cells, by activating the cellular death machinery, thus inducing apoptosis (Hilbi, Zychlinsky & Sansonetti, this volume). The host is not without all resources, however, and, with a last gasp, sends out cytokines such as IL-β and IL-18, whose activation relies on induction of apoptosis-inducing caspases (Gu *et al.* 1997). In turn, these molecules are instrumental in attracting and activating effector cells capable of restricting the spread of the pathogenic organisms.

Bacterial superantigens, such as staphylococcal food toxins and toxic shock syndrome toxin, remain informative examples of the mischief that ensues when otherwise specific T cell antigen receptors become indiscriminately engaged (Maillard *et al.* this volume). Although their initial activation results in the elaboration of numerous inflammatory mediators, including cytokines, these activated T cells slowly become anergized and deleted, perhaps due to their activation in the absence of fully functional costimulatory signals. Such kinetics are best understood with the mouse mammary tumour viruses. Integrated throughout the mouse genome, the MMTVs, which express superantigens that underlie the mouse minor lymphocyte stimulatory, or Mls loci, rely on expansion of T cells to provide expansion of the infected B cell pool and the eventual infection of mammary epithelial cells in order to complete their life cycle through passage in milk to the lactating offspring. In turn, the host deletes the requisite T cell pool, thus providing protection against subsequent infection by these oncogenic viruses. Although the role of endogenous retroviruses remains less understood in human biology,

the recent incrimination of a viral superantigen encoded by an endogenous MMTV-like human retrovirus in type 1 diabetes mellitus may open new avenues of research in autoimmune disease of man (Conrad *et al.* 1997).

Finally, as indicated above, latency and chronicity are the hallmarks of most of the parasitic diseases discussed here. Understanding the mechanisms by which these organisms blatantly resist immunological destruction by the host will undoubtedly prove fruitful in devising therapies against other devastating latent and chronic diseases – HIV infection and cancer – as discussed in two manuscripts that round out this volume (Füst; Stewart & Heppner, this volume). Parasites have coevolved with the human immune system, and have much to teach us. The temptation to brush such research aside as romanticized studies of diseases of underdeveloped societies should be resisted energetically and forcefully. As attested by the papers in this volume, nowhere will be found such insightful, informative and potentially enlightening probes of the human immune system. We need be only clever enough to ask the right questions.

ACKNOWLEDGEMENTS

Supported by the National Institutes of Health grants AI26918 and AI30663. R. Locksley is a Burroughs Wellcome Fund Scholar in Molecular Parasitology and an Investigator in the Howard Hughes Medical Institute.

REFERENCES

BEVERLEY, S. M. (1996). Hijacking the cell: parasites in the driver's seat. *Cell* **87**, 787–789.

CONRAD, B., WEISSMAHR, R. N., BÖNI, J., ARCARI, R., SCHÜPBACH, J. & MACH, B. (1997). A human endogenous retroviral superantigen as candidate autoimmune gene in type 1 diabetes. *Cell* **90**, 303–313.

GU, Y., KEISUKE, K., TSUTSUI, H., KU, G., HSIAO, K., FLEMING, M. A., HAYASHI, N., HIGASHINO, K., OKAMURA, H., NAKANISHI, K., KURIMOTO, M., TANIMOTO, T., FLAVELL, R. A., SATO, V., HARDING, M. W., LIVINGSTONE, D. J. & SU, M. S.-S. (1997). Activation of interferon-γ inducing factor mediated by interleukin-β converting enzyme. *Science* **275**, 206–209.

KERSH, G. J. & ALLEN, P. M. (1996). Essential flexibility in the T cell recognition of antigen. *Nature* **300**, 495–498.

MARINO, M. W., DUNN, A., GRAIL, D., INGLESE, M., NOGUCHI, Y., RICHARDS, E., JUNGBLUTH, A., WADA, H., MOORE, M., WILLIAMSON, B., BASU, S. & OLD, L. J. (1997). Characterization of tumor necrosis factor-deficient mice. *Proceedings of the National Academy of Sciences, USA* **94**, 8093–8098.

MURPHY, P. M. (1993). Molecular mimicry and the generation of host defense protein diversity. *Cell* **72**, 823–827.

REINER, S. L. & LOCKSLEY, R. M. (1995). The regulation of

immunity to *Leishmania major*. *Annual Review of Immunology* **13**, 151–177.

SUTTERWALA, F. S., NOEL, G. J., CLYNES, R. & MOSSER, D. M. (1997). Selective suppression of interleukin-12 after macrophage receptor ligation. *Journal of Experimental Medicine* **185**, 1977–1985.

REINER, S. L., ZHENG, S., WANG, Z.-E., STOWRING, L. & LOCKSLEY, R. M. (1994). *Leishmania* promastigotes evade interleukin 12 (IL-12) induction by macrophages and stimulate a broad range of cytokines from CD4+ T cells during initiation of infection. *Journal of Experimental Medicine* **179**, 447–456.

Leishmania, macrophages and complement: a tale of subversion and exploitation

D. M. MOSSER* *and* A. BRITTINGHAM

Department of Microbiology and Immunology, Temple University School of Medicine, 3400 N. Broad St, Philadelphia, PA 19140, USA

SUMMARY

Leishmania are intracellular protozoan parasites which reside primarily, if not exclusively, in host mononuclear phagocytes. Several studies have demonstrated that infectious promastigotes rapidly and efficiently fix complement when they encounter serum components during their transmission to the mammalian host. Activation of the complement system by a microorganism can have 3 distinct biological effects. First, fixation of the terminal complement components can result in complement-mediated lysis. Second, fixation of the 3rd component of complement can lead to opsonization of the organism for uptake by phagocytic cells. Finally, the elaboration of the complement anaphylotoxins, C3a and C5a, can lead to inflammation. In the present chapter, we discuss the interaction of leishmania promastigotes with the complement system. We show that infectious promastigotes avoid the lytic effects of complement and resist fixation of the terminal complement components. At the same time, however, these organisms depend on fixation of opsonic complement to invade host mononuclear phagocytes efficiently. We discuss the mechanisms which allow metacyclic leishmania promastigotes to exploit the opsonic properties of complement and the receptors on macrophages involved in leishmania recognition. The role of complement mediated inflammatory processes in the host response to leishmania infection is an area which requires additional study.

Key words: *Leishmania*, macrophages, phagocytosis, complement, gp63.

INTRODUCTION

Upon entry into their vertebrate host *Leishmania* promastigotes immediately encounter 2 of the most ancient, yet effective immune mechanisms: (1) lysis by complement; and (2) destruction by phagocytes. How the promastigote deals with these 2 obstacles may be the difference between the successful and unsuccessful initiation of infection. In this review we will discuss the mechanisms by which promastigotes not only subvert components of the innate immune response, but exploit them in the establishment of infection.

THE *LEISHMANIA* PARASITE

Life-cycle

There are 2 morphological forms of *Leishmania*, the promastigote and the amastigote. The promastigote is found in the vectors of leishmaniasis, female sandflies of the generas *Lutzomyia* and *Phlebotomus*. Promastigote development in the insect vector has been extensively characterized using numerous parasite/vector combinations (Walters, 1993). Unlike parasites of the subgenus *Viannia*, which include a developmental phase in the hindgut of the fly, parasites of the subgenus *Leishmania* restrict development to the midgut and foregut. Within the gut of the fly, numerous developmental forms can be

observed. *Leishmania* parasites are acquired by the fly in the amastigote form during the taking of a bloodmeal. The first developmental event observed is the transformation of the amastigote to the promastigote form. This transformation event occurs within the bloodmeal itself, which in the sandfly is encased in a peritrophic membrane (matrix), a structure secreted by the midgut epithelium (Killick-Kendrick, 1979). These early promastigote forms are ellipsoid in shape, with a body length of 6–8 μm. This stage of promastigote development is one of replication and is referred to as the procyclic stage. Approximately 3 days after feeding, the promastigotes begin to escape from the peritrophic membrane and spread toward the anterior portion of the midgut (Bates, 1994). At this point, promastigotes are commonly seen attached to the midgut epithelium or with their flagella inserted between microvilli. From day 5 onward, increasing numbers of slender, non-replicating, rapidly moving promastigotes can be observed in the lumen of the anterior midgut and foregut. It is this highly infectious metacyclic form of promastigotes which is delivered to the vertebrate host (Sacks, 1989). During a bloodmeal, the sandfly generates a small pool of blood, into which infectious promastigotes are regurgitated (Schlein, Jacobson & Messer, 1992).

Upon delivery to the vertebrate host, the promastigotes are quickly taken up by tissue phagocytes, and by monocytes and neutrophils brought to the site due to the damage created by sandfly feeding (Wilson *et al.* 1987). Within macrophages, promastigotes lose

* Corresponding author.

Parasitology (1997) **115**, S9–S23. © 1997 Cambridge University Press

their flagellum and transform into non-motile amastigotes. The amastigote survives and replicates within the acidic environment of the phagolysosome, eventually lysing the infected cell, freeing amastigotes to infect nearby cells. When a sandfly acquires a bloodmeal from an infected host, it acquires either free amastigotes, or amastigote-infected mononuclear cells, and the life-cycle continues.

Metacyclogenesis

The transition from replicating, non-infectious procyclic promastigotes, to non-replicating, highly infectious metacyclic promastigotes is a crucial step in the life cycle of *Leishmania* parasites (Sacks, 1992). The occurrence of several morphological forms of promastigotes in sandflies suggested the possibility that different morphological forms may possess different capacities to cause disease. The first indications that parasite growth phase or morphology may impact on promastigote infectivity were established using *in vitro*-derived promastigotes. Both Giannini (1974) and Keithley (1976) demonstrated a correlation between the age of *Leishmania donovani* in culture and their infectivity for hamsters. The identification of a subpopulation of infectious metacyclic-stage promastigotes within stationary-phase cultures of promastigotes was later made in several species of *Leishmania* (Sacks, 1989). The first indication that this parasite morphogenesis also occurred during promastigote development within sandfly vectors was made using *Leishmania major* and *Lutzomyia anthopora* (Sacks & Perkins, 1984). These investigators demonstrated that promastigotes taken from sandflies 3 days after a bloodmeal were essentially avirulent, whereas promastigotes taken 7–8 days after feeding were highly virulent in mice. These observations were later extended to include the development of *Leishmania amazonensis* in *Lutzomyia longipalpis* (Sacks & Perkins, 1985).

Multiple morphological and ultrastructural modifications have been demonstrated to occur as promastigotes progress from non-infectious procyclic forms to infectious metacyclic forms. Metacyclic promastigotes are generally more elongated and thinner than procyclic forms. Additionally, metacyclic promastigotes demonstrate increased motility in culture and an elongated flagellum (Sacks, Hieny & Sher, 1985). Studies characterizing ultrastructural changes occurring during metacyclogenesis have largely utilized *L. major* (Sacks *et al.* 1985), and to a lesser extent *L. donovani* (Howard, Sayers & Miles, 1987). The most intensely studied ultrastructural modifications that occur during metacyclogenesis are those that occur on the parasites' surface. Using *L major*, Sacks (Sacks *et al.* 1985) demonstrated differences in the surface glycosylation patterns of promastigotes as they progressed through their growth phase. Whereas procyclic promastigotes are readily agglu-

tinated by the D-galactose-binding lectin, peanut agglutinin (PNA), the same concentration of lectin does not agglutinate metacyclic promastigotes. *L. donovani* promastigotes appear to demonstrate a similar pattern of PNA agglutination as they develop (Howard *et al.* 1987). An analogous change in surface glycosylation has been reported for *Leishmania braziliensis* using lentil lectin (Almeida *et al.* 1993). The molecular basis for the varying susceptibilities of procyclic and metacyclic promastigotes to lectin agglutination is due to changes in the glycosylation of the major surface structure on *Leishmania* promastigotes, lipophosphoglycan (McConville *et al.* 1992).

Lipophosphoglycan

The major cell surface glycoconjugate on *Leishmania* promastigotes is called lipophosphoglycan (LPG). It is localized over the entire surface of the parasite, including the flagella. LPG is linked to the parasites surface via a phosphotidylinositol linkage. The structure of LPG is composed of four domains: a phosphotidylinositol lipid anchor, a glycan core, a repeating saccharide–phosphate region, and an oligosaccharide cap (Turco & Descoteaux, 1992). The lipid anchor and glycan core of LPG are conserved among all *Leishmania* species studied; however, extensive variability exists in the carbohydrate content of the repeating saccharide–phosphate region and oligosaccharide cap (Sacks, 1992).

During the development of promastigotes from procyclic to metacyclic forms, two alterations occur to LPG structure. The first alteration is an elongation of LPG due to an increase in the number of phosphorylated saccharides. In *L. major* this has been shown to result in almost a doubling in length of the LPG molecule (Sacks, Brodin & Turco, 1990). The second modification is a change in the terminal carbohydrate moieties found on the repeating saccharide–phosphate region. On procyclic promastigotes the majority of terminal sugars are galactose. On metacyclic promastigotes these have been replaced by an arabinopyranose (McConville *et al.* 1992). This structural modification explains the loss of PNA agglutinability by metacyclic promastigotes. These changes have been documented to occur not only during *in vitro* cultivation of promastigotes, but also during *in vivo* development of promastigotes in their insect vector (Saraiva *et al.* 1995). An examination of the role that LPG plays during development of *L. major* promastigotes in the insect vector demonstrates that the LPG expressed by procyclic promastigotes can act as a ligand for receptors on midgut epithelial cells. This attachment is necessary to prevent the developing promastigote from being removed from the gut during passage of the digested bloodmeal (Pimenta *et al.* 1992). The modification of LPG during metacyclogenesis results in a loss of binding to midgut epithelial cells, suggesting that these modifications are necessary to allow the infectious form of the promastigote to be released from the midgut and move toward the foregut, in anticipation of delivery to the host (Pimenta *et al.* 1992). This mechanism of stage-specific midgut adhesion has also been demonstrated to occur during *L. donovani* development (Sacks *et al.* 1995). Recent studies have also demonstrated that vectorial competence (the ability of a particular species of sandfly to transmit a particular species of parasite) may be regulated by the variation of LPG structure amongst different species of *Leishmania* (Pimenta *et al.* 1994).

Because LPG is the most abundant surface structure on *Leishmania* promastigotes, its role as a virulence factor has been extensively investigated. These studies have identified roles for LPG in the attachment of promastigotes to macrophages (Handman & Goding, 1985), protection from oxidative damage (Chan *et al.* 1989), protection from digestion within the phagolysosome (Eilam, El-On & Spira, 1985; Handman *et al.* 1986), and down-regulation of macrophage functions, including chemotaxis and IL-1 production (Frankenberg *et al.* 1990) and protein kinase C activity (Descoteaux *et al.* 1991, 1992).

The major surface protease, gp63

In addition to changes in the glycosylation of surface molecules, metacyclic promastigotes have also been shown to possess differences in surface protein antigen profiles (Sacks *et al.* 1985), the most prominent difference being an increase in the expression of the major surface protein on promastigotes (Kweider *et al.* 1987). The major surface protein on *Leishmania* promastigotes is a 63 kDa glycoprotein, commonly known as gp63 (Lepay, Nogueria & Cohn, 1983; Colomer-Gould *et al.* 1985).

The first identification of gp63 on the surface of *Leishmania* promastigotes was made during an analysis of monoclonal antibodies reacting with *L. amazonensis* promastigotes (Fong & Chang, 1982). The majority of monoclonal antibodies generated in this study were found to immunoprecipitate the same 63 kDa protein from *Leishmania* lysates. Additionally, these antibodies were shown to react with the parasites' surface. Soon after this report, the 63 kDa protein was demonstrated to be the major surface antigen on all species of promastigotes analysed (Lepay *et al.* 1983; Colomer-Gould *et al.* 1985). GP63 was first purified to homogeneity in 1985 (Bouvier, Etges & Bordier, 1985) and characterized as a zinc-containing metalloproteinase the following year (Etges, Bouvier & Bordier, 1986). GP63 is attached to the parasite membrane via a phosphotidylinositol linkage, as are many surface molecules of trypanosomatid protozoa (Bordier *et al.*

1986). Surface iodination, immunostaining and biochemical analysis have demonstrated not only that gp63 is the most abundant protein on the surface of all *Leishmania* studied, but that its expression increases as promastigotes become more infectious, or progress into metacyclogenesis (Kweider *et al.* 1987; Grogl *et al.* 1987; Kweider *et al.* 1989; Ramamoorthy *et al.* 1992; Brittingham *et al.* 1995).

Biochemical and molecular analysis has identified gp63 as a Zn^{2+}-dependent endopeptidase (Bouvier *et al.* 1989). The proteolytic activity of gp63 is resistant to inhibition by most known proteinase inhibitors, except for 1,10-phenanthroline (Etges *et al.* 1989). gp63 has been shown to be capable of degrading numerous substrates including albumin, casein, immunoglobulin, complement proteins and haemoglobin (Chaudhuri & Chang, 1988). In 1988 Button & McMaster cloned the gp63 gene from *L. major* (Button & McMaster, 1988). Further studies on the genomic organization of this gene revealed that it belonged to a family of tandemly linked genes (Button *et al.* 1989). In *L. major*, there are 6 copies of the gene, 5 tandemly arranged and one additional copy 8 kb downstream of the tandem array (Button *et al.* 1989). *L. donovani* has 7 copies of the gp63 gene, 5 tandemly arranged, and 2 further downstream (Webb, Button & McMaster, 1991). *L. amazonensis* possesses a more complex array of gp63 genes, with at least 10 copies of the gene, split into 3 distinct complexes, all occurring on the same chromosome (Medina-Acosta, Beverley & Russell, 1993). The most complex and best characterized gp63 gene array is that of *L. chagasi*. In this species, there are at least 18 genes encoding gp63 contained in a 80 kb region (Roberts *et al.* 1993). Differences in the gp63 mRNA size between log phase and stationary phase promastigote cultures of *L. chagasi* have been reported (Ramamoorthy *et al.* 1992). These differences were due, primarily to unique 3′ untranslated regions (3′ UTR). The mechanisms of gp63 gene regulation remain unknown; however, differences in the 5′ and 3′ UTR may play a role in regulating the stability and processing of these transcripts (Ramamoorthy *et al.* 1995, 1996).

Due to the great abundance, surface location, and proteolytic activity of gp63, much work has been done to define a role for gp63 in *Leishmania* virulence. These include studies implicating a role for gp63 in the attachment of promastigotes to macrophages (Chang & Chang, 1986; Russell & Wilhelm, 1986), as well as a role in the survival of phagocytosed promastigotes (Chaudhuri *et al.* 1989).

Two other potentially important alterations occur as promastigotes progress from the procyclic to the metacyclic form. Metacyclic organisms express greater amounts of surface-associated acid phosphatase activity (Gottlieb & Dwyer, 1981) and cysteine proteinases (Robertson & Coombs, 1992). Neither of these activities has been as extensively characterized as LPG and gp63. However, targeted deletion of the genes encoding *Leishmania* cysteine proteinases resulted in decreased intracellular survival of the parasite in macrophages (Mottram *et al.* 1996).

LEISHMANIA PROMASTIGOTES AND HOST SERUM FACTORS

Leishmania promastigotes are delivered to their vertebrate hosts by the bite of an infected sandfly. During the feeding process metacyclic promastigotes are regurgitated into a blood pool (Schlein *et al.* 1992). Due to this method of delivery, promastigotes immediately encounter host serum and the lytic factors which are contained in serum. The ability of fresh serum to lyse promastigotes was observed as early as 1912 by W. S. Patton (in Hindle, Hou & Patton, 1926). In 1926, Hindle observed that serum from patients with Kala-azar as well as non-immune individuals was capable of killing *L. donovani* promastigotes (Hindle *et al.* 1926). Adler suggested that the ability of 2 individuals to resist experimental infection with *Leishmania* was due to lytic factors in their serum (Adler, 1940). Ulrich was the first to suggest that the complement system was responsible for the lysis of promastigotes, by observing that the lytic factor was heat labile (Ulrich, Ortiz & Convit, 1968). A more thorough examination of complement activation by promastigotes demonstrated that both *L. major* and *L. enrietti* were capable of activating complement via the alternative pathway, a process that proceeds in the absence of antibody (Mosser & Edelson, 1984). The activation of the alternative complement pathway was later extended to include *L. mexicana*, *L. amazonensis*, and *L. braziliensis* (Mosser *et al.* 1986). Complement activation by *L. donovani* (Pearson & Steigbigel, 1980) and possibly metacyclic *L. major* promastigotes (Puentes *et al.* 1988) may involve components of the classical complement pathway as well as the alternative pathway. Early reports suggested the involvement of a naturally occurring antibody in complement activation by *L. donovani* (Pearson & Steigbigel, 1980; Mosser *et al.* 1986). More recent reports have demonstrated the binding of acute phase proteins, including C-reactive protein (CRP) (Pritchard *et al.* 1985) and mannan-binding proteins (MBP) (Green *et al.* 1994), to promastigotes. Both CRP and MBP are capable of activating the complement system via a 3rd activation pathway, sometimes referred to as the lectin pathway, which consumes some of the classical complement components in a manner similar to that of antibody mediated activation (Claus *et al.* 1977; Ikeda *et al.* 1987).

The deposition of C3b (the first step in the activation of the alternative complement pathway) onto any surface or structure requires the formation of an amide or ester linkage to be formed between the thioester of C3b and a free amino or hydroxyl group on the activating surface (Law & Levine, 1977). Surfaces that are activators of the alternative pathway support the interaction of C3b with Factor B, thereby preventing the inactivation of C3b to iC3b. Surfaces that are poor activators of the alternative pathway preferentially allow the binding of Factor H to C3b. Factor H is a cofactor for Factor I, the serine protease responsible for the inactivation of C3b to iC3b (Pangburn *et al.* 1980). Because of the highly reactive nature of the C3b thioester, and the relative random deposition of C3 upon activation to C3b, it is not surprising that the two most abundant surface structures on promastigotes, gp63 and LPG, have both been identified as C3 acceptor sites (Russell, 1987; Puentes *et al.* 1988). Using *L. amazonensis*, Russell (1987) demonstrated that the majority of radiolabelled C3 that was bound to the surface of promastigotes was immunoprecipitated with antibodies to gp63. Additionally, the incorporation of gp63 into liposomes converted them into efficient activators of the alternative complement pathway. A similar observation has been made using Chinese hamster ovary (CHO) cells stably expressing gp63 on their surface (Brittingham *et al.* 1995). In contrast to this work, Puentes demonstrated that the majority of C3 on the surface of *L. major* promastigotes was bound to LPG (Puentes *et al.* 1988). One of us demonstrated that LPG-coated beads could consume complement from normal serum (Mosser & Handman, 1992). Regardless of the site of fixation or mechanism of complement activation, deposition of C3 onto the surface of promastigotes has been shown to be a crucial step in the interaction of the promastigote with the innate immune system (Mosser & Rosenthal, 1994). Surface bound C3 can act as ligand for macrophage receptors, thereby mediating the attachment of promastigotes to macrophages. C3b can also lead to formation of the C5 convertase and subsequently to assembly of the complement membrane attack complex (MAC).

Early studies on the interaction of promastigotes with the complement system demonstrated that these parasites were quite susceptible to complement-mediated lysis (Ulrich *et al.* 1968; Mosser & Edelson, 1984). Subsequent observations that metacyclic organisms taken from the stationary phase of growth were more infectious than procyclic log phase promastigotes prompted an examination of parasites taken from different growth phases, with respect to their susceptibilities to complement-mediated lysis (Franke *et al.* 1985). This work demonstrated that stationary phase promastigotes of *Leishmania panamensis* and *L. donovani* were more resistant to complement-mediated lysis than were organisms taken from the log phase of growth (Franke *et al.* 1985). Using *L. major*, Puentes and colleagues demonstrated that PNA-selected metacyclic promas-

Table 1. Mechanisms by which microorganisms avoid complement attack

Failure to activate complement		
Escherichia coli K1	Factor H binding	(Pluschke *et al.* 1983)
Haemophilus influenzae	Factor H binding	(Quinn *et al.* 1977)
Neisseria meningitidis	Factor H binding	(Jarvis & Vedros, 1987)
Group A streptococci	M proteins	(Hortsmann *et al.* 1988)
Schistosoma mansoni	C1q binding	(Laclette *et al.* 1992)
	DAF	(Pearce *et al.* 1990)
Taenia solium	C1q binding	(Laclette *et al.* 1992)
Trypanosoma cruzi	C1q binding/degradation	(Rimoldi *et al.* 1989)
	Sialidase/trans-sialidase	(Tomlinson *et al.* 1994)
	DAF homolog	(Joiner *et al.* 1988; Norris *et al.* 1991)
Trypanosoma brucei	Antigenic variation	(Borst & Cross, 1982)
Consumption/degradation of complement components		
Porphymonas gingivalis	C3 degradation	(Fletcher *et al.* 1994)
Pseudomonas aeruginosa	C3 degradation	(Hong & Ghebrehiwet, 1992)
Serratia marcescens	C3/C5 degradation	(Oda *et al.* 1990)
Aspergillus fumigatus	C3 degradation	(Sturtevant & Latge, 1992)
Herpes virus	C3 binding/consumption	(McNearney *et al.* 1987)
Vaccinia virus	C4 binding	(Kotwal *et al.* 1990)
Entamoeba histolytica	C3 activation/consumption	(Reed *et al.* 1989; Reed & Gigli, 1990)
L. major	C3 degradation	(Brittingham *et al.* 1995)
Schistosoma mansoni	C3/C9 degradation	(Marikovsky *et al.* 1988)
Trichomonas vaginalis	C3 degradation	(Alderete *et al.* 1995)
Inhibition of MAC formation/function		
E. coli	MAC shedding	(Joiner *et al.* 1984)
Salmonella minnesota	MAC shedding	(Joiner *et al.* 1982*a*)
Klebsiella pneumoniae	MAC shedding	(Merino *et al.* 1992)
Entamoeba histolytica	CD59 homologue	(Braga *et al.* 1992)
L. major	MAC shedding	(Puentes *et al.* 1990)
Naegleria fowleri	Membrane blebbing	(Toney & Marciano-Cabral, 1994)
Schistosoma mansoni	CD59 homologue/SCIP-1	(Parizade z*et al.* 1994)
	No MAC insertion	(McLaren & Hockley, 1977)

tigotes were more resistant to complement-mediated lysis than were procyclic organisms, and correlated this increased resistance with modifications of LPG on metacyclic promastigotes (Puentes *et al.* 1988). Based on previous work identifying LPG as the major C3 acceptor site on promastigotes (Puentes *et al.* 1988), and the spontaneous release of C5b-C9 complexes from the surface of metacyclic promastigotes (Puentes *et al.* 1990), it was proposed that metacyclic promastigotes resist serum lysis in a manner analogous to certain strains of *Salmonella* (Table 1). Strains of *Salmonella* with long chain O-polysaccharides are resistant to serum lysis due to the assembly of the MAC at a site too distant from the cell membrane to allow insertion (Joiner *et al.* 1982*a*). Similar to the observations made with *Salmonella* (Joiner *et al.*, 1982*b*), the MAC complexes that were formed on metacyclic *leishmania* were incapable of attaching to the parasite membrane, and were eventually released from the microbes' surface (Puentes *et al.* 1990).

Previous reports had suggested that the release of C3 from the surface of metacyclic promastigotes may be due to the proteolytic cleavage of C3 by an endogenous parasite protease (Puentes *et al.* 1989). Since several groups reported the increased ex-

pression of gp63 on metacyclic promastigotes, relative to procyclic organisms (Grogl *et al.* 1987; Kweider *et al.* 1987, 1989; Ramamoorthy *et al.* 1992), we sought to define a role for gp63 in the resistance of promastigotes to complement-mediated lysis. Using gp63-transfected variants of a gp63-deficient strain of *L. amazonensis*, we demonstrated a correlation between gp63 expression and resistance to complement-mediated lysis (Brittingham *et al.* 1995; Fig. 1). Organisms expressing wild-type gp63 fixed reduced levels of terminal complement components and were much more resistant to complement-mediated lysis then were organisms lacking gp63. Furthermore, transfected organisms expressing a proteolytically inactive form of gp63, called E265D, also fixed the terminal complement components and were as susceptible to complement-mediated lysis as were gp63-deficient organisms. These data demonstrate that gp63 can render promastigotes resistant to complement-mediated lysis, and that the mechanism of this resistance is dependent on the proteolytic activity of gp63.

The mechanism by which proteolytically active gp63 confers resistance to complement-mediated lysis is revealed by the observation that C3b on the surface of parasites expressing wild-type gp63 was

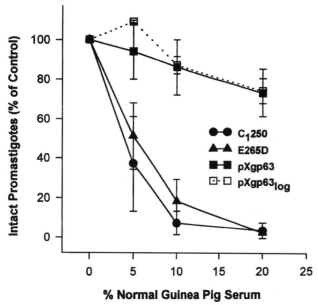

Fig. 1. Complement-mediated lysis of *L. amazonensis*
gp63 variants. The parental gp63-deficient strain,
designated C_1250 and described elsewhere (McGwire &
Chang, 1994) (●), was transfected with constructs
encoding either wild-type gp63, designated pXgp63 (■),
or gp63 containing a point mutation rendering it
proteolytically inactive, designated E265D (▲).
Promastigotes were resuspended in Hanks' balanced salt
solution containing increasing amounts of normal
guinea-pig serum, as a source of lytic complement.
Intact promastigotes were counted on a
haemocytometer. Organisms used in these assays were
taken from stationary growth phase cultures, except
$pXgp63_{log}$ (□), which were taken from cultures in
logarithmic growth. Values are expressed as a percentage
of the control, which is the number of intact
promastigotes present following an incubation in the
absence of serum. (Reproduced, with permission, from
Brittingham *et al.* 1995).

rapidly converted to a form antigenically similar to
that of iC3b. This inactive form of C3, while
remaining opsonic, is unable to support formation of
the C5 convertase and subsequent deposition of the
MAC. Using a cell-free system, we demonstrate that
purified gp63 degrades C3b to a form physically
similar to iC3b (Brittingham & Mosser, 1996). These
data demonstrate that gp63 affords the parasite
protection from lysis by directly cleaving C3b to a
form which can no longer promote the fixation of the
MAC.

As evidenced by numerous microbes, a successful
pathogen possesses more than one mechanism to
avoid destruction by complement (for review see
Moffitt & Frank, 1994; Jokiranta, Jokipii & Meri,
1995 and Table 1). Our work (Brittingham *et al.*
1995; Brittingham & Mosser, 1996) and the work of
Puentes and colleagues (Puentes *et al.* 1988, 1990)
suggest that the 2 most abundant surface structures
on promastigotes, gp63 and LPG, are both involved
in the resistance of parasites to complement-mediated

lysis. The relative contributions of these 2 factors
have yet to be determined. In a previous publication,
Camara and colleagues demonstrated that resistance
to complement-mediated lysis correlated with protease activity (Camara *et al.* 1995). We demonstrated
that logarithmic phase organisms expressing high
levels of recombinant gp63 were resistant to lysis
(Fig. 1), indicating that metacyclic LPG is not
required for resistance to complement-mediated
lysis. It should be noted that the LPG phenotype of
the strain of *L. amazonensis* used in these studies is
unknown. In the future, genetic manipulation of the
parasite to delete gp63 encoding genes, or enzymes
involved in LPG biosynthesis, may allow a more
thorough investigation of the relative roles of these 2
molecules in parasite virulence, as well as resistance
to complement-mediated lysis.

In addition to opsonization and lysis, activation of
complement also leads to the generation of the
chemotactic peptides C3a and C5a. These peptides
have been shown not only to be potent inducers of
leukocyte migration, but also to up-regulate the
expression of complement receptors on mononuclear
phagocytes (Yancey *et al.* 1985). Bray demonstrated
that macrophages show no chemotaxis toward promastigotes themselves, but following the activation
of complement by *Leishmania*, macrophages show a
directional migration toward the products of complement activation (Bray, 1983). Thus, by activating
complement, promastigotes can recruit a fresh
supply of newly migrated host cells (macrophages) to
the area of infection. In the visceral model of
leishmaniasis, newly recruited macrophages are
thought to be the effector cells that eliminate *L.
donovani* (Murray, 1994). In the cutaneous model,
however, newly migrated macrophages may be
necessary for the propagation of infection (Murray,
1994). Using C3 deficient mice, we have begun to
examine the importance of complement-mediated
inflammation in mice infected with cutaneous and
visceralizing species of *Leishmania*.

INTERACTION OF PROMASTIGOTES WITH MACROPHAGES

The interaction of promastigotes with macrophages
has been an extensively studied aspect of *Leishmania*
biology (reviewed in Mosser & Rosenthal, 1994).
Although early reports suggested a specific orien-
tation of the parasite (flagella first) was necessary for
macrophage invasion, implying active penetration
by the parasite (Pulvertaft & Hoyle, 1960; Miller &
Twohy, 1967), these reports were not substantiated
(Akiyama & Haight, 1971; Chang & Dwyer, 1978).
Promastigote entry into macrophages has been
shown to be a passive process on the part of the
parasite, which can be inhibited by treating macrophages with cytochalasin B, an inhibitor of phagocytosis (Chang, 1979). Heat-killed promastigotes are

Table 2. Mechanisms of promastigote attachment to macrophages

Parasite ligand	Macrophage receptor	Reference
Direct binding		
gp63	Mac-1	(Russell & Wright, 1988; Van Strijp et al. 1993)
	Fibronectin receptors	(Rizvi et al. 1988; Soteriadou et al. 1992)
LPG	Mac-1	(Van Strijp et al. 1993)
	p150/95	(Talamas-Rohana et al. 1990)
	???	(Handman & Goding, 1985)
???	Mannose receptor	(Channon et al. 1984; Wilson & Pearson, 1986)
	β-glucan receptor	(Mosser & Handman, 1992)
	Advanced glycosylation end product (AGE) receptor	(Mosser et al. 1987)
Opsonized binding		
C3b	CR1	(Da Silva et al. 1989)
iC3b	Mac-1	(Mosser & Edelson, 1985; Blackwell et al. 1985)
Fibronectin	???	(Wyler et al. 1985)
C-reactive protein	???	(Culley et al. 1996)

efficiently phagocytosed by macrophages (Chang, 1979). Early studies (Zenian, 1981; Wyler, 1982) demonstrated that promastigote binding and phagocytosis are receptor-mediated events. Multiple macrophage receptors, parasite ligands, and host opsonins have been implicated in the binding of promastigotes to macrophages (Mosser & Rosenthal, 1993; Table 2). Promastigotes encounter serum components early in the infectious process and therefore presumably encounter macrophages in a serum-opsonized state. However, this does not negate the importance of parasite surface ligands in the binding of promastigotes to macrophages. In fact, efficient attachment to and internalization by macrophages is probably dependent on multiple receptor–ligand interactions. To simplify the discussion of these receptor–ligand interactions we will characterize them as either (1) the direct binding of promastigotes to macrophages, or (2) the opsonin facilitated binding of promastigotes to macrophages (Table 2).

Direct adhesion of parasites to macrophages

The recognition of parasite carbohydrates by macrophage lectin-like receptors appears to be an important component of the direct recognition of parasites by macrophages. The binding of L. donovani promastigotes to murine (Channon, Roberts & Blackwell, 1984; Blackwell et al. 1985) or human (Wilson & Pearson, 1986) macrophages was inhibited by mannan and other inhibitors of the mannose receptor. The direct binding of L. major promastigotes to murine macrophages was not inhibited by mannan, but was inhibited by presumed ligands for the β-glucan receptor, laminarin and zymocel (Mosser & Handman, 1992). An alternative carbohydrate receptor that was implicated in the attachment of

promastigotes to macrophages is the receptor for advanced glycosylation end products (AGE) (Mosser et al. 1987). AGEs arise from the time-dependent, non-enzymic adduction of glucose to proteins (Brownlee, Vlassara & Cerami, 1984). The AGE receptor may be involved in the removal of senescent proteins and cells (Vlassara, Brownlee & Cerami, 1985).

The identification of an abundant glycoconjugate (LPG) on the surface of promastigotes (Handman, Greenblatt & Goding, 1984; Turco, Wilkerson & Clawson, 1984) initiated experiments to define a role for this molecule in the direct attachment of promastigotes to macrophages. Handman & Goding (1985) demonstrated that affinity-purified LPG from L. major bound specifically to murine macrophages. Promastigote binding could also be blocked by incubating the promastigotes with F(ab) fragments of antibodies against LPG (Handman & Goding, 1985). Further analysis of L. major LPG structure demonstrated that a PO_4-6[Gal(β1-3)Gal(β1-3)Gal(β1-3)]Gal(β1-4)Man α1-region of LPG was the region of LPG that bound to macrophages and inhibited L. major promastigote binding (Kelleher, Bacic & Handman, 1992). This region of LPG is unique to L. major and not found in the LPG of other species of Leishmania. The specificity of this interaction was demonstrated by the fact that this phospho-oligosaccharide repeat was capable of inhibiting L. major binding, but had no effect on L. donovani binding (Kelleher et al. 1992). The macrophage receptor which recognizes this region of LPG remains undetermined.

One group of receptors which has been implicated in recognizing LPG is the β2 (CD18) family of integrins. Using LPG-coated silica beads, and monoclonal antibodies, Talamas-Rohana and colleagues presented evidence suggesting that LPG could bind directly to Mac-1 (CD11b/CD18) and p150/95 (CD11c/CD18) (Talamas-Rohana *et al.* 1990). LPG-coated beads also bound to affinity-purified Mac-1 (Van Strijp *et al.* 1993). All these studies were performed using LPG coupled to an inert particle. Using viable intact promastigotes, however, several reports have demonstrated no direct binding of promastigotes to Mac-1 (Mosser, Springer & Diamond, 1992; Brittingham *et al.* 1995; Rosenthal *et al.* 1996). Using affinity-purified Mac-1 and Mac-1 transfected fibroblasts, the absolute requirement for opsonic complement in the binding of promastigotes to Mac-1 was observed (Mosser *et al.* 1992), suggesting that the direct interaction of LPG with Mac-1 that was previously observed was not of sufficient avidity to mediate the binding of viable promastigotes to Mac-1.

Like LPG, the other major surface molecule on promastigotes, gp63, has also been implicated in the direct adhesion of promastigotes to macrophages. Purified protein, as well as antibodies against gp63 have both been shown to inhibit promastigote adhesion (Chang & Chang, 1986; Russell & Wilhelm, 1986). gp63-Coated particles have been demonstrated to bind to macrophages (Russell & Wilhelm, 1986). Restoration of gp63 expression on a gp63-deficient variant of *L. amazonensis* was shown to improve promastigote binding to murine macrophages (Liu & Chang, 1992). The mechanisms of gp63-mediated adhesion to macrophages, as well as the macrophage receptors involved in gp63 recognition remain somewhat controversial. Original reports (Russell & Wright, 1988) suggested that gp63 contained the amino acid sequence Arg-Gly-Asp (RGD), a sequence which is recognized by many receptors of the integrin family (Ruoslahti, 1991). Further analysis of the gp63 sequence, however, demonstrated that gp63 does not possess an RGD region (Miller, Reed & Parsons, 1990), and that the RGD region of complement protein C3, the normal ligand for Mac-1, was not necessary for receptor binding (Taniguchi-Sidle & Isenman, 1992). The majority of work implicating an interaction between gp63 and Mac-1 utilized purified gp63 coupled to either liposomes or other inert particles (Russell & Wright, 1988; Van Strijp *et al.* 1993). Using viable promastigotes, however, no direct binding of promastigotes to substrates coated with purified Mac-1 or fibroblasts transfected with constructs encoding Mac-1 was observed (Mosser *et al.* 1992; Brittingham *et al.* 1995; Rosenthal *et al.* 1996. Additionally, CHO cells stably expressing leishmania gp63 on their surface exhibited no direct binding to purified Mac-1-coated substrates (Brittingham *et al.* 1995). Once again, the reason for these apparent discrepancies may be due to the low avidity of gp63 for Mac-1 and to subtle differences in the assays employed in these studies.

A second set of receptors which have been implicated in the gp63-mediated attachment of promastigotes to macrophages are the cellular receptors for fibronectin. The interaction of fibronectin with cellular receptors is often dependent on the Arginine-Glycine-Aspartic acid-Serine (RGDS) region of fibronectin (Pierschbacher & Ruoslahti, 1984). The gp63 molecule contains a conserved amino acid sequence, Serine-Arginine-Tyrosine-Aspartic acid (SRYD), which in *L. major* is found at amino acids 252–255 (Button & McMaster, 1988). Antibodies against fibronectin have been shown to cross-react with gp63 (Rizvi *et al.* 1988; Soteriadou *et al.* 1992) and also to inhibit the binding of promastigotes to macrophages (Wyler, Sypek & McDonald, 1985; Rizvi *et al.* 1988). Soteriadou *et al.* (1992) localized the cross-reactive epitopes of fibronectin and gp63 and demonstrated that the SRYD region of gp63 was antigenically similar to the RGDS region of fibronectin. Wyler *et al.* (1985) had previously demonstrated that the tetrapeptide RGDS could inhibit the immunoprecipitation of gp63 by antibodies to fibronectin. These data suggest that receptors for fibronectin present on macrophages may be capable of recognizing the 'fibronectin-like' region of gp63 and mediate promastigote attachment to macrophages. Despite the antigenic similarities of fibronectin and gp63, and the ability of anti-fibronectin antibodies and fibronectin-derived peptides to inhibit promastigote adhesion to macrophages, formal proof of the involvement of fibronectin receptors in parasite adhesion remains to be determined.

Using *L. amazonensis* gp63-transfected variants, we demonstrated that the presence of gp63 on the surface of promastigotes caused a modest but significant enhancement in the direct binding of promastigotes to human macrophages. This enhanced binding was not dependent on the presence of the SRYD sequence of gp63, because *L. amazonensis* variants expressing gp63 containing a point mutation, converting SRYD to SRDD, on their surface bound to macrophages as well as did those parasites expressing wild-type gp63 (Brittingham, unpublished). An examination of the interaction of promastigotes with CHO cells expressing Mac-1, or the fibronectin receptor ($\alpha5\beta1$), demonstrated no direct binding of promastigotes to these cellular receptors. The presence or absence of gp63, or its SRYD region, had no effect on the binding of parasites to these cells (A. Brittingham *et al.*, unpublished observations).

The increased attachment and spreading of fibroblasts to surfaces coated with gp63 has been reported (Rizvi *et al.* 1988). We measured the attachment of promastigotes to human and murine fibroblasts. The presence of gp63 on the surface of promastigotes significantly enhanced their binding to fibroblasts. Unlike our observation with macrophages, the

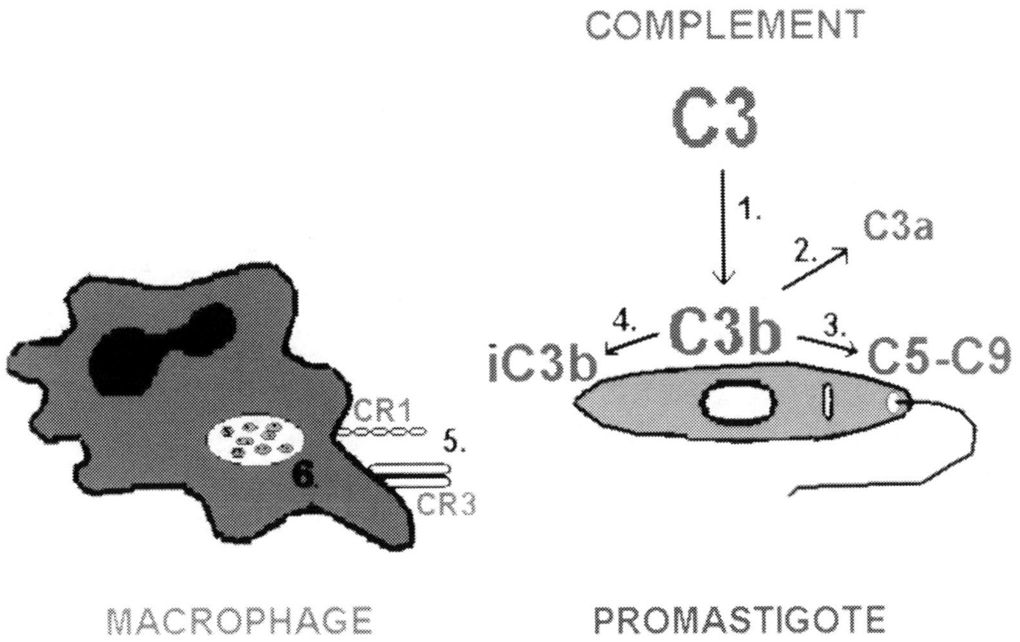

Fig. 2.

attachment of promastigotes to fibroblasts was dependent on the presence of an intact SRYD region (A. Brittingham *et al.*, unpublished). The receptors involved in the gp63-mediated adhesion of promastigotes to fibroblasts is unknown; however, it is probably not the α5β1, fibronectin receptor, since there was no direct binding of promastigotes to transfected CHO cells expressing the α5β1 receptor.

In summary, there appear to be multiple mechanisms for the direct adhesion of promastigotes to macrophages. All of these mechanisms are inefficient in promoting parasite adhesion, relative to the serum-dependent mechanisms which will be described below. However, the observation that mice lacking C3 can be infected with either visceral or cutaneous strains of leishmania (unpublished observation) indicates that these direct binding mechanisms can eventually lead to a productive infection *in vivo*.

Serum-dependent adhesion of promastigotes to macrophages

Numerous serum opsonins and cellular receptors for these opsonins have been implicated in the attachment of promastigotes to macrophages (Mosser & Rosenthal, 1993). The most well characterized of these interactions is the binding of complement-opsonized promastigotes to macrophage complement receptors (Mosser & Edelson, 1984; Wozencraft & Blackwell, 1987). Initial studies demonstrated that complement opsonization of promastigotes enhanced their attachment to macrophages (Mosser & Edelson, 1984). The amount of this enhancement depended on the species of *Leishmania* studied as well as the species and type of macrophage used (Mosser &

Rosenthal, 1993). Additional studies demonstrated that opsonization of promastigotes by serum not only affected their attachment to macrophages, but also their intracellular fate following phagocytosis (Mosser & Edelson, 1987; Sacks, 1992). Complement fixation by *L. major* promastigotes resulted in increased intracellular survival of the parasites in macrophages. These results indicate that, *in vitro*, the survival of *L. major* in macrophages is potentiated by the fixation of serum complement. The increased survival of complement opsonized promastigotes in macrophages may be due in part to the reduced elicitation of the respiratory burst by complement-opsonized promastigotes, relative to unopsonized organisms (Mosser & Edelson, 1987).

The receptors involved in the binding and phagocytosis of serum-opsonized promastigotes are the macrophage receptors for the complement protein C3 (Blackwell *et al.* 1985; Mosser & Edelson, 1985; Da Silva *et al.* 1989). The 2 major complement receptors on mononuclear phagocytes are Mac-1 (CD11b/CD18), the receptor for iC3b, and CR1 (CD35), the receptor for C3b and C4b. The CR4, p150,95, appears to play only a minor role in promastigote adhesion (Mosser *et al.* 1992). Initial studies examining the interaction of complement-opsonized promastigotes with macrophages were done with murine macrophages, which express Mac-1 but not CR1, and demonstrated the importance of this receptor in promastigote adhesion (Blackwell *et al.* 1985; Mosser & Edelson, 1985). Working with human macrophages, Da Silva and colleagues demonstrated a role for CR1 in the binding of metacyclic *L. major* promastigotes to human macrophages (Da Silva *et al.* 1989). Recently, using defined particles such as erythrocytes opsonized with C3 fragments, it has been demonstrated that CR1 and Mac-1 co-operate in a unique manner to mediate the complement-dependent adhesion of particles to macrophages (Sutterwala, Rosenthal & Mosser, 1996). CR1 mediates a transient adhesion of complement-opsonized particles to macrophages. Because of the Factor I-cofactor activity of CR1, bound C3b is rapidly converted to iC3b, the ligand for Mac-1. Mac-1, in turn, mediates a stable adhesion of complement-opsonized particles to leukocytes. This stable adhesion is a prerequisite for phagocytosis. These results were extended to include *Leishmania* promastigotes (Rosenthal *et al.* 1996). In these studies, both procyclic and metacyclic *L. major* promastigotes bound to both CR1 and Mac-1, similar to the observations with defined particles. The stable adhesion of complement-opsonized *Leishmania* was mediated primarily by Mac-1, and antibodies to Mac-1, but not CR1, inhibited the phagocytosis of serum-opsonized metacyclic *L. major* promastigotes (Rosenthal *et al.* 1996). Thus, the complement-dependent adhesion of *Leishmania* to macrophages is mediated by both Mac-1 and CR1, but Mac-1 is the primary receptor involved in the complement-dependent phagocytosis of *Leishmania* promastigotes.

CONCLUSION

The complement system is one of the first lines of host defence against microbial invasion. Therefore, it is not surprising that successful pathogens have evolved multiple mechanisms to avoid the destructive effect of complement activation (Moffitt & Frank, 1994; Jokiranta *et al.* 1995; Table 1). In the case of *Leishmania* promastigotes this avoidance has been taken to a higher level. Not only are the detrimental effects of complement activation avoided by the parasite, but the beneficial effects of opsonization and chemotaxis are exploited. In this way *Leishmania* promastigotes assure their successful entry into mononuclear phagocytes.

The final figure (Fig. 2) summarizes some of the observations made by several different laboratories throughout the world concerning the interaction of *Leishmania* promastigotes with complement and macrophage complement receptors. (1) Complement can be fixed by any of the 3 activation pathways, but all *Leishmania* species can activate complement via the alternative pathway, a process that proceeds in the absence of antibody. (2) The complement anaphylatoxin, C3a, is generated during complement fixation. This molecule is chemotactic for macrophages. (3) The lysis of *Leishmania* promastigotes by the C5-9 membrane attack complex of complement is inhibited by gp63 and LPG. (4) C3b on the surface of promastigotes is rapidly converted to iC3b by Factor I and gp63. (5) Two complement receptors, CR1 and Mac-1 (CR3), cooperate to mediate the

adhesion of complement-opsonized promastigotes. CR1 acts as a binding site for complement-opsonized *Leishmania*, whereas Mac-1 mediates internalization. (6) The opsonization of some species of *Leishmania* promastigotes by complement improves their intracellular survival in macrophages.

REFERENCES

ADLER, S. (1940). Attempts to transmit visceral leishmaniasis to man. *Transactions of the Royal Society of Medicine and Hygiene* **33**, 419–437.

AKIYAMA, H. J. & HAIGHT, R. D. (1971). Interaction of *Leishmania donovani* and hamster peritoneal macrophages. A phase-contrast microscopical study. *American Journal of Tropical Medicine and Hygiene* **20**, 539–545.

ALDERETE, J. F., PROVENZANO, D. & LEHKER, M. W. (1995). Iron mediates trichomonas resistance to complement lysis. *Microbial Pathogenesis* **19**, 93–103.

ALMEIDA, M. C., CUBA CUBA, C. A., DE SA, C. M., PHAROAH, M. M., HOWARD, K. M. & MILES, M. A. (1993). Metacyclogenesis of *Leishmania (Viannia) braziliensis in vitro*: evidence that lentil lectin is a marker of complement resistance and enhanced infectivity. *Transactions of the Royal Society of Tropical Medicine and Hygiene* **87**, 325–329.

BATES, P. A. (1994). The developmental biology of *Leishmania* promastigotes. *Experimental Parasitology* **79**, 215–218.

BLACKWELL, J. M., EZEKOWITZ, R. A. B., ROBERTS, M. B., CHANNON, J. Y., SIM, R. B. & GORDON, S. (1985). Macrophage complement and lectin-like receptors bind *Leishmania* in the absence of serum. *Journal of Experimental Medicine* **162**, 324–331.

BORDIER, C., ETGES, R. J., WARD, J., TURNER, M. J. & CARDOSO DE ALMEIDA, M. L. (1986). *Leishmania* and *Trypanosoma* surface glycoproteins have a common glycophospholipid membrane anchor. *Proceedings of the National Academy of Sciences, USA* **83**, 5988–5991.

BORST, P. & CROSS, G. M. (1982). Molecular basis for trypanosome antigenic variation. *Cell* **29**, 291–303.

BOUVIER, J., BORDIER, C., VOGEL, H., REICHELT, R. & ETGES, R. (1989). Characterization of the promastigote surface protease as a membrane-bound zinc endopeptidase. *Molecular and Biochemical Parasitology* **37**, 235–245.

BOUVIER, J., ETGES, R. J. & BORDIER, C. (1985). Identification and purification of membrane and soluble forms of the major surface protein of *Leishmania* promastigotes. *Journal of Biological Chemistry* **260**, 15504–15509.

BRAGA, L. L., NINOMIYA, H., McCOY, J. J., EACKER, S., WIEDMER, T., PHAM, C., WOOD, S., SIMS, P. J. & PETRI, W. A., JR (1992). Inhibition of the complement membrane attack complex by the galactose-specific adhesion of *Entamoeba histolytica*. *Journal of Clinical Investigation* **90**, 1131–1137.

BRAY, R. S. (1983). *Leishmania*: chemotaxic response of promastigotes and macrophages *in vitro*. *Journal of Protozoology* **30**, 322–329.

BRITTINGHAM, A. & MOSSER, D. M. (1996). Exploitation of the complement system by *Leishmania* promastigotes. *Parasitology Today* **12**, 444–447.

BRITTINGHAM, A., MORRISON, C. J., McMASTER, W. R., McGWIRE, B. S., CHANG, K. P. & MOSSER, D. M. (1995). Role of the *Leishmania* surface protease gp63 in complement fixation, cell adhesion, and resistance to complement-mediated lysis. *Journal of Immunology* **155**, 3102–3111.

BROWNLEE, M., VLASSARA, H. & CERAMI, A. (1984). Nonenzymatic glycosylation and the pathogenesis of diabetic complications. *Annals of Internal Medicine* **101**, 527–537.

BUTTON, L. L. & McMASTER, W. R. (1988). Molecular cloning of the major surface antigen of *Leishmania*. *Journal of Experimental Medicine* **167**, 724–729.

BUTTON, L. L., RUSSELL, D. G., KLEIN, H. L., MEDINA-ACOSTA, E., KARESS, R. E. & McMASTER, W. R. (1989). Genes encoding the major surface glycoprotein in *Leishmania* are tandemly linked at a single chromosomal locus and are constitutively transcribed. *Molecular and Biochemical Parasitology* **32**, 271–284.

CAMARA, M., ORTIZ, G., VALERO, P. L., MOLINA, R., NAVARRO, I., CHANCE, M. L. & SEGOVIA, M. (1995). Complement-mediated lysis and infectivity for mouse macrophages and sandflies of virulent and attenuated *Leishmania major* promastigotes varying in expression of the major surface protease and lipophosphoglycan. *Annals of Tropical Medicine and Parasitology* **89**, 243–251.

CHAN, J., FUJIWARA, T., BRENNAN, P., McNEIL, M., TURCO, S. J., SIBILLE, J. C., SNAPPER, M., AISEN, P. & BLOOM, B. R. (1989). Microbial glycolipids: possible virulence factors that scavenge oxygen radicals. *Proceedings of the National Academy of Sciences, USA* **86**, 2453–2457.

CHANG, C. S. & CHANG, K. P. (1986). Monoclonal antibody affinity purification of a *Leishmania* membrane glycoprotein and its inhibition of leishmania–macrophage binding. *Proceedings of the National Academy of Sciences, USA* **83**, 100–104.

CHANG, K. P. (1979). *Leishmania donovani*: promastigote–macrophage surface interactions *in vitro*. *Experimental Parasitology* **48**, 175–189.

CHANG, K. P. & DWYER, D. M. (1978). *Leishmania donovani*–hamster macrophage interactions *in vitro*: cell entry, intracellular survival and multiplication. *Journal of Experimental Medicine* **147**, 515–529.

CHANNON, J. Y., ROBERTS, M. B. & BLACKWELL, J. M. (1984). A study of the differential respiratory burst ellicited by promastigotes and amastigotes of *Leishmania donovani* in murine peritoneal macrophages. *Immunology* **53**, 345–355.

CHAUDHURI, G. & CHANG, K. P. (1988). Acid protease activity of a major surface membrane glycoprotein (gp63) from *Leishmania mexicana* promastigotes. *Molecular and Biochemical Parasitology* **27**, 43–52.

CHAUDHURI, G., CHAUDHURI, M., PAN, A. & CHANG, K. P. (1989). Surface acid proteinase (gp63) of *Leishmania mexicana*: a metalloenzyme capable of protecting liposome-encapsulated proteins from phagolysosomal degradation by macrophages. *Journal of Biological Chemistry* **264**, 7483–7489.

CLAUS, D. R., SIEGEL, J., PETRAS, K., OSMAND, A. P. & GEWURZ, H. (1977). Interaction of C-reactive protein with the first component of human complement. *Journal of Immunology* **119**, 187–196.

COLOMER-GOULD, V., QUINTAS, L. G., KEITHLEY, J. & NOGUEIRA, N. (1985). A common major surface antigen on amastigotes and promastigotes of *Leishmania* species. *Journal of Experimental Medicine* **162**, 902–916.

CULLEY, F. J., HARRIS, R. A., KAYE, P. M., McADAM, K. P. W. J. & RAYNES, J. G. (1996). C-reactive protein binds to a novel ligand on *Leishmania donovani* and increases uptake into human macrophages. *Journal of Immunology* **156**, 4691–4696.

DA SILVA, R. P., HALL, F. B., JOINER, K. A. & SACKS, D. L. (1989). CR1, the C3b receptor, mediates binding of infective *Leishmania major* promastigotes to human macrophages. *Journal of Immunology* **143**, 617–622.

DESCOTEAUX, A., MATLASHEWSKI, G. & TURCO, S. J. (1992). Inhibition of macrophage protein kinase C-mediated protein phosphorylation by *Leishmania donovani* lipophosphoglycan. *Journal of Immunology* **149**, 3008–3015.

DESCOTEAUX, A., TURCO, S. J., SACKS, D. L. & MATLASHEWSKI, G. (1991). *Leishmania donovani* lipophosphoglycan selectively inhibits signal transduction in macrophages. *Journal of Immunology* **146**, 2747–2753.

EILAM, Y., EL-ON, J. & SPIRA, D. T. (1985). *Leishmania major*: excreted factor, calcium ions, and the survival of amastigotes. *Experimental Parasitology* **59**, 161–168.

ETGES, R., BOUVIER, J. & BORDIER, C. (1986). The major surface protein of *Leishmania promastigotes* is a protease. *Journal of Biological Chemistry* **261**, 9098–9101.

ETGES, R., BOUVIER, J. & BORDIER, C. (1989). The promastigote surface protease of *Leishmania*: pH optimum and effects of protease inhibitors. In *Leishmaniasis: Current Status and New Strategies for Control* (ed. Hart, D.), pp. 627–633. NATO ASI Series A, vol. 163. New York, Plenum Press.

FLETCHER, H. M., SCHENKEIN, H. A. & MARCINA, F. L. (1994). Cloning and characterization of a new protease gene (prtH) from *Porphyromonas gingivalis*. *Infection and Immunity* **62**, 4279–4286.

FONG, D. & CHANG, K. P. (1982). Surface antigenic change during differentiation of a parasitic protozoan, *Leishmania mexicana*: identification by monoclonal antibodies. *Proceedings of the National Academy of Sciences, USA* **79**, 7366–7370.

FRANKE, E. D., McGREEVY, P. B., KATZ, S. P. & SACKS, D. L. (1985). Growth cycle dependent generation of complement-resistant *Leishmania* promastigotes. *Journal of Immunology* **134**, 2713–2718.

FRANKENBURG, S., LEIBOVICIC, V., MANSBACH, N., TURCO, S. J. & ROSEN, G. (1990). Effect of glycolipids of *Leishmania* parasites on human monocyte activity. Inhibition by lipophosphoglycan. *Journal of Immunology* **145**, 4284–4289.

GIANNINI, M. S. (1974). Effect of promastigote growth phase, frequency of subculture, and host age on promastigote-initiated infections in *Leishmania donovani* in the golden hamster. *Journal of Protozoology* **21**, 521–527.

GOTTLIEB, M. & DWYER, D. M. (1981). *Leishmania donovani*: surface membrane acid phosphatase activity of promastigotes. *Experimental Parasitology* **52**, 117–128.

GREEN, P. J., FEIZI, T., STOLL, M. S., THIEL, S., PRESCOTT, A. & McCONVILLE, M. J. (1994). Recognition of the major cell surface glycoconjugates of *Leishmania* parasites by the human serum mannan-binding protein. *Molecular and Biochemical Parasitology* **66**, 319–328.

GROGL, M., FRANKE, E. D., McGREEVY, P. B. & KUHN, R. E. (1987). *Leishmania braziliensis*: protein, carbohydrate, and antigen differences between log-phase and stationary-phase promastigotes *in vitro*. *Experimental Parasitology* **63**, 352–359.

HANDMAN, E. & GODING, J. W. (1985). The *Leishmania* receptor for macrophages is a lipid-containing glycoconjugate. *EMBO Journal* **4**, 329–336.

HANDMAN, E., GREENBLATT, C. L. & GODING, J. W. (1984). An amphipathic sulfated glycoconjugate of *Leishmania*: characterization with monoclonal antibodies. *EMBO Journal* **3**, 2301–2306.

HANDMAN, E., SCHNUR, L. F., SPITHILL, T. W. & MITCHELL, G. F. (1986). Passive transfer of *Leishmania* lipopolysaccharide confers parasite survival in macrophages. *Journal of Immunology* **137**, 3608–3613.

HINDLE, E., HOU, P. C. & PATTON, W. S. (1926). Serological studies in Chinese Kala Azar. *Proceedings of the Royal Society of London (Biology)* **100**, 368–373.

HONG, Y. Q. & GHEBREHIWET, B. (1992). Effect of *Pseudomonas* elastase and alkaline protease on serum complement and isolated components C1q and C3. *Clinical Immunology and Immunopathology* **62**, 133–138.

HORTSMANN, R. D., SIEVERTSEN, H. J., KNOBLOCH, J. & FISCHETTI, V. A. (1988). Antiphagocytic activity of streptococcal M protein: selective binding of complement control protein factor H. *Proceedings of the National Academy of Sciences, USA* **85**, 1657–1661.

HOWARD, M. K., SAYERS, G. & MILES, M. A. (1987). *Leishmania donovani* metacyclic promastigotes: transformation *in vitro*, lectin agglutination, complement resistance, and infectivity. *Experimental Parasitology* **64**, 147–156.

IKEDA, K., SANNOH, T., KAWASAKI, N., KAWASAKI, T. & YAMASHINA, I. (1987). Serum lectin with known structure activates complement through the classical pathway. *Journal of Biology Chemistry* **262**, 7451–7454.

JARVIS, G. A. & VEDROS, N. A. (1987). Sialic acid of group B *Neisseria meningitidis* regulates alternative complement pathway activation. *Infection and Immunity* **55**, 174–180.

JOINER, K. A., DA SILVA, W. D., RIMOLDI, M. T., HAMMER, C. H., SHER, A. & KIPNIS, T. L. (1988). Biochemical characterization of a factor produced by trypomastigotes of *Trypanosoma cruzi* that accelerates the decay of complement C3 convertase. *Journal of Biological Chemistry* **263**, 11327–11335.

JOINER, K. A., HAMMER, C. H., BROWN, E. J., COLE, R. J. & FRANK, M. M. (1982*a*). Studies on the mechanism of bacterial resistance to complement mediated killing. I. Terminal complement components are deposited and released from *Salmonella minnesota* S218 without causing bacterial death. *Journal of Experimental Medicine* **155**, 797–804.

JOINER, K. A., HAMMER, C. H., BROWN, E. J. & FRANK, M. M. (1982*b*). Studies on the mechanism of bacterial resistance to complement mediated killing. II. C8 and C9 release C5b67 from the surface of *Salmonella minnesota* S218 because the terminal complex does not insert into the bacterial outer membrane. *Journal of Experimental Medicine* **155**, 809–815.

JOINER, K. A., SCHMETZ, M. A., GOLDMAN, R. C., LEIVE, L. & FRANK, M. M. (1984). Mechanism of bacterial resistance to complement-mediated killing: inserted C5b-9 correlates with killing for *Escherichia coli* 0111B4 varying in o-antigen capsule and O-polysaccharide coverage of lipid A core oligosaccharide. *Infection and Immunity* **45**, 113–117.

JOKIRANTA, T. S., JOKIPII, L. & MERI, S. (1995). Complement resistance of parasites. *Scandinavian Journal of Immunology* **42**, 9–20.

KEITHLEY, J. S. (1976). Infectivity of *Leishmania donovani* amastigotes and promastigotes for golden hamsters. *Journal of Protozoology* **23**, 244–248.

KELLEHER, M., BACIC, A. & HANDMAN, E. (1992). Identification of a macrophage-binding determinant on lipophosphoglycan from *Leishmania major* promastigotes. *Proceedings of the National Academy of Sciences, USA* **89**, 6–10.

KILLICK-KENDRICK, R. (1979). Biology of *Leishmania* in phlebotomine sandflies. In *Biology of the Kinetoplastida* (ed. Lumsden, W. H. R. & Evans, D. A.), pp. 395–460. New York, Academic Press.

KOTWAL, G. J., ISAACS, S. N., McKENZIE, R., FRANK, M. M. & MOSS, B. (1990). Inhibition of the complement cascade by the major secretory protein of vaccinia virus. *Science* **250**, 827–830.

KWEIDER, M., LEMESRE, J., DARCY, F., KUSNIERZ, J. P., CAPRON, A. & SANTORO, F. (1987). Infectivity of *Leishmania braziliensis* promastigotes is dependent on the increased expression of a 65 000-dalton surface antigen. *Journal of Immunology* **138**, 299–305.

KWEIDER, M., LEMESRE, J. L., SANTORO, F., KUSNIERZ, J. P., SADIGURZKY, M. & CAPRON, A. (1989). Development of metacyclic *Leishmania* promastigotes is associated with the increasing expression of GP65, the major surface antigen. *Parasite Immunology* **11**, 197–209.

LACLETTE, J. P., SHOEMAKER, C. B., RICHTER, D., ARCOS, L., PANTE, N., COHEN, C., BING, D. & NICHOLSON-WELLER, A. (1992). Paramyosin inhibits complement C1. *Journal of Immunology* **148**, 124–128.

LAW, S. K. & LEVINE, R. P. (1977). Interaction between the third complement protein and cell surface molecules. *Proceedings of the National Academy of Sciences, USA* **74**, 2701–2705.

LEPAY, D. A., NOGUERIA, N. & COHN, Z. (1983). Surface antigens of *Leishmania donovani* promastigotes. *Journal of Experimental Medicine* **157**, 1562–1572.

LIU, X. & CHANG, K. P. (1992). Extrachromosomal genetic complementation of surface metalloproteinase (gp63)-deficient *Leishmania* increases their binding to macrophages. *Proceedings of the National Academy of Sciences, USA* **89**, 4991–4995.

MARIKOVSKY, M., ARNON, R. & FISHELSON, Z. (1988). Proteases secreted by transforming schistosomula of *Schistosoma mansoni* promote resistance to killing by complement. *Journal of Immunology* **141**, 273–278.

McCONVILLE, M. J., TURCO, S. J., FERGUSON, M. A. J. & SACKS, D. L. (1992). Developmental modification of lipophosphoglycan during the differentiation of *Leishmania major* promastigotes to an infectious stage. *EMBO Journal* **11**, 3593–3600.

McGWIRE, B. S. & CHANG, K. P. (1994). Genetic rescue of surface metalloproteinase (gp63) deficiency in *Leishmania amazonensis* variants increases their infection of macrophages at the early phase. *Molecular and Biochemical Parasitology* **66**, 345–347.

McLAREN, D. J. & HOCKLEY, D. J. (1977). Blood flukes have a double membrane. *Nature* **269**, 147–149.

McNEARNEY, T. A., ODELL, C., HOLERS, V. M., SPEAR, P. G. & ATKINSON, J. P. (1987). Herpes simplex virus glycoproteins gC-1 and gC-2 bind to the third component of complement and provide protection against complement-mediated neutralization of viral infectivity. *Journal of Experimental Medicine* **166**, 1525–1535.

MEDINA-ACOSTA, E., BEVERLEY, S. M. & RUSSELL, D. G. (1993). Evolution and expression of the *Leishmania* surface proteinase (gp63) gene locus. *Infectious Agents and Disease* **2**, 25–34.

MERINO, S., CAMPRUBI, S., ALBERTI, S., BENEDI, V. J. & THOMAS, J. M. (1992). Mechanism of *Klebsiella pneumoniae* resistance to complement-mediated killing. *Infection and Immunity* **60**, 2529–2535.

MILLER, H. C. & TWOHY, D. W. (1967). Infection of macrophages in culture by leptomonas of *Leishmania donovani*. *Journal of Protozoology* **14**, 781–789.

MILLER, R. A., REED, S. G. & PARSONS, M. (1990). *Leishmania* gp63 molecule implicated in cellular adhesion lacks an Arg-Gly-Asp sequence. *Molecular and Biochemical Parasitology* **39**, 267–274.

MOFFITT, M. C. & FRANK, M. M. (1994). Complement resistance in microbes. *Springer Seminars in Immunopathology* **15**, 327–344.

MOSSER, D. M., BURKE, S. K., COUTAVAS, E. E., WEDGEWOOD, J. F. & EDELSON, P. J. (1986). *Leishmania* species: mechanisms of complement activation by five strains of promastigotes. *Experimental Parasitology* **62**, 394–404.

MOSSER, D. M. & EDELSON, P. J. (1984). Activation of the alternative complement pathway by *Leishmania* promastigotes: parasite lysis and attachment to macrophages. *Journal of Immunology* **132**, 1501–1505.

MOSSER, D. M. & EDELSON, P. J. (1985). The mouse macrophage receptor for C3bi (CR3) is a major mechanism in the phagocytosis of leishmania promastigotes. *Journal of Immunology* **135**, 2785–2789.

MOSSER, D. M. & EDELSON, P. J. (1987). The third component of complement (C3) is responsible for the intracellular survival of *Leishmania major*. *Nature* **327**, 329–331.

MOSSER, D. M. & HANDMAN, E. (1992). Treatment of murine macrophages with interferon-γ inhibits their ability to bind *Leishmania* promastigotes. *Journal of Leukocyte Biology* **52**, 369–376.

MOSSER, D. M. & ROSENTHAL, L. A. (1993). *Leishmania*–macrophage interactions: multiple receptors, multiple ligands, and diverse cellular responses. *Seminars in Cell Biology* **4**, 315–322.

MOSSER, D. M. & ROSENTHAL, L. A. (1994). Divergent strategies used by the promastigote and amastigote forms of *Leishmania* to invade mammalian cells. In *Baillière's Clinical Infectious Disease*, vol. 1, *Strategies for Intracellular Survival of Microbes* (ed. Russell, D. G.), pp. 191–212. London, Baillière Tindall.

MOSSER, D. M., SPRINGER, T. A. & DIAMOND, M. S. (1992). *Leishmania* promastigotes require opsonic complement to bind to the human leukocyte integrin Mac-1 (CD11b/CD18). *Journal of Cell Biology* **116**, 511–520.

MOSSER, D. M., VLASSARA, H., EDELSON, P. J. & CERAMI, A. (1987). *Leishmania* promastigotes are recognized by the macrophage receptor for advanced glycosylation endproducts. *Journal of Experimental Medicine* **165**, 140–145.

MOTTRAM, J. C., SOUZA, A. E., HUTCHISON, J. E., CARTER, R., FRAME, M. & COOMBS, G. H. (1996). Evidence from disruption of the *lmcpb* gene array of *Leishmania mexicana* that cysteine proteinases are virulence factors. *Proceedings of the National Academy of Sciences, USA* **93**, 6008–6013.

MURRAY, H. W. (1994). Blood monocytes: differing effector role in experimental visceral versus cutaneous leishmaniasis. *Parasitology Today* **10**, 220–223.

NORRIS, K. A., BRADT, B., COOPER, N. R. & SO, M. (1991). Characterization of a *Trypanosoma cruzi* C3 binding protein with functional and genetic similarities to the human complement regulatory protein, decay-accelerating factor. *Journal of Immunology* **147**, 2240–2247.

ODA, T., KOJIMA, Y., AKAIKE, T., IJIRI, S., MOLLA, A. & MAEDA, H. (1990). Inactivation of chemotactic activity of C5a by the serratial 56-kilodalton protease. *Infection and Immunity* **58**, 1269–1272.

PANGBURN, M. K., MORRISON, D. C., SCHREIBER, R. D. & MÜLLER-EBERHARD, H. J. (1980). Activation of the alternative complement pathway: recognition of surface structures on activators by bound C3b. *Journal of Immunology* **124**, 977–982.

PARIZADE, M., ARNON, R., LACHMANN, P. J. & FISHELSON, Z. (1994). Functional and antigenic similarities between a 94 kDa protein of *Schistosoma mansoni* (SCIP-1) and human CD59. *Journal of Experimental Medicine* **179**, 1625–1636.

PEARCE, E. J., HALL, B. F. & SHER, A. (1990). Host-specific evasion of the alternative complement pathway by schistosomes correlates with the presence of phospholipase C-sensitive surface molecules resembling decay accelerating factor. *Journal of Immunology* **144**, 2751–2756.

PEARSON, R. T. & STEIGBIGEL, R. T. (1980). Mechanism of lethal effect of human serum upon *Leishmania donovani*. *Journal of Immunology* **125**, 2195–2201.

PIERSCHBACHER, M. D. & RUOSLAHTI, E. (1984). Cell attachment activity of fibronectin can be duplicated by small synthetic fragments of the molecule. *Nature* **309**, 30–33.

PIMENTA, P. F. P., SARAIVA, E. M. B., ROWTON, E., MODI, G. B., GARRAWAY, L. A., BEVERLEY, S. M., TURCO, S. J. & SACKS, D. L. (1994). Evidence that the vectorial competence of phlebotomine sand flies for different species of *Leishmania* is controlled by structural polymorphisms in the surface lipophosphoglycan. *Proceedings of the National Academy of Sciences, USA* **91**, 9155–9159.

PIMENTA, P. F. P., TURCO, S. J., McCONVILLE, M. J., LAWYER, P. G., PERKINS, P. V. & SACKS, D. L. (1992). Stage-specific adhesion of *Leishmania* promastigotes to the sandfly midgut. *Science* **256**, 1812–1815.

PLUSCHKE, G., MAYDEN, J., ACHTMAN, M. & LEVINE, R. P. (1983). Role of the O antigen is resistanced of O18:K1 *Escherichia coli* to complement-mediated killing. *Infection and Immunity* **42**, 907–913.

PRITCHARD, D. G., VOLANAKIS, J. E., SLUTSKY, G. M. & GREENBLATT, C. L. (1985). C-reactive protein binds leishmanial excreted factors. *Proceedings of the Society for Experimental Biology and Medicine* **178**, 500–503.

PUENTES, S. M., DA SILVA, R. P., SACKS, D. L., HAMMER, C. H. & JOINER, K. A. (1990). Serum resistance of metacyclic stage *Leishmania major* promastigotes is due to release of C5b-9. *Journal of Immunology* **145**, 4311–4316.

PUENTES, S. M., DWYER, D. M., BATES, P. A. & JOINER, K. A. (1989). Binding and release of C3 from *Leishmania donovani* promastigotes during incubation in normal human serum. *Journal of Immunology* **143**, 3743–3749.

PUENTES, S. M., SACKS, D. L., DA SILVA, R. P. & JOINER, K. A. (1988). Complement binding by two developmental stages of *Leishmania major* promastigotes varying in expression of a surface lipophosphoglycan. *Journal of Experimental Medicine* **167**, 887–902.

PULVERTAFT, R. J. V. & HOYLE, G. F. (1960). Stages in the life cycle of *Leishmania donovani*. *Transactions of the Royal Society of Tropical Medicine and Hygiene* **54**, 191–195.

QUINN, P. H., CROSSON, F. J., WINKELSTEIN, J. A. & MOXON, E. R. (1977). Activation of the alternative complement pathway by *Haemophilus influenzae* type B. *Infection and Immunity* **16**, 400–402.

RAMAMOORTHY, R., DONELSON, J. E., PAETZ, K. E., MAYBODI, M., ROBERTS, S. C. & WILSON, M. E. (1992). Three distinct RNAs for the surface protease gp63 are differentially expressed during development of *Leishmania donovani* chagasi promastigotes to an infectious form. *Journal of Biological Chemistry* **267**, 1888–1895.

RAMAMOORTHY, R., DONELSON, J. E. & WILSON, M. E. (1996). 5′ sequences essential for trans-splicing of msp (gp63) RNAs in *Leishmania chagasi*. *Molecular and Biochemical Parasitology* **77**, 65–76.

RAMAMOORTHY, R., SWIHART, K. G., McCOY, J. J., WILSON, M. E. & DONELSON, J. E. (1995). Intergenic regions between tandem gp63 genes influence the differential expression of gp63 RNAs in *Leishmania chagasi* promastigotes. *Journal of Biological Chemistry* **270**, 12133–12139.

REED, S. L. & GIGLI, I. (1990). Lysis of complement sensitive *Entamoeba histolytica* by activated terminal complement components. Initiation of complement activation by an extracellular neutral cysteine proteinase. *Journal of Clinical Investigation* **86**, 1815–1822.

REED, S. L., KEENE, W. E., McKERROW, J. H. & GIGLI, I. (1989). Cleavage of C3 by a neutral cysteine proteinase of *Entamoeba histolytica*. *Journal of Immunology* **143**, 189–195.

RIMOLDI, M. T., TENNER, A. J., BOBAK, D. A. & JOINER, K. A. (1989). Complement component C1q enhances invasion of human mononuclear phagocytes and fibroblasts by *Trypanosoma cruzi* trypomastigotes. *Journal of Clinical Investigation* **84**, 1982–1989.

RIZVI, F. S., OUAISSI, M. A., MARTY, B., SANTORO, F. & CAPRON, A. (1988). The major surface protein of *Leishmania* promastigotes is a fibronectin-like molecule. *European Journal of Immunology* **18**, 473–476.

ROBERTS, S. C., SWIHART, K. G., AGEY, M. W., RAMAMOORTHY, R., WILSON, M. E. & DONELSON, J. E. (1993). Sequence diversity and organization of the *msp* gene family encoding gp63 of *Leishmania chagasi*. *Molecular and Biochemical Parasitology* **62**, 157–172.

ROBERTSON, C. D. & COOMBS, G. H. (1992). Stage-specific proteinases of *Leishmania mexicana* promastigotes. *FEMS Microbiology Letters* **94**, 127–132.

ROSENTHAL, L. A., SUTTERWALA, F. S., KEHRLI, M. E. & MOSSER, D. M. (1996). *Leishmania major*-human macrophage interactions: cooperation between Mac-1 (CD11b/CD18) and complement receptor type 1 (CD35) in promastigote adhesion. *Infection and Immunity* **64**, 2206–2215.

RUOSLAHTI, E. (1991). Integrins. *Journal of Clinical Investigation* **87**, 1–5.

RUSSELL, D. G. (1987). The macrophage-attachment glycoprotein gp63 is the predominant C3-acceptor site on *Leishmania mexicana* promastigotes. *European Journal of Biochemistry* **164**, 213–221.

RUSSELL, D. G. & WILHELM, H. (1986). The involvement of the major surface glycoprotein (gp63) of *Leishmania* promastigotes in attachment to macrophages. *Journal of Immunology* **136**, 2613–2620.

RUSSELL, D. G. & WRIGHT, S. D. (1988). Complement receptor type 3 (CR3) binds to an Arg-Gly-Asp-containing region on the major surface glycoprotein, gp63, of leishmania promastigotes. *Journal of Experimental Medicine* **168**, 279–292.

SACKS, D. L. (1989). Metacyclogenesis in *Leishmania* promastigotes. *Experimental Parasitology* **69**, 100–103.

SACKS, D. L. (1992). The structure and function of the surface lipophosphoglycan on different developmental stages of *Leishmania* promastigotes. *Infectious Agents and Disease* **1**, 200–206.

SACKS, D. L., BRODIN, T. N. & TURCO, S. J. (1990). Developmental modification of the lipophosphoglycan from *Leishmania major* promastigotes during metacyclogenesis. *Molecular and Biochemical Parasitology* **42**, 225–234.

SACKS, D. L., HIENY, S. & SHER, A. (1985). Identification of cell surface carbohydrate and antigenic changes between noninfective and infective developmental stages of *Leishmania major* promastigotes. *Journal of Immunology* **135**, 564–569.

SACKS, D. L. & PERKINS, P. V. (1984). Identification of an infectious stage of *Leishmania* promastigotes. *Science* **223**, 1417–1419.

SACKS, D. L. & PERKINS, P. V. (1985). Development of infectious stage *Leishmania* promastigotes within Phlebotomine sand flies. *American Journal of Tropical Medicine and Hygiene* **343**, 456–459.

SACKS, D. L., PIMENTA, P. F. P., McCONVILLE, M. J., SCHNEIDER, P. & TURCO, S. J. (1995). Stage-specific binding of *Leishmania donovani* to the sand fly vector midgut is regulated by conformational changes in the abundant surface lipophosphoglycan. *Journal of Experimental Medicine* **181**, 685–697.

SARAIVA, E. M. B., PIMENTA, P. F. P., BRODIN, T. N., ROWTON, E., MODI, G. B. & SACKS, D. L. (1995). Changes in lipophosphoglycan and gene expression associated with the development of *Leishmania major* in *Phlebotomus papatasi*. *Parasitology* **111**, 275–287.

SCHLEIN, Y., JACOBSON, R. L. & MESSER, G. (1992). *Leishmania* infections damage the feeding mechanism of the sandfly vector and implement parasite transmission by bite. *Proceedings of the National Academy of Sciences, USA* **89**, 9944–9948.

SOTERIADOU, K. P., REMOUNDOS, M. S., KATSIKAS, M. C., TZINIA, A. K., TSIKARIS, V., SAKARELLOS, C. & TZARTOS, S. J. (1992). The Ser-Arg-Tyr-Asp region of the major surface glycoprotein of *Leishmania* mimics the Arg-Gly-Asp-Ser cell attachment region of fibronectin. *Journal of Biological Chemistry* **267**, 13980–13985.

STURTEVANT, J. E. & LATGE, J. P. (1992). Interactions between conidia of *Aspergillus fumigatus* and human complement component C3. *Infection and Immunity* **60**, 1913–1918.

SUTTERWALA, F. S., ROSENTHAL, L. A. & MOSSER, D. M. (1996). Cooperation between CR1 (CD35) and CR3 (CD11b/CD18) in the binding of complement-opsonized particles. *Journal of Leukocyte Biology* **59**, 883–890.

TALAMAS-ROHANA, P., WRIGHT, S. D., LENNARTZ, M. R. & RUSSELL, D. G. (1990). Lipophosphoglycan from *Leishmania mexicana* promastigotes binds to members of the CR3, p150,95 and LFA-1 family of leukocyte integrins. *Journal of Immunology* **144**, 4817–4824.

TANIGUCHI-SIDLE, A. & ISENMAN, D. E. (1992). Mutagenesis of the Arg-Gly-Asp triplet in human complement component C3 does not abolish binding of iC3b to the leukocyte integrin complement receptor type III (CR3, CD11b/CD18). *Journal of Biological Chemistry* **267**, 635–643.

TOMLINSON, S., PONTES DE CARVALHO, L. C., VANDEKERCKHOVE, F. & NUSSENZWEIG, V. (1994). Role of sialic acid in the resistance of *Trypanosoma cruzi* trypomastigotes to complement. *Journal of Immunology* **153**, 3141–3147.

TONEY, D. M. & MARCIANO-CABRAL, F. (1994). Alterations in protein expression and complement resistance of pathogenic *Naegleria* amoebae. *Infection and Immunity* **60**, 2784–2790.

TURCO, S. J. & DESCOTEAUX, A. (1992). The lipophosphoglycan of *Leishmania* parasites. *Annual Reviews in Microbiology* **46**, 65–94.

TURCO, S. J., WILKERSON, M. A. & CLAWSON, D. R. (1984). Expression of an unusual acid glycoconjugate in *Leishmania donovani*. *Journal of Biological Chemistry* **259**, 3883–3889.

ULRICH, M., ORTIZ, D. T. & CONVIT, J. (1968). The effect of fresh serum on leptomonads of leishmania. *Transactions of the Royal Society of Tropical Medicine and Hygiene* **62**, 825–830.

VAN STRIJP, J. A. G., RUSSELL, D. G., TOUMANEN, E., BROWN, E. J. & WRIGHT, S. D. (1993). Ligand specificity of purified complement receptor type three (CD11b/CD18, $\alpha m\beta 2$, Mac-1). *Journal of Immunology* **151**, 3324–3336.

VLASSARA, H., BROWNLEE, M. & CERAMI, A. (1985). High affinity receptor-mediated uptake and degradation of glucose-modified proteins: a potential mechanism for the removal of senescent macromolecules. *Proceedings of the National Academy of Sciences, USA* **82**, 5588–5592.

WALTERS, L. L. (1993). *Leishmania* differentiation in natural and unnatural sandfly hosts. *Journal of Eukaryotic Microbiology* **40**, 196–206.

WEBB, J. R., BUTTON, L. L. & McMASTER, W. R. (1991). Heterogeneity of the genes encoding the major surface glycoprotein of *Leishmania donovani*. *Molecular and Biochemical Parasitology* **48**, 173–184.

WILSON, M. E., INNES, D. J., SOUSA, A. Q. & PEARSON, R. D. (1987). Early histopathology of experimental infection with *Leishmania donovani* in hamsters. *Journal of Parasitology* **73**, 55–63.

WILSON, M. E. & PEARSON, R. D. (1986). Evidence that *Leishmania donovani* utilizes a mannose receptor on human mononuclear phagocytes to establish intracellular parasitism. *Journal of Immunology* **136**, 4681–4688.

WOZENCRAFT, A. O. & BLACKWELL, J. M. (1987). Increased infectivity of stationary-phase promastigotes of *Leishmania donovani*: correlation with enhanced C3 binding capacity and CR3-mediated attachment to host macrophages. *Immunology* **60**: 559–563.

WYLER, D. J. (1982). *In vivo* parasite-monocyte interactions in human leishmaniasis. Evidence for an active role of the parasite in attachment. *Journal of Clinical Investigation* **70**, 80–82.

WYLER, D. J., SYPEK, J. P. & McDONALD, J. A. (1985). *In vitro* parasite–monocyte interactions in human leishmaniasis: possible role of fibronectin in parasite attachment. *Infection and Immunity* **49**, 305–311.

YANCEY, K. B., O'SHEA, J., CHUSED, T., BROWN, E., TAKAHASHI, T., FRANK, M. M. & LAWLEY, T. J. (1985). Human C5a modulates monocyte Fc and C3 receptor expression. *Journal of Immunology* **135**, 465–470.

ZENIAN, A. (1981). *Leishmania tropica*: biochemical aspects of promastigotes' attachment to macrophages *in vitro*. *Experimental Parasitology* **51**, 175–188.

The fate and persistence of *Leishmania major* in mice of different genetic backgrounds: an example of exploitation of the immune system by intracellular parasites

P. LAUNOIS[1], J. A. LOUIS[1]* *and* G. MILON[2]

[1] *WHO Immunology Research & Training Center, Institute of Biochemistry, University of Lausanne, Chemin des Boveresses 155, CH-1066 Epalinges, Switzerland*
[2] *Immunophysiology Unit, Institut Pasteur, Paris, France*

SUMMARY

Leishmania spp. are intracellular protozoan parasites that are delivered within the dermis of their vertebrate hosts. Within this peripheral tissue and the draining lymph node, they find and/or rapidly create dynamic microenvironments that determine their ultimate fate, namely their more or less successful expansion, and favour their transmission to another vertebrate host though a blood-feeding vector. Depending on their genetic characteristics as well as the genetic make-up of their hosts, once within the dermis *Leishmania* spp. very rapidly drive and maintain sustained T cell-dependent immune responses that arbitrate their ultimate fate within their hosts. The analysis of the parasitism exerted by *Leishmania major* in mice of different genetic backgrounds has allowed us to recognize some of the early and late mechanisms driven by this parasite that lead to either uncontrolled or restricted parasitism. Uncontrolled parasitism by *Leishmania major* characterizing mice from a few inbred strains (e.g. BALB/c) is associated with the expansion of parasite reactive Th2 CD4 lymphocytes and results from their rapid and sustained activity. In contrast, restricted parasitism characteristic of mice from the majority of inbred strains results from the development of a polarized parasite-specific Th1 CD4 response. This murine model of infection has already been and will continue to be particularly instrumental in dissecting the rules controlling the pathway of differentiation of T cells *in vivo*. In the long run, the understanding of these rules should contribute to the rational development of novel immunotherapeutic interventions against severe infectious diseases.

Key words: *Leishmania major*, T cells, persistance, mice.

INTRODUCTION

Leishmania spp. are protozoan parasites that require the mononuclear phagocytes of their vertebrate hosts to achieve their life-cycle. Briefly, once inoculated within the dermis by a blood-feeding vector they enter rapidly in resident macrophages. In this first, peripheral site, *Leishmania* parasites initiate complex processes resulting in a micro-environment that favours their setting in. These complex processes involve both innate and adaptive immune mechanisms (Lanzavecchia, 1993; Marrack & Kapler, 1994; Abbas, Murphy & Sher, 1996; Fearon & Locksley, 1996).

It has been known for a long time that *Leishmania* spp.-driven parasitism in their vertebrate hosts (human beings, rodents or dogs) can result in either asymptomatic or pathological processes (leishmaniasis) (Sacks, Louis & Wirth, 1993). Such different outcomes are highly indicative of genetic polymorphism of either the parasites or their invertebrate/vertebrate hosts (Lanzaro & Warburg, 1995; Ivens & Blackwell, 1996; Wincker *et al.* 1996). The analysis of the contribution of the genetic component of the vertebrate hosts has been greatly facilitated by

devising experimental models based on the use of mice from genetically different inbred strains (Behin, Mauel & Sordat, 1979; Mitchell *et al.* 1983). In this context, the study of the murine model of parasitism (Russell, 1994) with *Leishmania major* has provided important insights in the understanding of the complex interactions between this intracellular parasite and the host genome, particularly the genes used by the host immune system (Blackwell, 1996).

In the present review, we will focus on recent approaches and results using this model pertaining to the characterization of both early and late immunological processes set in motion by *L. major*. Special emphasis will be given to mechanisms used by these parasites to exploit the host immune system, one of the end results being a persistence state probably favouring optimal transmission.

THE FATE OF *LEISHMANIA MAJOR* IN MICE OF DISTINCT GENETIC BACKGROUNDS: FROM THE CONCEPT OF RESISTANCE/SUSCEPTIBILITY TO THE NOTION OF THE EXPLOITATION OF THE IMMUNE SYSTEM

Following inoculation of *Leishmania major* promastigotes into the footpads, mice of the majority of inbred strains (CBA, C3H/He, C57BL/6, 129Sv/

* Corresponding author.

Ev, B10.D2) develop locally small lesions that spontaneously resolve after a few weeks (between 4 and 8 weeks). Since, once cured, these mice do not develop lesions after a second inoculation of *L. major*, they were designated as 'resistant' mice. In contrast, BALB/c mice develop locally severe lesions that do not resolve spontaneously; since BALB/c mice are not 'resistant' to secondary challenge (Behin *et al.* 1979; Handman, 1992) they were thus designated 'susceptible' mice.

The hypothesis that resistance to secondary challenge could be the consequence of the development of protective T lymphocytes was formulated in the late 1970s (Mitchell *et al.* 1981; Handman, 1992 for review). The effector functions of protective T lymphocytes were expected to be targeted to the host-cells of *L. major*, namely the mononuclear phagocytes, and to operate at both the level of invasion and intracellular growth (Nacy *et al.* 1983). The first hypothesis was validated by results showing that adoptive transfer of T lymphocytes from 'resistant' mice to otherwise very susceptible syngenic nu/nu mice allowed them to control the disease process (Mitchell *et al.* 1981). The demonstration that T cell-derived cytokines lead to activation of macrophages to a parasiticidal state validated the second hypothesis (Nacy *et al.* 1983; Titus *et al.* 1984*a*). It took a longer time to pinpoint the pro-parasite functions of T lymphocytes in 'susceptible' BALB/c mice (Titus *et al.* 1984*b*). The generation of monoclonal antibodies discriminating different T cell subsets (Fitch *et al.* 1993) permitted identification of CD4 T lymphocytes as the major subset paradoxically contributing to both resistance and susceptibility to *L. major* (Titus *et al.* 1987; Liew, 1989).

In 1986, the recognition of the existence in mice of functionally distinct CD4 subsets named Th1, Th2 (Mossman & Coffman, 1989 for review) led to the seminal observation that Th1 cytokine (IFNγ) and Th2 cytokine (IL-4) were associated with 'resistance' and 'susceptibility', respectively (Locksley *et al.* 1987). This observation favoured the hypothesis along which *L. major* triggers distinct CD4 T cell subsets in genetically different hosts. The monitoring, through Northern blot analysis, of those CD4 cytokine transcripts would be expected to reveal the expansion of Th1 (IFNγ and IL2) and Th2 (IL-4, IL5 and IL10) lymphocytes; it clearly showed the development of Th1 cells in lymph nodes of C57BL/6 mice and Th2 cells in lymph nodes of BALB/c mice (Heinzel *et al.* 1991). Other studies have confirmed that spontaneous healing of the lesion in 'resistant' mice correlated with reduced amounts of IL-4 mRNA and reduced numbers of IL-4 producing cells (Morris *et al.* 1993). However, several studies have shown that drastic differences between 'susceptible' and 'resistant' mice in IFNγ-mRNA and numbers of IFNγ producing cells are only observed at late stages of infection (Kopf *et al.*

1996; Stenger *et al.* 1995). Together with the demonstration that IL-4 can inhibit the interferony-mediated activation of macrophages (Liew *et al.* 1989; Bogdan *et al.* 1996), these observations illustrate the ability of *L. major* to exploit the immune system of certain hosts (such as BALB/c mice) to their benefit.

THE MATURATION OF FUNCTIONALLY DISTINCT T LYMPHOCYTES IN MICE OF DIFFERENT GENETIC BACKGROUNDS: A CRUCIAL ROLE FOR CYTOKINES DURING THE EARLY STAGE OF *L. MAJOR* INFESTATION?

The first approach to address this question relies on the neutralization of IL-4 either at the onset of infection or in a sustained manner throughout infection with the monoclonal antibody raised by Paul *et al.* (Ohara & Paul, 1985). Briefly, in BALB/c mice receiving injections of anti-IL-4 within the first week of *L. major* infection, the local lesions resolve, leading to a dominant Th1 response (Sadick *et al.* 1990). These results indicated a crucial role of IL-4 in the rapid and sustained differentiation of naïve CD4 T lymphocytes towards the Th2 phenotype after *L. major* inoculation. Even though recent results have questioned the importance of IL-4 in both susceptibility of BALB/c mice to infection and Th2 cell maturation (Noben-Trauth, Kropf & Muller, 1996), other results also using IL-4 deficient mice have clearly shown that these mice exhibit an impaired Th2 cell development and a state of resistance to infection with *L. major* (Kopf *et al.* 1996). In parallel, the early and/or sustained neutralization of IFNγ in C3H/HeN mice was shown to promote the development of Th2 T lymphocytes as assessed by the synthesis of IL-4 by CD4 T lymphocytes restimulated *in vitro* with soluble *L. major* extracts (Scott, 1991). These results suggested a role of IFNγ in the promotion of Th1 cell development and the parallel prevention of Th2 cell expansion.

In agreement with this contention, it has been shown that C57BL/6 mice with disruption of the IFNγ gene develop a Th2 cell response following inoculation of *L. major* (Wang *et al.* 1994). In contrast, administration of high doses of IFNγ to BALB/c mice did not impede Th2 cell development as assessed by the presence of IL-4 transcripts in lymphoid organs (Sadick *et al.* 1990). Furthermore, 129/Sv/Ev mice with disruption of the gene encoding the ligand-binding chain of the heterodimeric IFNγ receptor mounted a polarized Th1 CD4 response similar to that of control wild type mice (Swihart *et al.* 1995). These apparently conflicting results concerning the role of IFNγ in Th1 cell development could reflect genetic differences in the regulation of CD4 T cell subset differentiation. Indeed, recent results from our laboratory have shown a differential effect of neutralization of IFNγ at the initiation of the infection on the development of IL-4 transcription and secretion in 129/Sv/Ev and C57Bl/6 mice (P. Launois & J. A. Louis, unpublished).

Meanwhile, another cytokine, the heterodimeric IL-12, was identified as a triggering stimulus for IFNγ secretion by NK cells as well as CD4 and CD8 T cells (Chan *et al.* 1991). Using antibodies neutralizing the p35/p40 bioactive IL-12, and 129 Sv/Ev as well as C57Bl/6 mice with either p35 or p40 gene disruption, it was possible to establish the predominant role of this cytokine in the development of Th1 CD4 T lymphocytes in response to *L. major* (Mattner *et al.* 1996).

It is important to emphasize that the predominant role of both IL-4 and IL-12 at the initiation of antigenic stimulation in directing the differentiation of CD4 Naïve T lymphocytes towards the type 2 or type 1 functional phenotype has been most carefully established *in vitro* using naïve CD4 T cells from TCR αβ transgenic mice (Paul & Seder, 1994; O'Garra & Murphy, 1996; Swain & Cambier, 1996). Interestingly, when both exogenous IL-4 and IL-12 are added to cultures of TCR αβ transgenic in presence of antigen, but in absence of any other modulating reagents, the effect of IL-4 appears to predominate (Bradley & Watson, 1996; Swain & Cambier, 1996).

LEISHMANIA MAJOR RAPIDLY AND TRANSIENTLY INDUCES THE PRODUCTION OF CYTOKINES THAT SHAPE THE FUNCTIONAL CHARACTERISTICS OF THE T CELL RESPONSES, CONDITIONING THEIR ULTIMATE FATE/PERSISTENCE IN MICE

Using either neutralizing antibodies, exogenous recombinant cytokines and/or knock-out mice, four cytokines have been shown to play an early critical role in the differentiation of naïve T lymphocytes. Indeed, in addition to IL-4, IFNγ and IL-12 mentioned above, TGFβ also deserves attention (Locksley & Reiner, 1995). Therefore the subsequent steps were to compare directly, in mice of genetically different backgrounds, the production of critical cytokines in lymphoid organs during the initial period following *L. major* inoculation. This was possible to achieve due to the design of more quantitative methods by which to monitor directly mRNA transcripts or proteins, in cell suspensions, recovered from the lymphoid organs (Pannetier *et al.* 1993; Reiner *et al.* 1993; Openshaw *et al.* 1995).

Leishmania major *rapidly induces a burst of IL-4 transcripts in BALB/c mice*

The kinetics of IL-4 mRNA expression were carefully compared in BALB/c and C57BL/6 mice

during the first days of infection. In contrast to C57BL/6 mice, BALB/c mice exhibited (Fig. 1) in their draining lymph nodes, a burst of IL-4 transcripts peaking at 16 h after parasite inoculation (Launois *et al.* 1995). After returning to baseline level by 48 h, a second increase in IL-4 transcripts was observed in BALB/c mice from days 4 to 5 which remained elevated during the course of the parasitic process (Launois *et al.* 1997a) consistent with the development of the type 2 response. In C57BL/6 mice a small and gradual, but consistently 5 times lower than in BALB/c mice, production of IL-4 transcripts was observed peaking at day 4 and subsequently declining at day 7 to the baseline level of non-infected control mice (Launois *et al.* 1995). Another report has also documented the presence of IL-4 transcripts in CD4 lymphocytes recovered from lymph nodes 4 days after inoculation of *L. major*. Although in this study the *L. major* triggered a level of IL-4 transcripts which was usually higher in BALB/c than in C57BL/6 mice, the authors reported that they were similar in some experiments (Reiner *et al.* 1994). These differences in IL-4 transcript production 4 days after infection could be related to the use of *L. major* of different origins.

Is the production of IFNγ, IL-12, rapidly triggered by L. major *in mice of different genetic backgrounds? A still open issue*

Whatever the genetic background of mice inoculated with *L. major*, and in contrast to IL-4, it is difficult to document significant transcription of IFNγ in their lymph nodes during the first 5 days (Reiner *et al.* 1994; Launois *et al.* 1997a). However, the data documenting the detection of IFNγ at the protein level deserves comment (Scharton-Kersten *et al.* 1995). Lymph node cells recovered from C3H and BALB/c mice 2 days after inoculation of *L. major* were cultured in the absence of *L. major*-derived antigens, but in presence or absence of anti-IL-4, anti-IL10 and anti-TGFβ antibodies: 72 h later, IFNγ production was determined in the culture supernatants. While in the absence of anti-cytokine(s) neutralizing antibodies, C3H lymph node cells release 6000 pg/ml of IFNγ, BALB/c lymph node cells produce only 700 pg/ml. Interestingly, neutralization of TGFβ resulted in the release of larger amounts of IFNγ by BALB/c cells (≃ 13·000 pg/ml). IFNγ release was further increased in presence of the three antibodies neutralizing IL-4, IL10 and TGFβ. As far as IL-12 is concerned, the only available published data point out the delayed and low level of IL-12 p40 transcripts in both C57BL/6 and BALB/c lymph nodes as assessed from day 0 to day 7 following *L. major* inoculation (Reiner *et al.* 1994). Monitoring the 24 h production of IL-12 p40 immunoreactive protein in supernatant of BALB/c and C3H lymph node cells

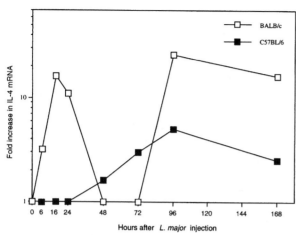

Fig. 1. Kinetics of IL-4 mRNA expression in popliteal lymph nodes following s.c. injection of *L. major* into the hind footpads. BALB/c and C57BL/6 mice were injected s.c. with 3×10^6 stationary phase *L. major* promastigotes. At various times following injection, the relative levels of IL-4 mRNA were determined by competitive RT–PCR (Reiner *et al.* 1993). Results in this figure are expressed as the increase in IL-4 mRNA in mice injected with *L. major* as compared to non-injected mice (reproduced with permission from *European Journal of Immunology* 1995, **25**, 3298).

obtained from day 1 to day 14 after parasite inoculation and cultured in absence of restimulation, a complex pattern was observed. While throughout the period under study the IL-12 p40 level increased by a factor 12 in C3H/mice, there was only a transient increase in BALB/c mice 1 day after infection ($\simeq 3$ fold) (Scharton-Kersten *et al.* 1995).

IL-12 prevents the early IL-4 burst triggered by Leishmania major

The administration of antibodies neutralizing the bioactive p40/p35 IL-12 to C57BL/6 mice not only allowed the occurrence of the first IL-4 burst, but also and as expected, the second and sustained IL-4 wave, reflecting the type 2 differentiation (Fig. 2). Together, these data suggest that IL-12 produced either constitutively or rapidly in response to *L. major* down-regulates the early burst of IL-4, thus preventing the subsequent Th2 differentiation of T lymphocytes.

Conversely, treatment of BALB/c mice with recombinant IL-12 abolished the burst of IL-4 transcription (Launois *et al.* 1995) preventing the second and sustained wave of IL-4 production (Launois *et al.* 1997*a*).

Experiments were designed to define the time during which treatment of BALB/c mice with IL-12 is efficient in preventing the development of a type 2 cytokine profile as assessed by monitoring IFNγ and IL-4 production. Results have shown that IL-12 has to be given sooner than 48 h after *L. major* inoculation (Launois *et al.* 1997*a*).

The importance of the early burst of IL-4 in directing the subsequent CD4 lymphocyte differentiation towards Th2 in BALB/c mice

The early burst of IL-4 transcripts detected in BALB/c mice lymph nodes was found to be independent of endogenous (i.e. neutralizable) IL-4. However, the subsequent and sustained IL-4 mRNA transcription by CD4 T lymphocytes reflecting Th2 commitment was found to be strictly dependent upon the IL-4 produced as the result of the early burst (16 h) of IL-4 mRNA transcription (Launois *et al.* 1997*a*).

Results from experiments designed to define the time during which the IL-4 must be present in order to drive the subsequent Th2 differentiation of CD4 T lymphocytes have revealed a narrow temporal window of less than 48 h after *L. major* inoculation (Launois *et al.* 1997*a*).

The early L. major-driven IL-4 burst rapidly induces unresponsiveness to IL-12 in BALB/c CD4 T lymphocytes

The rapid (< 48 h after *L. major* inoculation) loss of effectiveness of IL-12 in hampering Th2 commitment could be the result of the induction of IL-12 unresponsiveness in CD4 T lymphocytes. This hypothesis was validated by results from experiments showing that CD4 T lymphocytes recovered 72 hours after *L. major* inoculation do indeed loose their reactivity to IL-12 *in vitro* as assessed by IFNγ production. This acquired state of unresponsiveness to IL-12 by CD4 T lymphocytes was a direct consequence of the burst of IL-4 since it was prevented by treatment with anti-IL-4 neutralizing antibody at the initiation of infection (Launois *et al.* 1997*a*).

Using CD4 T cells specific for I-A-^{d}-restricted ovalbumin peptide obtained from α/β TCR transgenic BALB/c and B10.D2 mice, elegant studies have been performed with the aim of comparing the intrinsic tendencies of T cells from different genetic backgrounds to develop towards either Th1 or Th2 cells (Hsieh *et al.* 1995). Upon specific priming *in vitro* under neutral conditions, i.e. in the absence of exogenous cytokines, CD4$^+$ cells from α/β transgenic BALB/c mice develop a Th2 phenotype in contrast to similarly treated cells from B10.D2 mice. Interestingly, during priming, CD4$^+$ T cells from BALB/c mice lost the capacity to respond to IL-12 in terms of IFNγ production during secondary stimulation. In contrast, B10.D2 CD4$^+$ T cells remained responsive to IL-12 (Güler *et al.* 1996). Based on these results, Güler *et al.* have proposed a model of '*L. major* resistance' based on maintenance of the IL-12 signalling pathway rather than on a differential regulation of IL-4 regulation. However, our results rather suggest that the ability of *L. major* to set in,

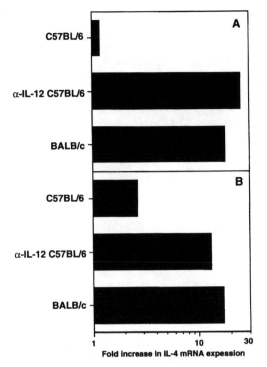

Fig. 2. Anti-IL-12 treatment results in an *L. major*-induced increase in IL-4 mRNA levels in resistant C57BL/6 mice. Mice were injected i.p. with 250 μg anti-IL-12 (polyclonal sheep anti-murine IL-12) 18 h and 2 h prior to s.c. injection of 3×10^6 *L. major*. Mice were killed 16 h (A) and 10 days (B) after injection of parasites, RNA was extracted from popliteal lymph nodes and the relative levels of IL-4 mRNA were determined by competitive RT–PCR (Reiner *et al.* 1993). For mice treated with anti-IL-12 the results are expressed as the fold increase in IL-4 mRNA in mice treated with anti-IL-12 and infected with *L. major* as compared to mice treated with the antibody but not infected with parasites.

in BALB/c mice, is primarily based on an up-regulation of IL-4 production which in turn induces IL-12 unresponsiveness in CD4 T lymphocytes. In a lymphoid environment devoid of IL-4, presumably as a result of the activity of IL-12, it is likely that CD4 T lymphocytes from C57BL/6 mice maintain their responsiveness to IL-12 resulting in their differentiation towards the Th1 phenotype.

LEISHMANIA MAJOR-DRIVEN EARLY CYTOKINE PRODUCTION: POSSIBLE CELLULAR ORIGIN

Although cell types other than CD4[+] T cells, i.e. mast cells, basophils, eosinophils, $\gamma\delta$T cells, have been reported to produce IL-4, recent results from our laboratory and the other available data (Bogdan *et al.* 1996) have clearly identified CD4[+] T cells themselves as the source of the IL-4 burst directing the subsequence Th2 cell differentiation (Launois *et al.* 1995). These CD4[+] T cells do not belong to the minor NK1.1 positive CD4[+] subset which has been

shown to produce IL-4 under other conditions (Yoshimoto & Paul, 1994; Yoshimoto *et al.* 1995). Indeed the IL-4 mRNA expression rapidly seen in BALB/c mice after infection with *L. major* did not occur in the CD4[+] T cell expressing either the Vβ8, Vβ7, or Vβ2 TCR β chains (Launois *et al.* 1995) which constitute the NK 1.1[+] CD4[+] subset (Lantz & Bendelac, 1994). Furthermore, cells from the NK1.1[+] subset were neither required for Th2 cell development nor progressive disease in BALB/c mice infected with *L. major* (Brown *et al.* 1996; von der Weid *et al.* 1996). More recent results have revealed this rapid IL-4 production, required for subsequent Th2 cell development, results from the activation, after recognition of a single antigen from *L. major*, of a restricted population of CD4[+] T cells expressing only the Vβ4-Vα8 TCR heterodimer (Launois *et al.* 1997*b*).

The other cytokines, namely IL-12 and IFNγ, could be rapidly produced and/or released by many different leucocytes once triggered by *L. major*-derived products. For IL-12, these include mononuclear phagocytes, dendritic leucocytes, neutrophils (Trinchieri, 1995). For IFNγ, these include non T leucocytes, such as NK cells, mast cells, as well as $\gamma\delta$ and $\alpha\beta$ T cells (Billiau, 1996).

CONCLUDING REMARKS

Since the first demonstration, in mice, of the crucial role of T lymphocytes in the healing or progression of lesions at the sites of *L. major* inoculation (Mitchell *et al.* 1983) critical reagents and quantitative readout assays have been developed allowing us to dissect the parameters controlling the T cell differentiation and functions. As a result, it is now possible to decipher when and how these *intracellular parasites* can exploit their host immune system to their benefit. We have described the available data indicating that *L. major* is able to trigger extremely rapidly the production of cytokines that determine the pathway of differentiation of naïve T cells of the host, ultimately conditioning the outcome of this parasite/host interplay.

The persistence of *Leishmania* spp. in their hosts, even at a time when cutaneous lesions are completely healed, represents another critical issue, particularly in relation to the transmission of these parasites. The persistence of *L. major* in resistant mice having recovered from a primary lesion has been well documented and, in this context, it has been recently shown that treatment of C57BL/6 mice with an inhibitor of the iNOS, at a time when the cutaneous lesion is resolved, results in an expansion, in the original lesion site, of those parasites which were either quiescent or replicating very slowly (Stenger *et al.* 1996). Although the subtle mechanisms allowing the persistence of *Leishmania* in immune mice are far from being elucidated, it is possible that the parasites persisting at the dermal sites in cured hosts reside in dermal macrophages continuously renewed from circulating bone-marrow-derived monocytes harbouring living parasites. Indeed, *Leishmania* spp., even those inducing pathological processes only at the level of the skin, have been detected within bone-marrow cells (Aebischer, Moody & Handman, 1993). Furthermore, among bone-marrow cells, stromal macrophages have been shown to become preferentially parasitized (Leclerq *et al.* 1996) in a murine model of infection with *L. infantum*. Since it has been shown that stromal bone marrow macrophages transiently bind monocyte precursors as well as neutrophils (Crocker *et al.* 1992), the monocyte precursors are therefore in the position of becoming randomly parasitized when contacting parasites harbouring stromal macrophages. These hypotheses will have to be carefully challenged with the proper tools and experimental approaches. If validated, such a pathway will indicate another subtle series of events pertaining to the ability of *Leishmania* spp. to exploit the immune system.

ACKNOWLEDGEMENTS

The authors' work was supported by grants from the Swiss National Science Foundation, the World Health Organization, the Roche Research Foundation, the European Union (No. TS3.CT.9403319), the Pasteur Institute and the DRET (G. Milon). We thank the Genetics Institute for providing polyclonal sheep anti-murine IL-12 antibodies.

REFERENCES

ABBAS, A. K., MURPHY, K. M. & SHER, A. (1996). Functional diversity of helper T lymphocytes. *Nature* **383**, 787–793.

AEBISCHER, T., MOODY, S. F. & HANDMAN, E. (1993). Persistence of virulent *Leishmania major* in murine cutaneous leishmaniasis: a possible hazard for the host. *Infection & Immunity* **61**, 220–226.

BEHIN, R., MAUEL, J. & SORDAT, B. (1979). *Leishmania tropica*: pathogenicity and *in vitro* macrophage function in strains of inbred mice. *Experimental Parasitology* **48**, 81–91.

BILLIAU, A. (1996). Interferon-gamma: biology and role in pathogenesis. *Advances in Immunology* **62**, 61–130.

BLACKWELL, G. G. (1996). Genetic susceptibility to leishmania infections: studies in mice and man. *Parasitology* **112**, 867–874.

BOGDAN, C., GESSNER, A., SOLBACH, W. & ROLLINGHOF, M. (1996). Invasion, control and persistence of *Leishmania* parasites. *Current Opinion in Immunology* **8**, 517–525.

BRADLEY, L. M. & WATSON, R. M. (1996). Lymphocyte migration into tissues: the paradigm derived from CD4 subsets. *Current Opinion in Immunology* **8**, 312–320.

BROWN, D. R., FOWELL, D. J., CORRY, D. B., WYNN, T. A., MOSKOWITZ, N. M., CHEEVER, A. W., LOCKSLEY, R. M. & REINER, S. L. (1996). β2-microglobulin-dependent NK1.1⁺ T cells are not essential for T helper cell 2 immune responses. *Journal of Experimental Medicine* **184**, 1295–1304.

CHAN, S. H., PERUSSIA, B., GUPTA, J. W., KOBAYASHI, M., POSPISIL, M., YOUNG, H. A., WOLF, S. F., YOUNG, D., CLARK, S. C. & TRINCHIERI, G. (1991). Induction of IFN-γ production by NK cell stimulation factor (NKSF): characterization of the responder cells and synergy with other inducers. *Journal of Experimental Medicine* **173**, 869–879.

CROCKER, P. R., KELM, S., MORRIS, L., BAINTON, D. F. & GORDON, S. (1992). Cellular interactions between stromal macrophages and haematopoietic cells. In *Mononuclear Phagocytes: Biology of Monocytes and Macrophages* (ed. R. Van Furth), pp. 55–69. Dordrecht, Kluwer Academic Publishers.

FEARON, D. T. & LOCKSLEY, R. M. (1996). The instructive role of innate immunity in the acquired immune response. *Science* **272**, 50–53.

FITCH, F. W., McKISIC, M. D., LANCKI, D. W. & GAJEWSKI, T. F. (1993). Differential regulation of murine T lymphocyte subsets. *Annual Review of Immunology* **11**, 29–48.

GULER, M. L., GORHAM, J. D., HSIEH, C.-S., MACKEY, A. J., STEEN, R. G., DIETRICH, W. F. & MURPHY, K. M. (1996). Genetic susceptibility to *Leishmania*: IL-12 responsiveness in Th1 cell development. *Science* **271**, 964–967.

HANDMAN, E. (1992). Host–parasite interactions in Leishmaniasis. *Advances in Molecular and Cellular Biology* **5**, 365–371.

HEINZEL, F. P., SADICK, M. D., MUTHA, S. S. & LOCKSLEY, R. M. (1991). Production of interferon-γ, IL-2 and IL-10 by CD4⁺ lymphocytes *in vivo* during healing and progressive murine leishmaniasis. *Proceedings of the National Academy of Sciences, USA* **88**, 7011–7015.

HSIEH, C.-S., MACATONIA, S. E., O'GARRA, A. & MURPHY, K. M. (1995). T cell genetic background determines default T helper phenotype development *in vitro*. *Journal of Experimental Medicine* **181**, 713–721.

IVENS, A. C. & BLACKWELL, J. M. (1996). Unravelling the *Leishmania* genome. *Current Opinion in Genetics and Development* **6**, 704–710.

KOPF, M., BROMBACHER, F., KOLHER, G., KIENZLE, G., WIDMANN, K.-H., LEFRANG, K., HUMBORG, C., LEDERMANN, B. & SOLBACH, W. (1996). IL-4 deficient BALB/c mice resist infection with *Leishmania major*. *Journal of Experimental Medicine* **184**, 1127–1136.

LANTZ, O. & BENDELAC, A. (1994). An invariant T cell receptor alpha chain is used by a unique subset of major histocompatibility complex class I-specific CD4⁺ and CD4⁻8-T cells in mice and humans. *Journal of Experimental Medicine* **180**, 1097–1106.

LANZARO, G. C. & WARBURG, A. (1995). Genetic variability in phlebotomine sandflies: possible implication for leishmaniasis epidemiology. *Parasitology Today* **11**, 151–154.

LANZAVECCHIA, A. (1993). Identifying strategies for immune interventions. *Science* **260**, 937–944.

LAUNOIS, P., MAILLARD, I., PINGEL, S., SWIHART, K., XENARIOS, I., ACHA-ORBEA, H., DIGGELMANN, H., LOCKSLEY, R. M. & LOUIS, J. A. (1997*b*). IL-4 rapidly produced by Vβ4Vα8 CD4⁺ T cells instructs Th2 cell development and susceptibility to *Leishmania major* in BALB/c mice. *Immunity* 6, 541–549.

LAUNOIS, P., OHTEKI, T., SWIHART, K., MacDONALD, H. R. & LOUIS, J. A. (1995). In susceptible mice, *Leishmania major* induce very rapid interleukin-4 production by CD4⁺ T cells which are NK 1.1⁻. *European Journal of Immunology* 25, 3298–3330.

LAUNOIS, P., SWIHART, K., MILON, G. & LOUIS, J. A. (1997*a*). Early production of IL-4 in susceptible mice infected with *Leishmania major* rapidly induces IL-12 unresponsiveness. *Journal of Immunology* 158, 3317–3324.

LECLERQ, V., LEBASTAD, M., BELKAID, Y., LOUIS, J. A. & MILON, G. (1996). The outcome of the parasitic process iniated by *Leishmania infantum* laboratory mice: a tissue-dependent pattern controlled by the *Lsh* and MHC loci. *Journal of Immunology* 157, 4537–4545.

LIEW, F.-Y. (1989). Functional heterogeneity of CD4⁺ T cells in leishmaniasis. *Immunology Today* 10, 40–45.

LIEW, F.-Y., MILLOTT, S., LI, Y., LELCHUCK, R., CHAN, W. L. & ZILTENER, H. (1989). Macrophage activation by interferon-γ from host protective T cells is inhibited by interleukin (IL)3 and IL-4 produced by disease-promoting T cells in leishmaniasis. *European Journal of Immunology* 19, 1227–1232.

LOCKSLEY, R. M., HEINZEL, F. P., SADICK, M. H., HOLADAY, B. J. & GARDNER, K. D. (1987). Murine cutaneous leishmaniasis: susceptibility correlates with different expansion of helper T cell subsets. *Annales de l'Institut Pasteur de Paris/Immunology* 138, 744–749.

LOCKSLEY, R. M. & REINER, S. L. (1995). Murine leishmaniasis and the regulation of CD4 T cell development. In *Molecular Approaches to Parasitology*, pp. 455–466. New York, Wiley-Liss Inc.

MARRACK, P. & KAPPLER, J. (1994). Subversion of the immune system by pathogens. *Cell* 76, 323–332.

MATTNER, F., MAGRAM, J., FERRANTE, J., LAUNOIS, P., DI PADOVA, K., BEHIN, R., GATELY, M. K., LOUIS, J. A. & ALBER, G. (1996). Genetically resistant mice lacking interleukin-12 are susceptible to infection with *Leishmania major* and mount a polarized Th2 cell response. *European Journal of Immunology* 26, 1553–1559.

MITCHELL, G. F. (1983). Murine cutaneous leishmaniasis: resistance in reconstituted nude mice and several F1 hybrids infected with *Leishmania tropica major*. *Journal of Immunogenetics* 10, 395–412.

MITCHELL, G. F., CURTIS, J. M., SCOLLAY, R. G. & HANDMAN, E. (1981). Resistance and abrogation of resistance to cutaneous leishmaniasis in reconstituted BALB/c nude mice. *Australian Journal of Experimental Biology and Medical Science* 59, 539–544.

MORRIS, L., AEBISCHER, T., HANDMAN, E. & KELSO, A. (1993). Resistance of BALB/c to *Leishmania major* infection is associated with a decrease in the precursor frequency of antigen-specific CD4⁺ cells secreting interleukin-4. *International Immunology* 5, 761–767.

MOSSMANN, T. R. & COFFMANN, R. L. (1989). Th1 and Th2

cells: different patterns of lymphokine secretion lead to different functional properties. *Annual Review of Immunology* 7, 145–173.

NACY, C. A., FORTIER, A. H., PAPPAS, M. G. & HENRY, R. R. (1983). Macrophage microbicidal activities. *Cellular Immunology* 77, 298–305.

NOBEN-TRAUTH, N., KROPF, P. & MULLER, I. (1996). Susceptibility to *Leishmania major* infection in interleukin-4-deficient mice. *Science* 271, 912–913.

O'GARRA, A. & MURPHY, K. M. (1996). Role of cytokines in development of Th1 and Th2. *Chemical Immunology* 63, 1–13.

OHARA, J. & PAUL, W. E. (1985). Production of a monoclonal antibody to and molecular characterization of B-cell stimulatory factor-1. *Nature* 315, 333–335.

OPENSHAW, P., MURPHY, E. E., HOSKEN, N. A., MAINO, V., DAVIS, K., MURPHY, K. & O'GARRA, A. (1995). Heterogeneity of intracellular cytokine synthesis at the single-cell level in polarized T helper 1 and T helper 2 populations. *Journal of Experimental Medicine* 182, 1357–1367.

PANNETIER, C., DELASSUS, S., DARCHE, S., SANCIER, C. & KOURILSKY, P. (1993). Quantitative titration of nucleic acids by enzymatic amplification reactions run to saturation. *Nucleic Acids Research* 21, 577–583.

PAUL, W. E. & SEDER, R. A. (1994). Lymphocyte responses and cytokines. *Cell* 76, 241–251.

REINER, S. L., ZHENG, S., CORRY, D. B. & LOCKSLEY, R. M. (1993). Constructing polycompetitor cDNAs for quantitative PCR. *Journal of Immunological Methods* 165, 37–46.

REINER, S. L., ZHENG, S., WANG, Z.-E., STOWRING, L. & LOCKSLEY, R. M. (1994). *Leishmania* promastigotes evade interleukin 12 (IL-12) induction by macrophages and stimulate a broad range of cytokines from CD4⁺ T cells during initiation of infection. *Journal of Experimental Medicine* 179, 447–456.

RUSSELL, D. G. (1994). Intracellular parasitism: the contrasting styles of *Leishmania* and *Mycobacterium*. In *Baillière's Clinical Infectious Diseases: Strategies for Intracellular Survival of Microbes* (Guest ed: Russell, D. G.), pp. 227–250. Baillière Tindall, London.

SACKS, D. L., LOUIS, J. A. & WRITH, D. F. (1993). Leishmaniasis. In *Immunology and Biology of Parasitic Infections* (ed. Waren, K. S.), pp. 71–86. Boston, Blackwell Scientific Publications.

SADICK, M. D., HEINZEL, F. P., HOLADAY, B. J., PU, R. T., DAWKINS, R. S. & LOCKSLEY, R. M. (1990). Cure of murine leishmaniasis with anti-interleukin 4 monoclonal antibody. *Journal of Experimental Medicine* 171, 115–127.

SCHARTON-KERSTEN, T., AFONSO, L. C. C., WYSOCKA, M., TRINCHIERI, G. & SCOTT, P. (1995). IL-12 is required for natural killer cell activation and subsequent T helper 1 cell development in experimental leishmaniasis. *Journal of Immunology* 154, 5320–5330.

SCOTT, P. (1991). IFN-γ modulates the early development of Th1 and Th2 responses in a murine model of cutaneous leishmaniasis. *Journal of Immunology* 147, 3149–3155.

STENGER, S., DONHAUSER, N., THURING, H., RÖLLINGHOFF, M. & BOGDAN, C. (1996). Reactivation of latent leishmaniasis by inhibition of inducible nitric oxide synthase. *Journal of Experimental Medicine* **183**, 1501–1514.

STENGER, S., THURING, H., RÖLLINGHOFF, M. & BOGDAN, C. (1995). Tissue expression of inducible nitric oxide synthase is closely associated with resistance to *L. major*. *Journal of Experimental Medicine* **180**, 783–793.

SWAIN, S. L. & CAMBIER, J. C. (1996). Orchestration of the immune response: multilevel regulation of divers regulatory processes. *Current Opinion in Immunology* **8**, 309–311.

SWIHART, K., FRUTH, U., MESSMER, N., HUG, K., BEHIN, R., HUANG, S., DEL GIUDICE, G., AGUET, M. & LOUIS, J. A. (1995). Mice from a genetically resistant background lacking the interferon γ receptor are susceptible to infection with *Leishmania major* but mount a polarized T helper cell 1-type CD4$^+$ T cell response. *Journal of Experimental Medicine* **181**, 961–971.

TITUS, R. G., KELSO, A. & LOUIS, J. A. (1984*a*). Intracellular destruction of *Leishmania tropica* by macrophages activated with macrophage activating factor/interferon. *Clinical and Experimental Immunology* **55**, 157–165.

TITUS, R. G., LIMA, G. C., ENGERS, H. D. & LOUIS, J. A. (1984*b*). Exacerbation of murine cutaneous leishmaniasis by adoptive transfer of parasite-specific helper T cell populations capable of mediating *L. major* specific delayed type hypersensitivity. *Journal of Immunology* **133**, 1594–1600.

TITUS, R. G., MILON, G., MARCHAL, G., VASSALI, P., CEROTTINI, J.-C. & LOUIS, J. A. (1987). Involvement of specific Lyt-2$^+$ T cells in the immunological control of experimentally induced murine cutaneous leishmaniasis. *European Journal of Immunology* **17**, 1429–1433.

TRINCHIERI, G. (1995). Interleukin-12: a proinflammatory cytokine with immunoregulatory functions that bridge innate resistance and antigen-specific adaptive immunity. *Annual Review of Immunology* **131**, 251–276.

VON DER WEID, T., BEEBE, A., ROOPENIAN, D. C. & COFFMANN, R. L. (1996). Early production of IL-4 and induction of Th2 responses in the lymph node originate from an MHC class I-independent CD4$^+$ NK1.1$^-$ T cell population. *Journal of Immunology* **157**, 4421–4427.

WANG, Z.-E., REINER, S. L., ZHENG, S., DALTON, D. K. & LOCKSLEY, R. M. (1994). CD4$^+$ effector cells default to the Th2 pathway in interferon-γ-deficient mice infected with *Leishmania major*. *Journal of Experimental Medicine* **179**, 1367–1371.

WINCKER, P., RAVEL, C., BLAINEAU, C., PAGES, M., JAUFFRET, Y., DEDET, J. P. & BASTIEN, P. (1996). The *Leishmania* genome comprises 36 chromosomes conserved across widely divergent human pathogenic species. *Nucleic Acids Research* **24**, 1688–1694.

YOSHIMOTO, T., BENDELAC, A., WATSON, C., HU-LI, J. & PAUL, W. E. (1995). Role of NK1.1$^+$ T cells in a th2 response and immunoglobulin E production. *Science* **270**, 1845–1847.

YOSHIMOTO, T. & PAUL, W. E. (1994). CD4^{pos}, NK1.1^{pos} T cells promptly produce interleukin 4 in response to *in vivo* challenge with anti-CD3. *Journal of Experimental Medicine* **179**, 1285–1295.

The relationship between immunological responsiveness controlled by T-helper 2 lymphocytes and infections with parasitic helminths

D. I. PRITCHARD[1]* C. HEWITT[1] *and* R. MOQBEL[2]

[1] *Centre for the Mechanisms of Human Toxicity, The Hodgkin Building, Lancaster Road, Leicester LE1 9HN, UK*
[2] *Pulmonary Research Group, University of Alberta, Edmonton, T6G 2S2, Canada.*

SUMMARY

It should have been difficult until relatively recently for immunologists to ascribe a sound biological reason for the continued possession of the allergic phenotype in human populations. Nevertheless, for the past 20 years or so textbooks of immunology have routinely exhibited fanciful and perhaps exaggerated diagrams as to how IgE and eosinophils killed all helminth parasites. These diagrams were largely based on perhaps selective *in vitro* observations, and it is only now that immunoparasitologists, working on human populations under arduous conditions in the field, are able to provide data to corroborate these findings, and perhaps ascribe a useful purpose for a generally pathological immune response termed Type I *hypersensitivity*. The present paper reviews much of this recent literature, and asks a number of pertinent questions relating to the relationship between what we now know to be T-helper 2 lymphocyte-driven immunological responsiveness and infections with parasitic helminths.

Key words: T-helper 2 lymphocyte, helminth infection, IgE, eosinophil, immunity, allergy, allergen, cytokine.

INTRODUCTION

For clinical reasons, and for the sake of simplifying a complex picture, the mammalian immune response was categorized decades ago into 4 types of 'hypersensitivity' (Coombs & Gell, 1963). The types of hypersensitivity (there are now 5) described the molecular and cellular components of immune responses which led in many cases to the clinical manifestation of disease, in which the immune response was over-reactive or hyper-sensitive, leading to tissue damage. The immune response which was categorized as a type 1 immediate hypersensitivity is characterized by the production of immunoglobulin ϵ or IgE following exposure to allergens, and the sensitization of mast cells or basophils for the release of vasoactive mediators on subsequent exposure to the allergenic insult. Later, T cells were shown to be pivotal in the control of IgE synthesis, and sub-sets called Th1 and Th2 counter regulate each other through the secretion of cytokines which either switch on (Th-2, IL4)† or switch off (Th1, IFN-γ) B cell IgE synthesis. B cells also require a second signal to differentiate into IgE-secreting plasma cells, and this signal is classically provided by the T cell, but can also be provided by cells of the

mast cell/basophil lineage, which also produce IL4 and IL13 (Ochensberger *et al.* 1996) (other sources of 'early' IL4 apparently include CD4+ NK 1·1+ cells, utilizing CD1 in the antigen recognition process), although there are also data to suggest that NK1·1+T cells are *not* involved in initiating Th2 responses (Brown *et al.* 1996).

This second signal is triggered by an interaction between the CD40 ligand, previously called gp (glycoprotein) 39, on the cooperative cell, and CD40 on the B cell. Once secreted, IgE attaches to the mast cell and basophil surface by interacting with its high-affinity receptor, FcϵRI. Cross-linkage of adjacent IgE molecules by allergen triggers mediator release. More recently a low-affinity receptor for IgE, called FcϵRII or CD23, was described and localized on a range of leucocyte types, including lymphocytes and eosinophils. It is through high and low affinity receptors that IgE mediates its biological and immunological activity, leading to the manifestation of allergic disease or, as is now believed, interference with the physiological activity of parasitic helminths. This latter point will be developed in the present article.

The allergic phenotype is considered to be of little benefit to humans in 'developed' countries; indeed, anti-allergic therapies targeted at the pivotal Th2 lymphocyte are justified on this basis. However, some believe that this type of immune response remains beneficial to a large proportion of the global population, people living in areas of high parasite endemicity. This statement is based on several reports in the literature which indicate that individuals best equipped to mount Th2-biased responses

---

* Corresponding author, on leave from Dept. of Life Science, University of Nottingham, University Park, Nottingham NG7 2RD.

† There is recent evidence, in mice, that class switching to IgE can be initiated, by retroviral infection, using IL-4 independent pathways (Morawetz *et al.* 1996). Helminths apparently do *not* use this pathway.

to infection are offered a degree of protection against helminths. In this context, allergy could be regarded as an evolutionary 'hangover from parasitism', and that those demonstrating a predisposition toward Type 1 hypersensitivity in relatively parasite-free societies (notwithstanding *Enterobius vermicularis*) would fare better than their non-hypersensitive neighbours if returned permanently to areas of high parasite endemicity. This would indeed be an interesting experiment to perform! However, in the absence of data from experiments of this sort, one is left to assess the current literature relating to atopy and parasitism, to decide whether allergic responses can be regarded as beneficial under certain circumstances. Indeed, the data supporting this concept are far stronger than those in favour of a parasite-protective role for Th2-driven immune responses. Thus a review based on the latter subject remains untenable in the continued presence of such sparse literature on the subject. However, in the context of host-protective Th2 responses, a number of questions can be asked, given the relative richness of the literature. (1) Do Th2 driven immune responses protect the host against parasitic helminths? (Alternatively do they protect the parasite?) (2) If Th2 responses protect the host, how are parasites damaged by these immune responses? (3) Can Th2 responses protect against tissue damage? (4) Why are parasitic helminths so allergenic? (5) How can study design be advanced/optimized to provide definitive answers?

IgE AND HUMAN HELMINTHIASES: EVIDENCE THAT TH2-DRIVEN RESPONSES ARE PROTECTIVE

The strongest evidence that allergic responses may be beneficial in areas of parasite endemicity comes from field studies of human schistosomiasis and necatoriasis (Hagan *et al.* 1991; Rihet *et al.* 1991; Dunne *et al.* 1992; Hagan, 1993; Pritchard, Quinnell & Walsh, 1995). These studies compellingly demonstrated a link between the ability to exhibit a Th2-biased response and resistance to infection or reduced parasite fitness and fecundity. For example, Hagan's studies of *S. haematobium* infection in The Gambia indicated that individuals with high levels of specific IgE against the parasite were less likely to become reinfected after successful chemotherapeutic intervention. This was an age-related phenomenon, with IgE levels and increased resistance to reinfection both associated with increasing age under conditions of equal exposure. Rihet's study of *S. mansoni* infections in Brazil, and Dunne's later study, support the link between high IgE responses and resistance to reinfection with schistosomes. Furthermore, undifferentiated Th0 cells derived from individuals with a resistant phenotype exhibited a high ratio of IL4 to IFN-γ production. This bias

may be maintained by IL10, a Th1-blocking cytokine which has been implicated in the modulation of T cell responsiveness in patients infected with *S. mansoni* (Araujo *et al.* 1996).

Additional supportive data come from a study of necatoriasis in Papua New Guinea, where significant correlations were seen between the Th2-mediated phenomena of high IgE levels and eosinophilia and reduced parasite weight and fecundity (Pritchard *et al.* 1995). Although other isotypes were shown to have anti-parasitic effects in this study (Quinnell *et al.* 1995), evidence is mounting to support the hypothesis that type-1 hypersensitivity is useful when it manifests itself in the context of parasitism. Intriguingly, there are also data which indicate that schistosomes may interfere with the activity of antigen-reactive IL4-producing T cells (Grogan *et al.* 1996), reinforcing the belief that this response is host-protective, and demonstrating that the co-evolution of host–parasite relationships continues. However, one should not ignore the fact that Th1 responses seem to be successful in controlling the invasive larval stages of some human parasites (Pritchard & Wilson, 1997). It is therefore likely that Th1 and Th2 responses act together in possibly a stage-specific manner and perhaps at different anatomical locations to control human parasitic infection.

IgE responses also seem to be more reliable and specific indicators of current infection status with hookworms (Ganguly *et al.* 1988; Pritchard & Walsh, 1995) and filarial parasites (Weiss, Hussain & Ottesen, 1982; Cabrera, Cooper & Parkhouse, 1986), indicating the additional value of this type of immune response to human medicine.

What is the biological significance of the polyclonal IgE response so often seen following helminth infection?

Some consider polyclonal IgE responses, induced by helminth parasites, to be parasite-protective (reviewed by Pritchard, 1993*a*). The concept that polyclonal IgE protects the parasite from potentially damaging specific IgE by saturating Fcε receptors with 'irrelevant' antibody is difficult to support experimentally (Moqbel & Pritchard, 1990), and explains the apparent bias of the present article, which is largely supportive of a host-protective role for Th2, and IgE responses. In fact, the existence of an excess of blocking polyclonal IgE could explain why adverse allergic reactions are rare following the chemotherapeutic treatment of systemic helminths such as schistosomes. In this instance, a polyclonal response would have to be seen as host-protective, preventing possible lethal episodes of systemic anaphylaxis following drug/immunologically-induced parasite destruction and concomitant allergen release.

Therefore, the true biological significance of polyclonal IgE is difficult to evaluate given the evidence available, and it can be seen as parasite- or host-protective, or both, depending on your viewpoint.

IF TH2-DRIVEN RESPONSES ARE HOST-PROTECTIVE, HOW IS THE PARASITE DAMAGED?

Mechanisms of damage in vitro

Studies in this area have largely concentrated on the role of eosinophils and antibodies (particularly IgE) in cytotoxicity reactions to larval stages. Initial *in vitro* experiments using the antibody-dependent cellular cytotoxicity (ADCC) reaction demonstrated that human peripheral blood granulocytes (particularly eosinophils) from patients infected with *Schistosoma mansoni* were able to adhere to and kill larvae of this trematode (Butterworth, 1984; Butterworth & Richardson, 1985; Butterworth & Thorne, 1993). These observations were confirmed using rat and human eosinophils against schistosome larvae (McLaren *et al.* 1977, 1978, 1981, 1984; Ramalho-Pinto, McLaren & Smithers, 1978). Similar antibody-dependent, eosinophil-mediated *in vitro* cytotoxicity was reported for a number of other helminth parasites. The targets employed in these systems included the eggs of *S. mansoni* using murine eosinophils (James & Colley 1976; Hsu *et al.* 1980; Feldman, Dannenberg & Seed, 1990), invading larvae of *Fasciola hepatica* (Duffus & Franks, 1980) and *M. corti* (Cook, Ashworth & Chernin, 1988). Newborn larvae of *T. spiralis*, nematode larvae that, although less susceptible because of their thick collagen-rich cuticle, were also killed following adherence to eosinophils (Kazura & Aikawa, 1980). These studies were extended by others to demonstrate eosinophil cytotoxicity *in vitro* against microfilariae of *Onchocerca volvulus* (Greene, Taylor & Aikawa, 1981; Mackenzie *et al.* 1981; Williams *et al.* 1987), larval stages of *Dictyocaulus viviparus* (Butterworth & Thorne, 1993), *Toxocara canis* (Fattah *et al.* 1986; Badley *et al.* 1987; Lombardi *et al.* 1990), *Brugia malayi* (Sim, Kwa & Mak, 1982; Chandrashekar *et al.* 1986) and *N. americanus* (Desakorn *et al.* 1987).

Eosinophils adhere readily and firmly to opsonized worms or larvae and release their granule contents which can be detected by electron microscopy as thick layers of electron-dense deposits on the surface of the organism (Mclaren *et al.* 1977, 1978, 1981; Glauert *et al.* 1978; Caulfield, 1980; Caulfield *et al.* 1980). Such deposition results in damage to the parasite in the vicinity of contact with the attached eosinophil. Damage was associated with the appearance of vacuoles in the syncytial tegument of the larvae and was followed by detachment of the tegumental membrane leading to the exposure of the underlying muscle layers. It has been suggested that tegumental membrane detachment may be mediated in part by worm-derived lysophospholipids (Golan *et al.* 1986; Furlong & Caulfield, 1989; Caulfield & Chiang, 1990) and by released eosinophil granule proteins including EPO, MBP and ECP. These products have the capacity to damage schistosomula directly when incubated as isolated proteins with parasitic larvae at very low molar concentrations. MBP and ECP both produced ballooning in the tegument in a similar pattern to that observed with whole eosinophils (Butterworth *et al.* 1979*a*, *b*; McLaren *et al.* 1984). ECP was 10 times more active on a molar basis than MBP. However, because MBP is present in the granule in larger amounts, it may account for a higher proportion of the toxicity observed (Ackerman *et al.* 1985). That eosinophil cationic proteins induce damage to schistosomula of *S. mansoni* through their basic charge was confirmed by use of a variety of synthetic polycations (Butterworth *et al.* 1979*b*; Jones, Helms & Kusel, 1988) and its inhibition by polyanions such as heparin (Young *et al.* 1986). Purified eosinophil cationic proteins were also toxic *in vitro* for the newborn larvae of *T. spiralis* (Wassom & Gleich, 1979; Hamann *et al.* 1987), the eggs of *S. mansoni* (Sher *et al.* 1980; Kephart, Andrade & Gleich, 1988) and larvae of both *B. malayi* and *B. Pahangi* (Hamann *et al.* 1990). These cytotoxic effects may, therefore, be due primarily to the intensely basic nature of these proteins rather than to other properties such as the ribonuclease activity of ECP (Barker *et al.* 1989). This is particularly relevant, since worm tegument and cuticle were shown to be strongly polyanionic (Pritchard *et al.* 1985). Recent evidence suggests that only 1 isoform of ECP, out of a total of 7, possesses any cytotoxic activity (P. Venge, personal communication). In contrast, EPO and EDN are relatively inactive on their own in causing direct damage to the parasite (Pincus *et al.* 1981; Ackerman *et al.* 1985). EPO action was enhanced in the presence of hydrogen peroxide and a halide. The effects of these polycations were inhibited by polyanions such as heparin (Gleich *et al.* 1980; Venge *et al.* 1983).

Although eosinophils can release higher concentrations of superoxide radicals than neutrophils in response to stimulation, Pincus *et al.* (1981) demonstrated that damage to schistosomula by normal eosinophils can occur under strictly nonaerobic conditions, suggesting that such oxidative metabolism is not necessary for *in vitro* killing. However, oxygen may be required for degranulation (Baskar & Pincus, 1988). Eosinoplasts (experimentally) generated granule-containing eosinophils devoid of a nucleus) elaborated oxygen metabolites and synergized with ECP in helminth toxicity (Yazdanbakhsh *et al.* 1987). Oxidative mechanisms also appear to be essential in killing of newborn larvae of *T. spiralis* by eosinophils (Bass & Szejada, 1979; Buys *et al.* 1981, 1984) while a reduction in oxygen tension limited the capacity of intact eosinophilic granulomas or isolated granuloma cells to kill eggs of *S. mansoni in vitro* (Feldman *et al.* 1990). Adherence of eosinophils to both live and fixed *S. mansoni* larvae induced *de novo* synthesis and release of lipid mediators including LTC$_4$ (Moqbel *et al.* 1990). The precise role of this mediator in parasite damage is not yet known. The ligands that mediate killing by both normal and activated eosinophils include immunoglobulins, particularly IgG, IgA and IgE. In addition, complement components C3b and C3bi were shown to facilitate adherence and killing in the absence of any immunoglobulin and this may be achieved through eosinophil adherence by CR1 and CR3 respectively (Anwar, Smithers & Kay, 1979; Moqbel *et al.* 1983; Fischer *et al.* 1986). Antibody-dependent killing may also be enhanced by LFA-1 associated mechanisms. Monoclonal antibodies against the α-chain of this β-1 integrin partly blocked the killing of *S. mansoni* larvae (Capron *et al.* 1987). That eosinophils utilize the receptor for IgA (Abu-Ghazaleh *et al.* 1989) to induce eosinophil degranulation (following incubation with either anti-IgA-coated sepharose beads, particularly sIgA (Fujisawa *et al.* 1990)) suggests that this ligand is an important receptor for mediator release. However, the involvement of sIgA in eosinophil-mediated cytotoxic response against parasitic helminthic targets is not yet established.

In vivo *correlates of* in vitro *damage*

The precise regulatory and functional roles of IgE and eosinophils in the progression of human helminthiases is unknown. Information in man is largely limited to measurements of blood and tissue eosinophilia and IgE during the migration of helminth(s) in various tissue sites (Wardlaw & Moqbel, 1992). There is some evidence of direct contact between eosinophils and adult worms during infections. Eosinophil-rich granulomas surrounding dead fragments of skin invading larvae of *Strongyloides ratti* were found in hyper-immune rats after challenge with infective larvae (Moqbel, 1980). Eosinophils were also found in close contact with the surface tegument of schistosomula of *S. haematobium* in the cutaneous tissue of immune monkeys. This was associated with the presence of large numbers of dead larvae in eosinophil-rich sites (Hsu *et al.* 1980). Similar observations were made in other host/parasite systems (Gleich & Adolphson, 1986; Butterworth & Thorne, 1993). Using appropriate antibodies, eosinophil-derived toxic proteins were identified on worm targets, *in vivo*. Immunofluorescent staining for MBP revealed the deposition of this eosinophil-derived product onto the surface of microfilariae of *Onchocerca volvulus* in skin biopsies of patients with onchocerciasis following treatment

with diethylcarbamazine (Kephart *et al.* 1988). The presence of an eosinophilic infiltrate in association with human onchocerciasis was shown to be correlated with microfilarial production from pregnant female adult worms but not with the host's immune status (Wildenburg *et al.* 1995). Adult *O. volvulus* elicited tissue eosinophilia only if microfilariae appear in the surrounding tissue. The levels of blood ECP were elevated in patients with filariasis suggesting the activation and degranulation of eosinophils (Spry, 1981). The rate of reinfection in African children with *S. haematobium* indicated that both IgE and eosinophils appear to influence resistance in that age group (Hagan *et al.* 1991; Woolhouse *et al.* 1991). Thus, much of our existing knowledge about the possible *in vivo* role of eosinophils in helminth-induced inflammation arises from studies in laboratory animals, although data are being gradually obtained on the sequence of events that may regulate human eosinophil-mediated responses to helminths (Mahanty *et al.* 1992).

While rat and human eosinophils have been shown to possess IgE-dependent anti-parasitic effector functions both *in vivo* and *in vitro*, BAL eosinophils from lungs of mice infected with *T. canis* were devoid of receptors for sIgM, sIgA, sIgE, but were positive for receptor for sIgG1, although FcϵRII and FcγRII were absent in mouse eosinophils (Jones *et al.* 1994). This suggests that there is a heterogeneity in the profile of eosinophils in different host species thus raising doubts concerning the role of eosinophils in helminth disease. It has been suggested that the presence of Th2-type responses (i.e. IgE-dependent mechanisms with its associated eosinophilia) to helminth infection may either contribute to host protection or lead to prolonged parasite survival. This appears particularly true in mouse models of schistosomiasis, in which the presence of IFN-γ and IgG have a more prominent protective role than IL-4 and IL-5, at least against larval stages of infection.

Reassessment of the leak–lesion hypothesis

Given that components of a Th2 driven response can be shown to damage parasite larvae *in vitro*, and that some *in vivo* correlates can be demonstrated, it is appropriate to reassess briefly the merits of the 'leak–lesion' hypothesis. This hypothesis was first mooted in the late 1960s, in an attempt to explain how type I hypersensitivity responses could effect parasite expulsion. Central to the hypothesis is the ability of IgE to sensitize mast cells and basophils specifically for the release of vasoactive amines, increasing vascular permeability and consequently flooding any site of parasitic challenge with blood rich in potentially host-protective cells and molecules. This phenomenon is most dramatically demonstrated following *Nippostrongylus brasiliensis* infection in rats, where the challenge of a sensitized small intestine results in 'gut shock', which is manifested by the engorgement of the tissues with vascular components. This type of reaction would certainly make the gut environment inhospitable to parasite colonization, and inflammatory responses are certainly implicated in worm expulsion in other systems, e.g. *Trichinella spiralis* (Wakelin, 1993). Also mast cell proteinases are known to digest nematode cuticle collagens (McKean & Pritchard, 1989) suggesting that mast cell degranulation and associated proteinase release could have a direct and physically damaging effect on the helminth. Such reactions would also be beneficial in anatomical locations such as the skin following challenge with hookworm larvae. Anti-larval IgG responses appear to be beneficial in necatoriasis, and occur in tandem with IgE responses (Quinnell *et al.* 1995). The leakage of plasma and inflammatory cells onto the site of parasite invasion, mediated by a type 1 response (the cause of 'ground itch'?) should be beneficial to the host; but, as we know, the parasite seems to have evolved evasion strategies (Hotez & Pritchard, 1995). The 'leak–lesion' hypothesis does not command overwhelming support (Ahmad, Wang & Bell, 1991). Nevertheless, the immunological components for leakage are certainly induced by infection and would be activated upon exposure to parasite allergens. It is certainly a hypothesis that warrants resurrection. Readers are directed to a recent review by Bell (1996) who expresses some interesting views on the role of IgE in helminth infection.

CAN TH2 RESPONSES PROTECT THE HOST FROM PATHOLOGY BY SUPPRESSING POTENTIALLY TISSUE-DAMAGING HOST IMMUNE RESPONSES?

As T cell responses can counter-regulate each other, depending on the balance of cytokines in the milieu surrounding the interactive components (Abbas, Murphy & Sher, 1996; Anderson & Coyle, 1994), it is conceivable that Th2-driven responses may suppress the potentially tissue-damaging Th1-driven responses, known to occur in some parasitic infections. The prime manifestation of this type of interaction would seem to occur in lymphatic filariasis, where microfilariaemic yet asymptomatic and hyporesponsive individuals demonstrate Th2 responses to crude antigen preparations, although the literature on this subject matter is highly confusing and apparently contradictory in places. Nevertheless, because these individuals are considered to be tolerating the infection, they are described as being in an antigen-specific anergic state. When anergy is overcome, and Th1 responses come to the fore, the state of hypersensitivity of the patient would appear to swing from the immediate (Type 1) to the delayed (Type IV) phenotype,

leading to the clearance of microfilariae, but the development of clinical disease (chronic lymphatic obstruction).

There is also evidence that IgG4, an isotype also under the control of Th2 cytokines and able to compete with IgE for both allergen epitopes and Fcε receptors, may play an important role in modulating IgE-mediated allergic responses *in vivo*. IgG4 is, thus, often described as a blocking antibody, and it is becoming apparent that it may act, in micro-filariaemic patients, to suppress the manifestation of TPE or tropical pulmonary eosinophilia (reviewed by King & Nutman, 1993). Consequently, individuals predisposed to mount Th2-biased responses in areas of high endemicity for lymphatic filariasis, particularly those with a high IgG4 to IgE ratio, could be considered to be less prone to the pathological sequelae of infection. In this sense, the Th2 response would be seen as beneficial to the individual.

WHY ARE HELMINTH INFECTIONS SO ALLERGENIC?

Helminth parasites are particularly adept at stimulating IgE synthesis, and a number of parasite allergens have recently been cloned and characterized (McReynolds, Kennedy & Selkirk, 1993). Do certain parasite antigens have molecular properties which support their allergenicity? This is a question which has puzzled immunologists for decades and a number of theories have been put forward. One popular theory at present is based on the fact that many allergens have enzymic activity (e.g. *Der p* I from *Dermatophagoides pteronyssinus*), and that enzymic activity somehow deviates antigen processing in a way that supports the generation of Th2 lymphocytes. It is, therefore, of interest that highly allergenic parasites such as schistosomes and hookworms are rich sources of secreted proteinases (Smith *et al.* 1994; Brown *et al.* 1995). How do these enzymes promote IgE synthesis? One clue to a possible mechanism comes from an observation made in a population in Papua New Guinea infected with the hookworm *N. americanus*. This population was noted to have exceedingly high levels (on average, 40 times that of normal) of soluble CD23 (the low-affinity receptor for IgE) and IgE in its plasma (up to 17000 IU/ml) (Pritchard, Kumar & Edmonds, 1993). This observation led to the hypothesis that parasite proteases might accelerate the natural proteolytic cleavage of CD23 from the leucocyte surface, leading to an upregulation of immediate-type hypersensitivity through the generation of soluble CD23 (a pro-allergic cytokine) and the removal of a negative feedback signal for IgE synthesis (Pritchard, 1993*a*). IgE-containing immune complexes bind to CD23, to send a feedback inhibition signal to the B cell, which reduces IgE synthesis (reviewed by Delespesse *et al.*

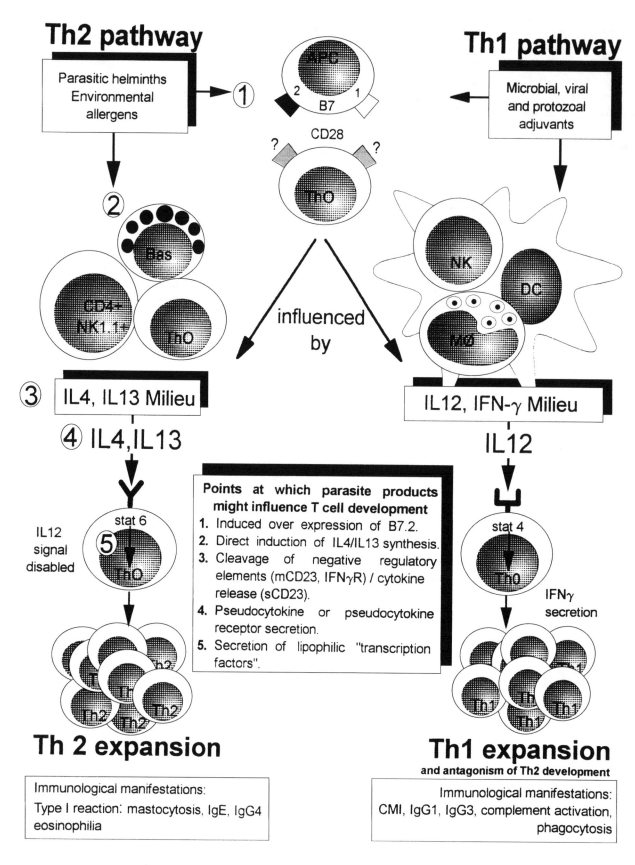

Fig. 1. T cell development can be influenced by a number of factors in the immunological environment. The points in T cell development where parasitic helminths and their ES products could conceivably exert an influence are shown here. For example, it is possible that helminth ES, particularly secreted proteinases and proteinase inhibitors (doubling as anticoagulants?), could influence the proteinase-rich antigen processing milieu to an extent that a pathway favouring Th2 development is selected. The mechanism by which CD28 controls IgE synthesis is sketchy. Some data support a positive role for CD28 (King *et al.* 1996) through an interaction with the B7 marker (Corry

1992). Consequently, allergens which selectively cleave CD23 would theoretically be pro-allergic. It is therefore of interest that *Der p* I, a proteolytically active allergen of the house mite *D. pteronyssinus*, selectivity cleaves CD23 from the human B cell surface (Hewitt *et al.* 1995). This is the first and most convincing experimental evidence that dust mites dysregulate the IgE network by directly interfering with the control of IgE synthesis; these observations have since been confirmed by other workers.

This explanation may not hold for all allergens. For example, many fatty acid binding proteins (FABP) are now proven allergens (McReynolds *et al.* 1993). However, it provides an experimental basis for testing the pro-allergic activity of at least a major group of common environmental toxins. Given the ability of FABPs to sequester vitamins, and the fact that some vitamins (e.g. D3) promote Th-2 development (Rook, Hernandez-Pando & Lightman 1994), it is possible that some allergens promote the allergic response through selective vitamin uptake. A résumé of recent developments in our understanding of the development of helper T cells at this point wold help to indicate other points in the development processes where parasites and their products could exert influence on Th-2 bias.

Recent advances in our understanding of the development of helper T cell subsets

The early cytokine milieu seems to be of critical importance to the differential development of T cells. Both Th1 and Th2 cells can develop from the same T cell precursor, with IL-12 from macrophages and activated dendritic cells the predominant Th1 promoting cytokine, and Il-4, possibly from basophils, T cells themselves or CD4$^+$ NK1·1$^+$, equivalents in man, promoting Th2 development. However, a number of caveats exist. Although IL-4-independent pathways of class-switching for IgE synthesis have been demonstrated, initiated by retroviral infection (Morawetz *et al.* 1996), nematode infection does not seem to influence this pathway.

Also there are data to suggest that β_2 microglobulin-independent NK1·1$^+$ T cells are not involved in Th2 responses (Brown *et al.* 1996).

The downstream effects of these cytokines on further T cell development would appear to be mediated through a number of transcription factors belonging to the *stat* family. Of these, *stat* 4 would appear to be activated by IL-12 (with IFNγ playing a supportive role), and *stat* 6 would appear to be activated by IL-4 (Kaplan *et al.* 1996 *a*, *b*). The other points in T cell development in which parasite products would be active are illustrated (in Fig. 1) and described below. Table 1 lists known and putative helminth secretions which could polarize the development of T cells.

Are there other possible explanations for the allergenicity of helminth parasites?

The CD23 effect described would be largely mediated through B cells committed to IgE synthesis and it has already been suggested that helminth parasites may promote their own survival by stimulating polyclonal IgE synthesis, although clear evidence for this hypothesis is still being found. As well as cleaving CD23 to promote IgE synthesis, it is also possible that parasitic helminths secrete molecules which act as pseudocytokines, to maintain the Th2 bias. Molecules with the activities of IL-4, Il-10 and sCD23 would fulfil these criteria.

Alternatively, if Th2 responses, in a specific sense, begin to damage the parasite, as has been inferred from a number of field studies described above, then it would be in the interest of the parasite to induce a Th1 bias. This could be done through the secretion of Th1-promoting pseudocytokines such as IFNγ and IL-12. This may indeed be the case in *Trichuris muris* infection, (see Grencis and Entwistle, this volume). Work by Birgit Helm and her colleagues (Dudler *et al.* 1995; Machado *et al.* 1996) would also suggest that enzymically active allergens induce IL-4 synthesis by cells of the basophil lineage in the absence of sensitization with IgE. This kind of

---

et al. 1994), but it may not always stimulate through B7 (Life *et al.* 1995). Other data suggest that co-stimulation of T cells via CD28 inhibits IgE production (Van der Pouwkraan *et al.* 1996). Interestingly, the latter effect was reversed by pertussis toxin. Proteinases could also effect IL-4 secretion directly, in a non-antigenic fashion, in a manner similar to that already described for other enzymically active allergens. It is also already known that another proteolytically active environmental allergen (*Der p* I) has the ability to dysregulate the IgE network; formal proof demonstrating a similar effect by helminth proteases is lacking, although the suggestion that they might potentiate IgE synthesis by cleaving molecules such as CD23 was first muted in the early 90s (Pritchard, 1993 *a*). The pro Th2 developmental pathway would then be propagated by the possible release of cytokines (IL-4, sCD23) or secretion of pseudocytokines or pseudocytokine receptors (e.g. IFN-γR) and even lipophilic homoserine lactones similar to those already shown to be active in chemical communication between bacteria (Williams, 1994). Another parasite secretion with the potential to modulate T cell development is acetylcholinesterase. Cholinergic influences on T cell development/activation have been recorded, yet it remains to be established whether extraneous sources of AChE, which would compete with T cell membrane AChE, exert any polarizing effect on T cell development (Pritchard, 1993 *b*). Abbreviations: APC, antigen presenting cell; BAS, basophil; DC, dendritic cell; ES, excretory/secretory product; MO, macrophage; NK, natural killer; ThO, undifferentiated T cell; STAT, signal transducer and activator of transcription.

Table 1. *Examples of known and putative parasite secretions with the potential to support the development and growth of Th2 lymphocytes*

ES product	Putative pro-Th2 effect
Proteinases e.g. cysteinyl proteinases, glycosidases, lipases (cf. phospholipase A2)	(i) cleavage of negative regulatory elements (CD23, IFN-γR/IL-12R); (ii) release of cytokines (sCD23) and 'anti-cytokines' (sIFN-γR/IL-12R) (iii) direct induction of IL-4 and IL-13 synthesis.
Proteinases/inhibitors e.g. serpins	Interference with cytosolic pathways of antigen processing/presentation.
Acetylcholinesterase	Interference with muscarinic (+ve) and nicotinic (−ve) influences on T cell function, and T-cell mAChE.
Pseudocytokines/receptors/ligands	Supportive of IL-4 milieu, sequestration of negative regulatory cytokines. Presentation of CD40 ligand-like molecules.
Superantigens	Induction of CD40 ligand expression and class switching to IgE.
Mitogens	Expansion of IL-4-receptive B (as found in *Ascaris* body fluid) and T cell sub-populations.
Protein kinases	Deviation of developmental pathways to Th-2 via intracellular signalling processes.
Lipophilic 'transcription factors'	Conversion of cytosolic regulating proteins into transcription factors.

direct activity would generate the early IL-4 milieu necessary for Th2 development. It is also intriguing that a number of pathogenic and non-pathogenic bacteria signal to each other through the secretion of homoserine lactones, which are lipophilic and enter cells to interact with proteins; the complex in turn becomes a transcription factor (Williams, 1994). Could parasitic helminths do the same? B cell mitogens which support the growth and division of ε B cells would also be important (Lee, 1995). Finally, superantigens can induce CD40 ligand expression, to modulate class-switching to IgE (Jabara & Geha, 1996). The possibility that helminths secrete superantigens should be fully explored. A full list of parasite molecules with the putative ability to control T cell development is shown in Table 1.

A PROSPECTIVE VIEW

When you ask the question, 'does a Th2 response protect against helminth infection?', you are almost asking whether atopics are protected against nematodes. These are thorny questions, which are difficult to answer in a definitive fashion (Moqbel & Pritchard, 1990) because many of the studies designed to answer these questions have been incomplete. How can study design be improved? The simplest experiments would involve the deliberate trickle infection of large numbers of human volunteers exhibiting a range of abilities to mount type 1 responses, or the transfer of such individuals from areas of low to areas of high parasite endemicity. Failing this, high-quality longitudinal studies should be conducted in which children of known atopic predisposition (judged by a combination of skin tests, IgE levels and symptomology) living in areas of high parasite endemicity and exhibiting similar behavioural profiles are followed over a 15-year period. For example, hookworm infection is acquired in a cumulative form from 3 years to adolescence, where infection intensity

levels off (Pritchard *et al.* 1990). The assessment of atopic status would be supported by genetic analysis. However it is likely that atopy/asthma has a polygenic basis, involving genes in the 5q31–q33 region, the 11q13 region and other genetic loci (Marsh *et al.* 1994; Herwerden *et al.* 1995; Levitt & Holroyd *et al.* 1995; Daniels *et al.* 1996). We currently have a collection of DNA samples from hookworm infected individuals under analysis in the context of total IgE levels induced by infection, but larger-scale individual studies will be required to produce definite answers.

ACKNOWLEDGEMENTS

The authors are indebted to Barbara Williams for her excellent secretarial support over the years, to the Wellcome Trust for their generous and consistent financial support, to an anonymous referee for a very useful critique, and to Ms Mari Nowell for the artwork. Redwan Moqbel is an Alberta Heritage Medical Senior Scholar.

REFERENCES

ABBAS, A. K., MURPHY, K. M. & SHER, A. (1996). Functional diversity of helper T lymphocytes. *Nature* **383**, 787–793.

ABU-GHAZALEH, R. I., FUJISAWA, T., MESTECKY, J., KYLE, R. A. & GLEICH, G. J. (1989). IgA-induced eosinophil degranulation. *Journal of Immunology* **142**, 2393–2400.

ACKERMAN, S. J., KEPHART, G. M., HABERMAN, T. M., GRIEPP, P. R. & GLEICH, G. J. (1985). Comparative toxicity of purified human eosinophil granule cationic proteins for schistosomula of *Schistosoma mansoni*. *American Journal of Tropical Medicine and Hygiene* **34**, 735–745.

AHMAD, A., WANG, C. H. & BELL, R. G. (1991). A role for IgE in intestinal immunity. Expression of rapid expulsion of *Trichinella spiralis* in rats transfused with IgE and thoracic duct lymphocytes. *Journal of Immunology* **146**, 3563–3570.

ANDERSON, G. P. & COYLE, A. J. (1994). T$_H$2 and 'T$_H$2-like' cells in allergy and asthma: pharmacological perspectives. *Trends in Pharmacological Sciences* **15**, 324–332.

ANWAR, A. R. E., SMITHERS, S. R. & KAY, A. B. (1979). Killing of schistosomula of *Schistosoma mansoni* coated with antibody and/or complement by human leucocytes *in vitro*: requirement for complement in preferential killing by eosinophils. *Journal of Immunology* **122**, 628–637.

ARAUJO, M. I., RIBEIRO DE JESUS, A., BACELLAR, O., SABIN, E., PEARCE, E. & CARVALHO, E. M. (1996). Evidence of a T helper type 2 activation in human schistosomiasis. *European Journal of Immunology* **26**, 1399–1403.

BADLEY, J. E., GRIEVE, R. B., ROCKY, J. G. & GLICKMAN, L. T. (1987). Immune mediated adherence of eosinophils to *Toxocara canis* infective larvae: the role of excretory–secretory antigens. *Parasite Immunology* **9**, 133–143.

BARKER, R. L., LOEGERING, D. A., TEN, R. M., AMANN, K. J., PEASE, L. R. & GLEICH, G. J. (1989). Eosinophil cationic protein cDNA. Comparison with other toxic cationic proteins and ribonucleases. *Journal of Immunology* **143**, 952–955.

BASKAR, P. & PINCUS, S. H. (1988). Oxidative requirements for degranulation of human peripheral blood eosinophils. *Infection and Immunity* **56**, 1907–1911.

BASS, D. A. & SZEJADA, P. (1979). Mechanisms of killing of newborn larvae of *Trichinella spiralis* by neutrophils and eosinophils. Killing by generators of hydrogen peroxide *in vitro*. *Journal of Clinical Investigation* **64**, 1558–1564.

BELL, R. G. (1996). IgE, allergies and helminth parasites: A new perspective on an old conundrum. *Immunology and Cell Biology* **74**, 337–345.

BROWN, A., BURLEIGH, J. M., BILLETT, E. E. & PRITCHARD, D. I. (1995). An initial characterization of the proteolytic enzymes secreted by the adult stage of the human hookworm *Necator americanus*. *Parasitology* **110**, 555–563.

BROWN, D. R., FOWELL, D. J., CORRY, D. B., DYNN, T. A., MOSKOQUITZ, N. H., CHEEVER, A. W., LOCKSLEY, R. M. & REINER, S. L. (1996). β_2 microglobulin independent NK1·1⁺ T cells are not essential for T helper cell and immune responses. *Journal of Experimental Medicine* **184**, 1295–1304.

BUTTERWORTH, A. E. (1984). Cell-mediated damage to helminths. *Advances in Parasitology* **23**, 143–235.

BUTTERWORTH, A. E. & RICHARDSON, B. A. (1985). Factors affecting the levels of antibody- and complement-dependent eosinophil-mediated damage to schistosomula of *Schistosoma mansoni in vitro*. *Parasite Immunology* **7**, 119–131.

BUTTERWORTH, A. E. & THORNE, K. J. I. (1993). Eosinophils and parasitic diseases. In *Immunopharmacology of Eosinophils* (ed. Smith, H. & Cook, R. M.), pp. 119–150. London: Academic Press.

BUTTERWORTH, A. E., WASSOM, D. L., GLEICH, G. J., LOEGERING, D. A. & DAVID, J. R. (1979a). Interactions between human eosinophils and schistosomules of *Schistosoma mansoni*. II. The mechanisms of irreversible eosinophil adherence. *Journal of Experimental Medicine* **150**, 1456–1471.

BUTTERWORTH, A. E., VADAS, M. A., WASSOM, D. L., DESSEIN, A., HOGAN, M., SHER, A. & GLEICH, J. (1979b).

Damage to schistosomula of *Schistosoma mansoni* induced directly by eosinophil major basic protein. *Journal of Immunology* **122**, 221–229.

BUYS, J. WEVER, R. & RUITENBERG, E. J. (1984). Myeloperoxidase is more efficient than eosinophil peroxidase in the *in vitro* killing of newborn larvae of *Trichinella spiralis*. *Immunology* **51**, 601–607.

BUYS, J., WEVER, R., VAN STIGT, R. & RUITENBERG, E. J. (1981). The killing of newborn larvae of *Trichinella spiralis* by eosinophil peroxidase, *in vitro*. *European Journal of Immunology* **11**, 843–845.

CABRERA, Z., COOPER, M. D. & PARKHOUSE, R. M. E. (1986). Differential recognition patterns of human immunoglobulin classes to antigens of *Onchocerca gibsoni*. *Tropical Medicine and Parasitology* **37**, 113–116.

CAPRON, M., KAZATHCHKINE, M. D., FISCHER, E., JOSEPH, M., BUTTERWORTH, A. E., KUSNIERZ, J.-P., PRIN, L. & PAPIN, J. (1987). Functional role of the alpha-chain of complement receptor type 3 in human eosinophil-dependent antibody-mediated cytotoxicity against schistosomes. *Journal of Immunology* **139**, 2059–2065.

CAULFIELD, J. P. (1980). The adherence of human neutrophils and eosinophils to schistosomula: evidence for membrane fusion between cells and parasites. *Journal of Cell Biology* **86**, 46–63.

CAULFIELD, J. P. & CHIANG, C. P. (1990). How does the schistosome evade host defenses? *Gastroenterology* **98**, 1712–1713.

CAULFIELD, J. P., KORMAN, G., BUTTERWORTH, A. E., HOGAN, M. & DAVID, J. R. (1980a). Partial and complete detachment of neutrophils and eosinophils from schistosomula: evidence for the establishment of continuity between a fused and normal parasite membrane. *Journal of Cell Biology* **86**, 64–76.

CHANDRASHEKAR, R., RAO, U. R., PARAB, P. B. & SUBRAHMANYAN, D. (1986). *Brugia malayi*: rat cell interactions with infective larvae mediated by complement. *Experimental Parasitology* **62**, 362–369.

COOK, R. M., ASHWORTH, R. F. & CHERNIN, J. (1988). Cytotoxic activity of rat granulocytes against *Mesocestoides corti*. *Parasite Immunology* **10**, 97–109.

COOMBS, R. R. A. & GELL, P. G. H. (1963). The classification of allergic reaction underlying disease. In *Clinical Aspects of Immunology* (ed. Gell, P. G. H. & Coombs, R. R. A.), pp. 317–337. Oxford: Blackwell Scientific Publications.

CORRY, D. B., REINER, S. L., LINSLEY, P. S. & LOCKSLEY, R. M. (1994). Differential effects of blockage of CD28–B7 on the development of Th1 and Th2 effector cells in experimental leishmaniasis. *Journal of Immunology* **153**, 4142–4148.

DANIELS, S. E., BHATTACHARRYA, S., JAMES, A., LEAVES, N. I., YOUNG, A., HILL, M. R., FAUX, J. A., RYAN, G. F., LE SÖUEF, P. N., LATHROP, G. M., MUSK, A. W. & COOKSON, W. O. C. M. (1996). A genome-wide search for quantitative trait loci underlying asthma. *Nature* **383**, 247–250.

DELESPESSE, G., SARFATI, M., WU, C. Y., FOURNIER, S. & LETELLIER, M. (1992). The low affinity receptor for IgE. *Immunology Review* **125**, 77–97.

DESAKORN, V., SUNTHARASAMAI, P., PUKPITTAYAKAMEE, S., MIGASENA, S. & BUNNAG, D. (1987). Adherence of human eosinophils to infective filariform larvae of *Necator americanus in vitro. Southeast Asian Journal of Tropical Medicine and Public Health* **18**, 66–72.

DUDLER, T., CANTARELLI-MACHADO, D., KOLBE, L., ANNAND, R. R., RHODES, N., GELB, M. H., KOELSCH, E., SUTER, M. & HELM, B. A. (1995). A link between catalytic activity, IgE-independent mast cell activation, and allergenicity of bee venom phospholipase A_2. *Journal of Immunology* **155**, 2605–2613.

DUFFUS, W. P. & FRANKS, D. (1980). *In vitro* effect of immune serum and bovine granulocytes on juvenile *Fasciola hepatica. Clinical and Experimental Immunology* **41**, 430–440.

DUNNE, D. W., BUTTERWORTH, A. E., FULFORD, A. J. C., KARIUKI, H. C., LANGLEY, J. G., OUMA, J. H., CAPRON, A., PIERCE, R. J. & STURROCK, R. F. (1992). Immunity after treatment of human schistosomiasis: association between IgE antibodies to adult worm antigens and resistance to reinfection. *European Journal of Immunology* **22**, 1483–1494.

FATTAH, D. I., MAIZELS, R. M., McLAREN, D. J. & SPRY, C. J. F. (1986). *Toxocara canis*: interaction of human blood eosinophils with the infective larvae. *Experimental Parasitology* **61**, 421–431.

FELDMAN, G. M., DANNENBERG, A. M. & SEED, J. L. (1990). Physiology oxygen tensions limit oxidant-mediated killing of schistosome eggs by inflammatory cells and isolated granulomas. *Journal of Leukocyte Biology* **47**, 344–354.

FISCHER, E., CAPRON, M., PRIN, L., KUSNIERZ, J.-P. & KAZATCHKINE, M. D. (1986). Human eosinophils express CR1 and CR3 complement receptors for cleavage fragments of C3. *Cellular Immunology* **97**, 297–306.

FUJISWA, T., ABU-GHAZALEH, R., KITA, H., SANDERSON, C. J. & GLEICH, G. J. (1990). Regulatory effect of cytokines on eosinophil degranulation. *Journal of Immunology* **144**, 642–646.

FURLONG, S. T. & CAULFIELD, J. P. (1989). *Schistosoma mansoni*: synthesis and release of phospholipids, lysophospholipids, and neutral lipids by schistosomula. *Experimental Parasitology* **69**, 65–77.

GANGULY, N. M., MAHAJAN, R. C., SEHGAL, R., SHETTY, P. & DILAWARI, J. B. (1988). Role of specific immunoglobulin E to excretory–secretory antigen in diagnosis and prognosis of hookworm infection. *Journal of Clinical Microbiology* **26**, 737–742.

GLAUERT, A. M., BUTTERWORTH, A. E., STURROCK, R. G. & HOUBA, V. (1978). The mechanism of antibody-dependent, eosinophil-mediated damage to schistosomula of *Schistosoma mansoni in vitro*: a study by phase-contrast and electron microscopy. *Journal of Cell Science* **34**, 173–192.

GLEICH, G. J. & ADOLPHSON, C. R. (1986). The eosinophil leukocyte: structure and function. *Advances in Immunology* **39**, 177–253.

GLEICH, G. J., LOEGERING, D. A., FRIGAS, F., WASSOM, D. L., SOLLEY, G. O. & MANN, K. (1980). The major basic protein of the eosinophil granule: physicochemical properties, localization and function. In *The Eosinophil in Health and Disease* (ed. Mahmoud, A. A. F. & Austen, K. F.); pp. 79–94. New York: Grune Stratton.

GOLAN, D. E., BROWN, C. S., GIANCI, C. M., FURLONG, S. T. & CAULFIELD, J. F. (1986). Schistosomula of *Schistosoma mansoni* use lysophosphatidylcholine to lyse adherent human red blood cells and immobilize red cell membrane components. *Journal of Cell Biology* **103**, 819–828.

GREENE, B. M., TAYLOR, H. R. & AIKAWA, M. (1981). Cellular killing of microfilariae of *Onchocerca volvulus*: eosinophil and neutrophil-mediated immune serum-dependent destruction. *Journal of Immunology* **127**, 1611–1618.

GROGAN, J. L., KREMSNER, P. G., DEELDER, A. M. & YAZDANBAKHSH, M. (1996). Elevated proliferation and interleukin-4 release from $CD4^+$ cells after chemotherapy in human *Schistosoma haematobium* infection. *European Journal of Immunology* **26**, 1365–1370.

HAGAN, P. (1993). IgE and protective immunity to helminth infections. *Parasite Immunology* **15**, 1–4.

HAGAN, P., BLUMENTHAL, U. J., DUNN, D., SIMPSON, A. J. G. & WILKINS, H. A. (1991). Human IgE, IgG4 and resistance to reinfection with *Schistosoma haematobium. Nature* **349**. 243–245.

HAMANN, K. J., BARKER, R. L., LOEGERING, D. A. & GLEICH, G. J. (1987). Comparative toxicity of purified human eosinophil granule proteins for newborn larvae of *Trichinella spiralis. Journal of Parasitology* **73**, 523–529.

HAMANN, K. J., BARKER, R. L., LOEGERING, D. A. & GLEICH, G. J. (1987). Comparative toxicity of purified human eosinophil granule proteins for newborn larvae of *Trichinella spiralis. Journal of Parasitology* **73**, 523–529.

HAMANN, K. J., TEN, R. M., LOEGERING, D. A., JENKINS, R. B., HEISE, M. T., SCHAD, C. R., PEASE, L. R. & GLEICH, G. J. (1990). Structure and chromosomal localization of the human eosinophil derived neurotoxin and eosinophil cationic protein genes: evidence for intron-less coding sequences in the ribonuclease gene superfamily. *Genomics* **7**, 535–546.

HERWERDEN, L., HARRAP, S. B., WONG, Z. Y. H., ABRAMSON, M. J., KUTIN, J. J., FORBES, A. B., RAVEN, J., LANIGAN, A. & WALTERS, E. H. (1995). Linkage of high-affinity IgE receptor gene with bronchial hyperactivity, even in absence of atopy. *Lancet* **346**, 1262–1265.

HEWITT, C. R. A., BROWN, A. P., HART, B. J. & PRITCHARD, D. I. (1995). A major house dust mite allergen disrupts the immunoglobulin E network by selectively cleaving CD23: innate protection by antiproteases. *Journal of Experimental Medicine* **182**, 1537–1544.

HOTEZ, P. & PRITCHARD, D. I. (1995). Hookworm infection. *Scientific American* **272**, 42–48.

HSU, S. Y., MITROS, F. A., HELMS, C. M. & SOLOMON, R. I. (1980). Eosinophils as effector cells in the destruction of *Schistosoma mansoni* eggs in granulomas. *Annals of Tropical Medicine and Parasitology* **74**, 179–183.

JABARA, H. H. & GEHA, R. S. (1996). The superantigen toxic shock syndrome toxin 1 induces CD40 ligand expression and modulates IgE isotype switching. *International Immunology* **10**, 1503–1510.

JAMES, S. L. & COLLEY, D. G. (1976). Eosinophil-mediated destruction of *Schistosoma mansoni* eggs. *Journal of the Reticuloendothelial Society* **20**, 359–374.

JONES, J. T., HELMS, C. N. & KUSEL, J. R. (1988). Variation in susceptibility of *Schistosoma mansoni* to damage by polycations. *Molecular and Biochemical Parasitology* **30**, 35–44.

JONES, R. E., FINKLEMAN, F. D., HESTER, R. B. & KAYES, S. G. (1994). *Toxocara canis*: failure to find IgE receptors (Fc epsilon R) on eosinophils from infected mice suggests that murine eosinophils do not kill helminth larvae by an IgE-dependent mechanism. *Experimental Parasitology* **78**, 64–75.

KAPLAN, M. H., SCHINDLER, U., SMILEY, S. T. & GRUSBY, M. J. (1996*a*). Stat 6 is required for mediating responses to IL-4 and for the development of Th-2 cells. *Immunity* **4**, 313–319.

KAPLAN, M. H., SUN YA-LIN, HOEY, T. & GRUSBY, M. J. (1996*b*). Impaired IL-12 responses and enhanced development of Th2 cells in *Stat* 4-deficient mice. *Nature* **382**, 174–177.

KAZURA, J. W. & AIKAWA, M. (1980). Host defense mechanisms against *Trichinella spiralis* infection in the mouse: eosinophil-mediated destruction of newborn larvae *in vitro*. *Journal of Immunology* **124**, 355–361.

KEPHART, G. M., ANDRADE, Z. A. & GLEICH, G. J. (1988). Localization of eosinophil major basic protein onto eggs of *Schistosoma mansoni* in human pathologic tissue. *American Journal of Pathology* **133**, 389–396.

KING, C. L. & NUTMAN, T. B. (1993). Cytokines and immediate hypersensitivity in protective immunity to helminth infections. *Infectious Agents and Disease* **2**, 103–108.

KING, C. L., XIANLI, J., JUNE, C. H., ABE, R. & LEE, K. P. (1996). CD28 deficient mice generate an impaired Th2 response to *Schistosoma mansoni* infection. *European Journal of Immunology* **26**, 2448–2455.

LEE, T. D. G. (1995). IgE regulation by nematodes: the body fluid of *Ascaris* contains a B cell mitogen. *Journal of Allergy and Clinical Immunology* **95**, 1246–1254.

LEVITT, R. C. & HOLROYD, K. J. (1995). Fine-structure mapping of genes providing susceptibility to asthma on chromosome 5q31–q33. *Clinical and Experimental Allergy* **25**, supplement 2, 119–123.

LIFE, P., AUBRY, J. P., ESTOPPEY, S., SCHNURIGER, V. & BONNEFOY, J. Y. (1995). CD28 functions as an adhesion molecule and is involved in the regulation of human IgE synthesis. *European Journal of Immunology* **25**, 333–339.

LOMBARDI, S., VEGNI TALLURI, M., BANCHIERI, L. & ESPOSITO, F. (1990). The *in vitro* adherence of murine eosinophils, neutrophils and non-induced and induced macrophages to infective larvae of *Toxocara canis* (Nematoda, Ascarididae). *International Journal for Parasitology* **20**, 603–613.

MACHADO, D. C., HORTON, D., HARROP, R., PEACHELL, P. T. & HELM, B. A. (1996). Potential allergens stimulate the release of mediators of the allergic response from cells of mast cell lineage in the absence of sensitisation with antigen-specific IgE. *European Journal of Immunology* **26**, 2972–2980.

MACKENZIE, C. D., JUNGERY, M., TAYLOR, P. M. & OGILVIE, B. M. (1981). The *in vivo* interaction of eosinophils, neutrophils, macrophages and mast cells with nematode surfaces in the presence of complement or antibodies. *Journal of Pathology* **133**, 161–175.

MAHANTY, S., ABRAMS, J. S., KING, C. L., LIMAYE, A. P. & NUTMAN, T. B. (1992). Parallel regulation of IL-4 and IL-5 in human helminth infections. *Journal of Immunology* **148**, 3567–3571.

MARSH, D. G., NEELY, J. D., BREAZEALE, D. R., GHOSH, B., FRIEDHOFF, L. R., EHRLICH-KAUTZKY, E., SCHOU, C., KRISHZAWAMY, G. & BEATY, T. H. (1994). Linkage analysis of IL-4 and other chromosome 5q31·1 markers and total serum immunoglobulin E concentrations. *Science* **264**, 1152–1155.

McKEAN, P. G. & PRITCHARD, D. I. (1989). The action of a mast cell protease on the cuticular collagens of *Necator americanus*. *Parasite Immunology* **11**, 293–297.

McLAREN, D. J., MACKENZIE, C. D. & RAMALHO-PINTO, F. J. (1977). Ultrastructural observations on the *in vitro* interaction between rat eosinophils and some parasitic helminths (*Schistosoma mansoni*, *Trichinella spiralis* and *Nippostrongylus brasiliensis*). *Clinical and Experimental Immunology* **30**, 105–118.

McLAREN, D. J., McKEAN, J. R., OLSEN, I., VENGE, P. & KAY, A. B. (1981). Morphological studies on the killing of schistosomula of *Schistosoma mansoni* by human eosinophil and neutrophil cationic proteins *in vitro*.; *Parasite Immunology* **3**, 359–373.

McLAREN, D. J., PETERSON, C. G. & VENGE, P. (1984). *Schistosoma mansoni*: further studies of the interaction between schistosomula and granulocyte-derived cationic proteins, *in vitro*. *Parasitology* **88**, 491–503.

McLAREN, D. J., RAMALHO-PINTO, F. J. & SMITHERS, S. R. (1978). Ultrastructural evidence for complement and antibody-dependent damage to schistosomula of *Schistosoma mansoni* by rat eosinophils *in vitro*. *Parasitology* **77**, 313–324.

McREYNOLDS, L. A., KENNEDY, M. W. & SELKIRK, M. E. (1993). The polyprotein allergens of nematodes. *Parasitology Today* **9**, 403–406.

MOQBEL, R. (1980). Histopathological changes in rats following primary, secondary and repeated infections with *Strongyloides ratti*, with special reference to tissue eosinophils. *Parasite Immunology* **2**, 11–27.

MOQBEL, R., MACDONALD, A. J., CROMWELL, O. & KAY, A. B. (1990). Release of leukotriene C₄ (LTC₄) from human eosinophils following adherence to IgE- and IgG-coated schistosolula of *Schistosoma mansoni*. *Immunology* **69**, 435–442.

MOQBEL, R. & PRITCHARD, D. I. (1990). Parasites and allergy: evidence for a 'cause and effect' relationship. *Clinical and Experimental Allergy* **20**, 611–618.

MOQBEL, R., SASS-KUHN, S. P., GOETZL, E. J. & KAY, A. B. (1983). Enhancement of neutrophil- and eosinophil-mediated complement-dependent killing of schistosomula of *Schistosoma mansoni in vitro* by leukotriene B₄. *Clinical and Experimental Immunology* **52**, 519–527.

MORAWETZ, R. A., GABRIELE, L., RIZZO, I. V., NOBENTRAUTH, N., KUHN, R., RAJEWSKY, K., MULLER, W., DOHERTY, T. M., FINKELMAN, F., COFFMANN, R. L. & MORSE, H. C. (1996). Interleukin (IL 4-independent immunoglobulin class switch to immunoglobulin (Ig) E in the mouse. *Journal of Experimental Medicine* **184**, 1651–1661.

OCHENSBERGER, B., DAEPP, G. C., RIGS, S. & DAHUNDEN, C. A. (1996). Human blood basophils produce interleukin 13 in response to IgE receptor dependent and independent activation. *Blood* **88**, 3028–3037.

PINCUS, S. H., BUTTERWORTH, A. E., DAVID, J. R., ROBBINS, M. & VADAS, M. A. (1981). Antibody-dependent eosinophil-mediated damage to schistosomula of *Schistosoma mansoni*: lack of requirement for oxidative metabolism. *Journal of Immunology* **126**, 1794–1799.

PRITCHARD, D. I. (1993 *a*). Immunity to helminths: is too much IgE parasite – rather than host-protective? *Parasite Immunology* **15**, 5–9.

PRITCHARD, D. I. (1993 *b*). Why do some parasites need to secrete acetylcholinesterase (AChE)? *International Journal for Parasitology* **3**, 549.

PRITCHARD, D. I., CRAWFORD, C. R., DUCE, I. R. & BEHNKE, J. M. (1985). Antigen stripping from the nematode epicuticle using the cationic detergent cetyltrimethylammonium bromide (CTAB). *Parasite Immunology* **7**, 575–585.

PRITCHARD, D. I., KUMAR, S. & EDMONDS, P. (1993). Solubles(s) CD23 levels in the plasma of a parasitized population from Papua New Guinea. *Parasite Immunology* **15**, 205–208.

PRITCHARD, D. I., QUINNELL, R. J., SLATER, A. F. G., McKEAN, P. G., DALE, D. D. S., RAIKO, A. & KEYMER, A. (1990). Epidemiology and immunology of *Necator americanus* infection in a community in Papua New Guinea: humoral responses to excretory–secretory and cuticular collagen antigens. *Parasitology* **100**, 317–326.

PRITCHARD, D. I., QUINNELL, R. J. & WALSH, E. A. (1995). Immunity in humans to *Necator americanus*: IgE, parasite weight and fecundity. *Parasite Immunology* **17**, 71–75.

PRITCHARD, D. I. & WALSH, E. A. (1995). The specificity of the human IgE response to *Necator americanus*. *Parasite Immunology* **17**, 605–607.

PRITCHARD, D. I. & WILSON, R. A. (1997). Helminth infections. In *Pulmonary Defences* (ed. Stockley, R. A.), pp. 305–319. Chichester: John Wiley and Sons Ltd.

QUINNELL, R. J., WOOLHOUSE, M. E. J., WALSH, E. A. & PRITCHARD, D. I. (1995). Immunoepidemiology of human necatoriasis: correlations between antibody responses and parasite burdens. *Parasite Immunology* **17**, 313–318.

RAMALHO-PINTO, F. J., McLAREN, D. J. & SMITHERS, S. R. (1978). Complement mediated killing of schistosomula of *Schistosoma mansoni* by rat eosinophils *in vitro*. *Journal of Experimental Medicine* **147**, 147–156.

RIHET, P., DEMEURE, C. E., BOURGOIS, A., PRATA, A. & DESSEIN, A. J. (1991). Evidence for an association between human resistance to *Schistosoma mansoni* and high anti-larval IgE levels. *European Journal of Immunology* **21**, 2679–2686.

ROOK, G. A. W., HERNANDEZ-PANDO, R. & LIGHTMAN, S. L. (1994). Hormones, peripherally activated prohormones and regulation of the Th1/Th2 balance. *Immunology Today* **15**, 301–303.

SHER, A., WADEE, A. A., MASON, P. R. & FRIPP, P. J. (1980). Cytotoxic effect of human eosinophils and eosinophil lysate on *Schistosoma haematobium* ova. *South African Journal of Science* **76**, 266–268.

SIM, B. K., KWA, B. H. & MAK, J. W. (1982). Immune response in human *Brugia malayi* infections: serum-

dependent cell-mediated destruction of infective larvae *in vitro*. *Transactions of the Royal Society of Tropical Medicine and Hygiene* **76**, 362–370.

SMITH, A. M., DALTON, J. P., CLOUGH, K. A., KILBANE, C. L., HARROP, S. A., HOLE, N. & BRINDLEY, P. J. (1994). Adult *Schistosoma mansoni* express cathepsin L proteinase activity. *Molecular and Biochemical Parasitology* **67**, 11–19.

SPRY, C. J. F. (1981). Alterations in blood eosinophil morphology, binding capacity for complexed IgG and kinetics in patients with tropical (filarial) eosinophilia. *Parasite Immunology* **3**, 1–11.

VAN DER POUWKRAAN, C. T. M., RENSINK, H. J. A. M., RAPPUOLI, R. & AARDEN, L. A. (1996). Co-stimulation of T cells via CD28 inhibits human IgE production, reversal by pertussis toxin. *Clinical and Experimental Immunology* **99**, 473–478.

VENGE, P., DAHL, R., FREDENSK, K., HAUGREN, R. & PETERSON, C. (1983). Eosinophil cationic proteins (ECP and EPX) in health and disease. In *Immunobiology of the Eosinophil* (ed. Yoshida, T., Torisu, M.) pp. 163–179. New York: Elsevier.

WAKELIN, D. (1993). Allergic inflammation as a hypothesis for the expulsion of worms from tissues. *Parasitology Today* **9**, 115–116.

WARDLAW, A. J. & MOQBEL, R. (1992). The eosinophil in allergic and helminth-related inflammatory responses. In *Allergy and Immunity to Helminths. Common Mechanisms or Divergent Pathways?* pp. 154–186. (London: Taylor & Francis

WASSOM, D. L. & GLEICH, G. J. (1979). Damage to *Trichinella spiralis* newborn larvae by eosinophil major basic protein. *American Journal of Tropical Medicine and Hygiene* **28**, 860–863.

WEISS, N., HUSSAIN, R. & OTTESEN, E. A. (1982). IgE antibodies are more species-specific than IgG antibodies in human onchocerciasis and lymphatic filariasis. *Immunology* **45**, 129–137.

WILDENBURG, G., KRÖMER, M., BONOW, I. & BÜTTNER, D. W. (1995). Distribution of effector cells in nodules of *Onchocerca volvulus*. *Zentralblatt für Bakteriologie* (Gustav Fischer, Verlag) **282**, 221 (abstr.).

WILLIAMS, J. F., CHALIB, H. W., MACKENZIE, C. D., ELDHALIFA, M. Y., AYUYA, J. M. & KRON, M. A. (1987). Cell adherence to microfilariae of *Onchocerca volvulus*: a comparative study. *Ciba Foundation Symposium* **127**, 146–163.

WILLIAMS, P. (1994). Compromising bacterial communication skills. *Journal of Pharmacology and Pharmacological Science* **46**, 252–260.

WOOLHOUSE, M. E. J., TALOR, P., MATANHIRE, D. & CHANDIWANA, S. K. (1991). Acquired immunity and epidemiology of *Schistosoma haematobium*. *Nature* **351**, 757–758.

YAZDANBAKHSH, M., TAI, P.-C., SPRY, C. J. F., GLEICH, G. J. & ROOS, D. (1987). Synergism between eosinophil cationic protein and oxygen metabolites in killing of schistosomula of *Schistosoma mansoni*. *Journal of Immunology* **138**, 3443–3447.

YOUNG, J. D. E., PETERSON, C. G. B., VENGE, P. & COHN, Z. A. (1986). Mechanism of membrane damage mediated by human eosinophil cationic protein. *Nature* **321**, 613.

Modulation of host cell receptors: a mechanism for the survival of malaria parasites

M. HOMMEL

Liverpool School of Tropical Medicine, Molecular Biology and Immunology Division, Pembroke Place, Liverpool L3 5QA, UK

SUMMARY

Intra-erythrocytic stages of malaria parasites can alter the surface of their host cells and release toxins which induce the production of cytokines, which in turn can up- or down-regulate the expression of adhesion receptors on the surface of microvascular endothelial cells. New adhesion receptors on endothelial cells provide the parasite with increased chances of survival despite an increasing level of host immunity. In order to take advantage of these new opportunities for survival, the parasite itself needs to make best use of its considerable ability to vary its surface antigens and adherent molecules. The paper describes the various players in this survival game and articulates a working hypothesis to explain how it may all fit together.

Key words: *Plasmodium falciparum*, *P. berghei*, *P. chabaudi*, cytoadherence, sequestration, cytokines, cerebral malaria, tumour necrosis factor, intercellular adhesion molecule-1, microvascular endothelial cells, malaria toxins, polar lipids, antigenic variation, haemozoin, chondroitin-sulphate A.

INTRODUCTION

In the course of their development, malaria parasites interact in a variety of ways with receptor molecules on the surface of host cells. Apart from the fact that intracellular stages of the parasite induce changes in the surface membrane of their host cells, a variety of molecules are released by the parasite some of which have a systemic effect on the host and either lead to an ineffective immune response and improve the chances of parasite survival, or contribute to the pathophysiology of malaria. This paper concentrates on the erythrocytic cycle of malaria parasites and examines, in particular, the interaction between molecules at the surface of infected erythrocytes and molecules at the surface of endothelial cells. Although, at first sight, this appears to be a relatively simple receptor–ligand interaction, it is becoming increasingly obvious from recent published literature that this represents a very dynamic form of interaction, where each side possesses a very considerable capacity to change not only its own set of surface receptors but also to induce receptor switching on the other. While previous reviews on this topic have concentrated mostly on whether or not cytoadherence could be a determining factor for virulence and disease severity (Goldring & Hommel, 1992; Berendt, Turner & Newbold, 1994), it is suggested here that the modulation of receptor expression may represent a mechanism evolved by the parasite to adapt to a continuously changing cellular environment and, thus, improve its chances of survival.

The players in this survival game are: the sticky molecules expressed on the surface of infected erythrocytes, the molecules on the surface of endothelial cells, the molecules (or 'toxins') released by the parasite, the cytokines released in response to toxins which, in turn, can induce an up- and a down-regulation of receptors on endothelial cells. By modulating the expression of endothelial adhesion molecules and, at the same time, switching the expression of its own molecules on the surface of infected erythrocytes to adapt to this changing environment, the parasite improves its means of survival despite increasing immune pressure. Since the mechanism by which the parasites modulate host receptor expression (via the release of toxin-inducing cytokines) is relatively random and difficult to control, depending as it does on the host's genetic make-up and its immune status, the parasite needs to make use of its own considerable biodiversity and its capability for rapid switching from one set of sticky molecules to another to adapt best to the changing environment. From such a viewpoint, what had been described earlier as 'antigenic variation' (Hommel, David & Oligino, 1983; Su *et al.* 1995; Smith *et al.* 1995), may in fact represent an elegant process of host–parasite adaptation rather than a mere 'escape mechanism'.

RECEPTORS ON THE SURFACE OF INFECTED ERYTHROCYTES

After infection by the malarial parasite, the red cell surface undergoes structural changes, which may substantially alter its function, appearance and antigenicity (Hommel & Semoff, 1988). The alterations identified so far on *Plasmodium falciparum*-infected cells include: a visible change of shape and reduced deformability; the presence of electron-dense protrusions or 'knobs' (Trager, Rudzinska & Bradbury, 1966); the expression of new sugar moieties, particularly galactose (David, Hommel & Oligino, 1981); the cytoadherence to endothelial cells or rosetting with normal erythrocytes (Wahlgren, Carlson & Udomsangpetch, 1987); the presence of new metabolic channels (Ginsburg et al. 1985); the evidence of new parasite-specific antigens associated with the red cell membrane (Hommel et al. 1983) and the reorganization of normal erythrocyte components (e.g. dimerization of erythrocyte Band 3 to form 'Pfalhesin') (Winograd & Sherman, 1989). The precise molecular organisation of the surface of infected erythrocytes is not yet known, but is generally believed to consist of a combination of red cell membrane alterations and the insertion of parasite molecules into the altered membrane (including PfEMP-1, sequestrin, rosettins, HRP-1 and PfEMP-3) (Leech et al. 1984; Rock et al. 1988; Ockenhouse et al. 1991; Helmby et al. 1993; Pasloke et al. 1993). As a result of this complex reorganization of the infected red cell membrane, new receptors are expressed on the surface, whose antigenic and cytoadherence properties are highly variable not only from one parasite isolate to another, but also within a given isolate due to a rapid phenotypic switching.

It is usually accepted that the severe forms of malaria in humans are due to ability of mature forms of *P. falciparum* to adhere to microvascular endothelium, a process referred to as 'sequestration' and which does not occur in the mild forms of malaria due to *P. vivax* or *P. malariae*. A crucial issue has been whether sequestration causes severe malaria pathology directly (e.g. by reducing blood flow) or indirectly (by producing a concentrated toxin release which influences the local cytokine environment), or whether sequestration is altogether a side-effect secondary to the up-regulation of endothelial cells receptors. The major difference between severe falciparum malaria and the severe murine malaria model is that, although both produce coma, sequestration of infected red cells occurs only in falciparum malaria, a feature which has raised questions about the relevance of the murine model for the study of the pathophysiology of human cerebral malaria (Porta et al. 1993). The observation that mixed infections *P. chabaudi/P. berghei* may be able to re-create the conditions for cerebral malaria with brain sequestration not only of leukocytes but also of infected erythrocytes (Dennison & Hommel, 1993) offers interesting prospects.

A number of adhesion receptors has been identified on the surface of endothelial cells and incriminated as cytoadherence receptors for *P. falciparum*-infected cells, including thrombospondin, CD36, ICAM-1, VCAM-1, chondroitin sulphate A, thrombomodulin, E-selectin, P-selectin and a N-linked glycosaminoglycan) (see reviews by Hommel, 1993; Rogerson & Brown, 1997). Pfalhesin has strong binding affinities for CD36 on endothelial cells (Crandall, Land & Sherman, 1994), sequestrin appears to have a specific affinity for CD36 (Ockenhouse et al. 1991), while rosettins probably recognize thrombospondin (Wahlgren et al. 1987). Of the parasite molecules expressed at the surface of infected red cells, PfEMP-1 is the best known: a polymorphic, high molecular weight protein (250–300 kDa), encoded by the *var* genes (Su et al. 1995; Smith et al. 1995); each parasite clone has approximately 50 genes of this family and is able to switch rapidly from one to another (Roberts et al. 1992). Variants of PfEMP-1 are capable of adherence to all the above endothelial cell receptors, but also capable of producing rosettes of normal erythrocytes via a binding to complement receptor-1 (Newbold et al. 1997; Rowe et al. 1997). Although PfEMP-3 and HRP-1 are not expressed on the surface, these molecules contribute to the structure of knobs and a change in the balance between these two molecules may affect the expression of other surface ligands (Le Scanf et al. 1997). Whether a single infected red cell can adhere to only one or to a variety of different endothelial receptors has not been finally established since most experiments have been performed on populations of parasites rather than single cells. Thus, the finding of a differential sensitivity to proteolytic enzymes has been interpreted as suggesting either that more than one red cell receptor may be involved (Chaiyaroj et al. 1994) or that a given PfEMP-1 molecule may possess different binding domains (Gardner et al. 1996).

ANTIGENIC VARIATION AND INHIBITION OF CYTOADHERENCE

Antigenic variation was first described in the *P. knowlesi*/rhesus monkey model (Brown & Brown, 1965), but has since been shown to exist in *P. chabaudi*, *P. falciparum* and *P. fragile* (McLean, Pearson & Phillips, 1982; Hommel et al. 1983; Handunetti, Mendis & David, 1987); and it is reasonable to assume that the phenomenon is widespread among malaria parasites. Many of the studies on antigenic variation have concentrated on the molecules expressed on the surface of infected erythrocytes (e.g. SICA in *P. knowlesi* and PfEMP-1 in *P. falciparum*). Switching from one variant antigenic type to another is a random event, which apparently occurs without any external stimulus or pressure. The rate of switching from one antigenic phenotype to another is thought to be very fast. In a study of *P. chabaudi* in mice, 40% of the parasites present at the peak of infection were already different from the original inoculum (Brannan, Turner & Phillips, 1994), while in an *in vitro* study of *P. falciparum* it was estimated that 2% of parasites switched to a new antigen in every erythrocytic cycle (Roberts *et al.* 1992), which is comparable to the *in vivo* results with *P. chabaudi*. The rapid switching of the expression of PfEMP-1 may be a consequence of the sub-telomeric position of *var* genes on almost all the chromosomes of the parasites, a particularly unstable chromosomal area, where genes are frequently reorganized in these parasites (Thompson *et al.* 1997). Switching from one antigenic phenotype to another is associated with a change in the cytoadherence properties of the infected cells and there is an undeniable relationship between antigenic phenotype and adherent phenotype, even if only based on circumstantial evidence. The consequence of antigenic variation is that, even in a cloned population of parasites *in vitro*, the parasites present at any particular time always constitute a mixture of different variant populations with different cytoadherence properties. Over time in culture, the relative proportion of the different variant populations present continuously change. If the situation is complex in a cloned population *in vitro*, it is even more complex in a natural infection, when parasites injected by a single mosquito consist usually of more than one clone and where an individual may receive more than one infective bite each night (Thaithong *et al.* 1984; Paul *et al.* 1995).

The molecules expressed on the surface of infected erythrocytes (whether pfEMP-1, rosettins, sequestrin or Pfalhesin) are all immunogenic and capable of inducing a potent, long-lasting immune response in infected individuals (Hommel, 1985; Marsh & Howard, 1986; Treutiger *et al.* 1992; Crandall *et al.* 1995). After primo-infection, only antibodies to the homologous isolate of parasites can be detected and the repertoire of antibodies to different variants increases with each further infection. What is important here is that variant-specific antibodies recognize the sticky molecules on infected cells, and can inhibit or reverse cytoadherence to the corresponding endothelial cell receptors. This had first been demonstrated as inhibition/reversal of cytoadherence to amelanotic melanoma cells (David *et al.* 1983) and later to more purified ligands (e.g. purified thrombospondin or CD36 attached to plastic, or ICAM-1-transfected CHO cells) (Roberts *et al.* 1985; Berendt *et al.* 1989; Oquendo *et al.* 1989). It may be of interest to note that *in vitro* cytoadherence to monocytes/macrophages can neither be inhibited nor reversed using antimalarial hyperimmune serum (Goldring & Hommel, 1992), since this may explain the absence of effect of passive transfer of hyperimmune serum in humans on sequestration (Taylor *et al.* 1992) which contrasts to the reversal of cytoadherence that had been observed *in vivo* in a primate model (David *et al.* 1983). The immune response to surface antigens on infected erythrocytes represents, therefore, a potent selection pressure for cytoadherence

molecules and, after many years of exposure to malaria, an individual would eventually be protected from further re-infection when the repertoire of cytoadherent molecules of the endemic parasite population has been exhausted. While infection still occurs in adults living in endemic areas (i.e. no 'sterile immunity'), the level of parasitaemia always remains low and clinical manifestations are unusual. It looks as if, by continuously changing the adherence receptors expressed on infected erythrocytes, the parasite can extend its survival despite immunity; by the same token, this means that, over a period of time, the parasite needs to recognize a series of different endothelial cell molecules, but it also means that it will eventually run out of molecules to bind to. For parasites like *P. falciparum*, there is no possible survival in the absence of sequestration, with the exception of infection in splenectomized individuals, where sequestration does not take place (Israeli, Shapiro & Ephros, 1987), or infection during pregnancy, where new adherence receptors become available in the placenta (Fried & Duffy, 1996; Maubert, Guilbert & Deloron, 1997).

PARASITE 'TOXINS'

Various molecules may be released by malarial parasites during the intra-erythrocytic stage of their life cycle which can influence the function of host cells, alter the development of immunity or be, directly or indirectly, responsible for pathological events. Here we use the term 'malarial toxins' to describe this entire group of bio-active molecules, although some authors prefer to restrict the use of the term to molecules capable of inducing fever (Kwiatkowski, 1995). The release of 'toxins' is most likely to occur when a mature meront (or segmenter) opens up to allow new merozoites to escape, but it is conceivable that true excretion of parasite-made molecules (or 'exo-antigens') through the erythrocyte-membrane may occur via the parasitophorous vacuole, the tubo-vesicular membrane network and a transient 'duct' (Haldar, 1994), while the parasite develops within the red cell.

Mitogens

Malarial antigens can induce polyclonal proliferation of B cells (Greenwood & Vick, 1975; Rosenberg, 1978) and of $\gamma\delta$ T cells (Ho *et al.* 1994). While the molecular nature of the former has not yet been identified, it appears that the malarial 'super-antigens' responsible for the proliferation of $V\gamma9/V\delta2$ $\gamma\delta$ T cells are phosphorylated molecules, similar to isopentenol pyrophosphate from *Mycobacterium tuberculosis* (Behr *et al.* 1996). The polyclonal activation of B cells is generally considered to represent one cause of immunosuppression during malaria; the stimulation of $\gamma\delta$ T cells, which is particularly intense during malaria in naive individuals, is believed to contribute to create a Th1-type environment, which may be one of the prerequisites for the development of severe forms of malaria (Grau & Behr, 1994).

Apoptosis

The ultimate modulating effect the malarial parasite can exert on the cells of its host is the induction of apoptosis, but there is no evidence that the parasite uses this process as a mechanism of defence. Extracts of *P. falciparum* schizonts can induce apoptosis in human lymphocytes (Touré-Balde *et al.* 1996) and unusually high levels of spontaneous apoptosis are observed in short-term cultures of lymphocytes from individuals with clinical malaria, akin to what has been described in asymptomatic HIV-infected individuals. The nature of apoptosis-inducing malarial toxins has not yet been identified.

Toxic proteins

Early work on exo-antigens of *P. falciparum* had concentrated on proteins found in culture supernatants capable of inducing the secretion of inflammatory cytokines, including the Ag7 complex, PfMSP-1, RAP-1 and RESA (Jakobsen *et al.* 1993; Picot *et al.* 1993), very little can be concluded from such work based on crude preparations and probably contaminated with the far more potent polar lipids of malaria parasites.

Polar lipids

Polar lipids extracted from malarial parasites and malaria culture supernatant are responsible for a variety of effects on host cells, including the production of inflammatory cytokines by monocytes and macrophages (Bate, Taverne & Playfair, 1988; Jakobsen *et al.* 1995) and the induction of lipogenesis by adipocytes (Taylor *et al.* 1992). The biological activity is heat stable and pronase resistant and has, for the most, been assigned to phospholipids particularly phosphatidylinositol (Bate *et al.* 1992). The glycosylphosphatidylinositol (GPI) anchor of merozoite surface antigens MSP-1 and MSP-2 has been claimed to account for a large portion of the biological activity ascribed to malarial phospholipids (Schofield & Hackett, 1993; Schofield *et al.* 1996), but this view is controversial and others believe that different molecules may be responsible for the various biological effects observed (Taverne *et al.* 1995). The phospholipids that induce lipogenesis by adipocytes *in vitro* also stimulate the release of insulin *in vivo* (Elased & Playfair, 1994) and cause a rapid drop in blood glucose when injected into mice, but these effects seem distinguishable from the TNF-inducing effects of malarial polar lipids.

Haemozoin

Malarial pigment is found in the parasite food vacuole and represents the residue of haemoglobin digestion (crystalline β-haematin) combined with a variety of compounds of parasite and host origin (including aggregated proteins, lipids and phospholipids) (see review by Arese & Schwarzer, 1997). On schizont rupture, the residual body containing haemozoin is released and is rapidly taken up by phagocytic cells where it remains undigested for long periods of time. The belief that haemozoin was responsible for fever in malaria dates back to a paper by Brown in 1912, but the concept was later challenged and forgotten, only to be revived in recent years. It was observed that malarial pigment was not only responsible for the release of cytokines by monocytes (Pichyangkul, Saengkrai & Webster, 1994), but also seriously impaired the functionality of phagocytic cells (reduced phagocytic activity, failure to produce oxidative burst, failure to kill invading microorganisms) (Arese, Turrini & Ginsburg, 1991; Schwarzer *et al.* 1992).

Endotoxin and Mycoplasma *contamination*

Early work on malaria toxins was confused by the belief that the biological effects observed were the consequence of a contamination of samples by bacterial lipopolysaccharide endotoxin; some studies succeeded in detecting endotoxin or endotoxin-like molecules by using the *Limulus* amoebocyte lysate (Tubbs, 1980; Jakobsen, Baeck & Jepsen, 1988), while others failed to do so (Greenwood, Evans-Jones & Stratton, 1975). The recent finding of a frequent contamination by *Mycoplasma* of *Plasmodium* lines maintained *in vitro* (Turrini *et al.* 1997) has raised different questions about the possible artefactual nature of some or all work describing so-called 'malaria toxins' and published over the past ten years. Since the *Mycoplasma* themselves secrete highly potent 'toxic' molecules (including glycolipids, lipoglycans and proteins) with a wide range of effects, from the induction of inflammatory cytokine secretion by macrophages to mitogenic effects on B cells and immunoglobulin production, it will take considerable efforts to disentangle what, in the published literature on 'malarial toxins' can truly be ascribed to *Plasmodium*.

CYTOKINES

In the course of malaria, cytokines are produced as a result of the action of malarial toxins on host cells, particularly macrophages/monocytes and $\gamma\delta$ T cells, but also as a result of the recognition of malarial antigens by T cells. Although these different modes of cytokine production are described separately here, the two are obviously linked and one may either synergise or antagonise the other; macrophages will, for example, respond very differently to malaria toxins if they have previously been stimulated by T cell cytokines or not.

Effect of malaria toxins on cytokine production

A feature common to most malaria toxins is the ability to induce the production of inflammatory cytokines (TNFα, IL-1 and IL-6) by macrophages and, to a lesser extent, monocytes (as reviewed by Jakobsen *et al.* 1995). The identification of TNF-inducing malaria toxins was first performed using murine models (Bate, Taverne & Playfair, 1988) and the *in vivo* effects have been particularly well analysed in the murine malaria models using CBA/A mice infected with *P. berghei* Anka (Grau *et al.* 1988). By extension of the murine model of severe malaria many conclusions have been drawn on the pathophysiology of severe falciparum malaria in man and, although there are substantial differences between the two, there appears at least to be a clinical correlation between increased levels of circulating TNFα and disease severity (Grau *et al.* 1989). The most interesting feature about inflammatory cytokines induced by malarial toxins is that this pathway may actually be interrupted by antibodies against the toxins; this was first demonstrated with the serum of mice vaccinated against *P. yoelii* which was shown to block the induction of TNFα release (Bate, Taverne & Playfair, 1988). Immunization experiments with crude *P. yoelii* toxin have interesting features, in as much as they have shown that such immunity is T-independent, short-lived and predominantly IgM, but also that it is not *P. yoelii*-specific but cross-reacts with the toxins of other malaria parasites (e.g. *P. vivax*). Such experiments have formed the basis for the concept of an anti-toxic or 'anti-disease' vaccine against malaria (Playfair *et al.* 1990). Together with the notion of the existence of an anti-toxin immunity, the variability of different parasites isolates in their ability to induce TNFα (Allan, Rowe & Kwiatkowski, 1993) may represent an important feature for our understanding of the diversity of parasite–host interactions.

T cell cytokines

As immunity to malaria develops, a variety of cytokines is produced (Kumaratilake & Ferrante, 1994) and studies in murine models have shown that the nature of anti-malarial immunity changes over time, switching from an initial Th1 response to an essential Th2 response. Evidence that a Th1 to Th2 switch occurs in man is best demonstrated by studies performed in Gabon, where young children with acute malaria have high levels of IFN-γ, while the levels of IFN-γ decrease drastically in school-age children in whom IL-4 then becomes detectable (Mshana *et al.* 1994). This is important since Th1 cytokines will act in synergy with inflammatory cytokines while Th2 cytokines will essentially have an 'anti-inflammatory' effect. Studies with the *P. berghei* ANKA model had shown that if TNFα had a central role in the pathological events, this only occurs in the presence of IFNγ (Grau *et al.* 1988; Grau & Behr, 1994) and that no cerebral malaria could be induced in IFNγ-R-deficient mice (Rudin *et al.* 1997). The Th1 environment that prevails early in the disease may, to some extent, be the result of a considerable stimulation of γδ T cell activity and the related increase of IFNγ and TNF (Grau & Behr, 1994; Ho *et al.* 1994). Conversely, levels of IL-4 and TGF-β are significantly reduced in cases of severe malaria (Wenisch *et al.* 1995), but in situations where these cytokines are experimentally increased this correlates with a reduced pathology.

UP- AND DOWN-REGULATION OF RECEPTORS BY CYTOKINES

The expression of host cell receptors on microvascular endothelial cells is, to a large extent, modulated by the micro-environment generated by the prevailing local balance of different cytokines. Inflammatory cytokines, IFNγ, GM-CSF and IL-3 have all been shown to be capable of significantly increasing the *in vitro* cytoadherence of *P. falciparum*, particularly when acting in synergy with one another (Ringwald, Le Bras & Savel, 1992). This increased cytoadherence reflects a change in the receptors expressed at the surface of the cells used in such *in vitro* cytoadherence assays. It has been shown that, depending on the cytokines present, there could either be an up-regulation of ICAM-1, V CAM-1 and E-selectin (in the presence of TNFα, IFNγ or IL-1, as well as a direct effect of certain malaria toxins, e.g. GPI) (Berendt *et al.* 1989; Schofield *et al.* 1996) or an up-regulation of CD36 and thrombospondin (in the presence of IL-4 and GM-CSF) (Yesner *et al.* 1996). The relative cytokine balance prevailing in the microvascular bed is more difficult to assess, but *in vivo* up- and down-regulations of host cell receptors clearly affect the level of sequestration and the degree of malaria severity. In IFNγ-R-deficient mice, for example, the absence of cerebral malaria has been associated with an absence of ICAM-1 up-regulation (Rudin *et al.* 1997). In humans who died of cerebral malaria, the study of *P. falciparum* cytoadherence and relative expression of adherence receptors has shown an increased expression of ICAM-1 and E-selectin in brain capillaries where cytoadherence occurred (Turner *et al.* 1994). This type of study is fraught with difficulties since the up-regulation of receptor molecules may be a consequence rather than a cause of parasite cytoadherence. Thus, while there is no suggestion that TNF receptors on endothelial cells play any role in cytoadherence *per se*, there is a clear correlation between the expression of TNFR2 (but not TNFR1) in microvascular endothelial cells and cerebral malaria (Lucas *et al.* 1997) with the suggestion that the expression of TNFR2 is directly implicated in the up-regulation of ICAM-1. The topic of adhesion molecules on endothelial cells and the up- and down-regulation of these molecules in response to local stimulation by cytokines is a fast-evolving one, where many questions still need to be answered. We know that large vessel endothelial cells behave very differently from microvascular endothelial cells in response to cytokines, but how different are the brain microvascular endothelial cells from those of other tissues which are easier to study? Are we concentrating on certain adhesion molecules because we have the reagents, while we should be looking for less promiscuous receptors which may explain unique cytoadherence characteristics? We must not lose sight of the fact that in every *in vitro* model of cytoadherence comparing large numbers of wild *P. falciparum* isolates, there is always a percentage of isolates that do not bind to any of the cytoadherence receptors presented in the assay (while they had clearly been sequestered *in vivo*) (Goldring *et al.* 1992), which suggests that our inventory of host adhesion receptor is still not yet complete.

HOST FACTORS

The genetic make-up of the host will influence the constitutional expression of adhesion receptors on endothelial cells, the ability to respond to malarial toxins, the relative ability to produce certain cytokines and the intrinsic ability to respond to certain malarial antigens and develop immunity. Host differences have been clearly established in murine models (e.g. CBA/A, but not BALB/C, mice develop severe malaria with *P. berghei* ANKA (Grau *et al.* 1988) and some studies suggest that comparable overall differences in susceptibility to severe malaria or to the development of immunity to malaria may exist in humans (Hill, 1992; McGuire *et al.* 1994; Riley, 1996).

MODULATION OF HOST RECEPTORS AND PARASITE SURVIVAL

The concept that the dynamic relationship that exists between the parasite and its host could actually improve the chances of survival of the parasite may be articulated in the following ways. (1) The parasites continuously switch the expression of their erythrocyte-associated antigens (e.g. by switching *var* gene expression) and such a switching is random, independent of any external pressure. In consequence, at each new erythrocytic cycle, a small percentage of the parasite population present will express a new phenotype of variant adhesive molecules. In order to survive *in vivo*, all the variant adhesive molecules expressed will need to bind to a corresponding adhesion receptor on endothelial cells; any parasite that expresses a variant adhesive molecule that corresponds to an endothelial receptor not expressed at that particular moment will be eliminated. (2) Host microvascular endothelial cells express constitutively a given repertoire of adhesive receptors which provides the parasite with a 'basic' vascular bed for sequestration; this would, for instance, be used by parasites early in infection, before the development of inflammatory responses or immunity. Basic receptors may include CD36 and thrombospondin. (3) As the malaria infection progresses, the excretion of malarial toxins will bring the inflammatory cytokines into play, which will induce a series of up- and down-regulations of receptors and, hence, produce an opportunity for different variant populations to cytoadhere (and thus survive). Since the total parasite load is finite, the availability of new receptors will cause a redistribution of the infection to new sites of sequestration. Within these new sites of 'upregulated' sequestration, the presence of cytoadherence itself will further stimulate upregulation. ICAM-1 is probably the model for such receptors. (4) Once immunity to malarial antigens has been initiated, the essentially Th1 environment that is produced will enhance the production of inflammatory cytokines and Th1 cytokines will act in synergy with inflammatory cytokines for an enhanced upregulation of adhesion receptors. This will offer further opportunities for a selection of new variant parasite populations to cytoadhere; at this point, parasites that only adhere to molecules which are not constitutionally expressed on 'resting' endothelial cells (e.g. E-selectin, VCAM-1, P-selectin) will now be able to survive. Both in terms of intensity and specific location this is probably the time during infection when cerebral forms of the disease are most likely to occur. (5) With repeated malarial infections, immunity against erythrocyte-associated surface antigens will start to develop and variant-specific antibodies will inhibit the cytoadherence of more and more variants; this form of immunity is solid and long-lasting and those variants which can no longer cytoadhere will be eliminated. It is conceivable, in the natural evolution of immunity, that antibodies to variants adherent to 'upregulated' receptors may develop earlier than against variants adherent to constitutional receptors, or that the repertoire of variants of adherent to constitutional receptors is substantially larger than the upregulated type, or that there may be different degrees of binding affinity (to lowest affinity remaining operative the longest). The result of this, and regardless of the actual mechanism involved, would be an early reduction of severe complications, while mild infections may continue to occur. (6) After long-time exposure to serial malaria infections, such as would occur in an individual living in a malaria-endemic area, there will be an effective immunity to most of the variant populations to which the host has been exposed and parasites will no longer be able to take a hold and develop above a certain threshold, presumably related to the lag-time required for boosting immune memory. Of importance may be the observation that similar *var* gene transcripts may be shared, different parasites having a particular adherence phenotype in common (Smith *et al.* 1995); this may help to explain why immunity eventually transcends strain-specificity. (7) In an adult living in a hyperendemic area, this state of immunity will effectively persist for life, except for malaria occurring during pregnancy. From a cytoadherence point of view, pregnancy is a situation where new adhesion receptors become available, not on endothelial cells but on syncytiotrophoblasts, and these new receptors offer a new opportunity for cytoadherence of malarial parasites, despite an otherwise functional immunity. Hence in pregnancy, it would be rational to expect to find only variants capable of adherence to unusual receptors, unique to the placenta (therefore new to the host), since immunity to previously experienced variants would prevent them from developing. This hypothesis has the advantage over the older immunosuppression hypothesis (Rasheed *et al.* 1993; Matteeli *et al.* 1997) to provide an explanation for the differences in malaria severity between primigravida and multigravida, since the host becomes less and less susceptible to the 'placenta adhesive' variants with each of the subsequent pregnancies. The finding of a high frequency of parasites cytoadherent to chondroitin-sulphate A and thrombomodulin produces the first evidence to support this hypothesis (Fried & Duffy, 1996; Maubert, Guilbert & Deloron, 1997; Rogerson & Brown, 1997).

REFERENCES

ALLAN, R. J., ROWE, A. & KWIATKOWSKI, D. (1993). *Plasmodium falciparum* varies in its ability to induce tumor necrosis factor. *Infection and Immunity* **61**, 4772–4776.

ARESE, P. & SCHWARZER, E. (1997). Malarial pigment (haemozoin): a very active 'inert' substance. *Annals of Tropical Medicine and Parasitology* **91**, 501–516.

ARESE, P., TURRINI, F. & GINSBURG, H. (1991). Erythrophagocytosis in malaria: host defence or menace to the macrophage. *Parasitology Today* **7**, 25–28.

BATE, C. A. W., TAVERNE, J. & PLAYFAIR, J. H. L. (1988). Soluble malarial antigens are toxic and induce the production of tumour necrosis factor *in vivo*. *Immunology* **66**, 600–605.

BATE, C. A. W., TAVERNE, J., ROMAN, E., MORENO, C. &

PLAYFAIR, J. H. L. (1992). TNF induction by malaria exoantigens depends upon phospholipids. *Immunology* **75**, 129–135.

BEHR, C., POUPOT, R., PEYRAT, M. A., POQUET, Y., CONSTANT, P., DUBOIS, P., BONNEVILLE, M. & FOURNIE, J. J. (1996). *Plasmodium falciparum* stimuli for human γδ T cells are related to phosphorylated antigens of Mycobacteria. *Infection and Immunity* **64**, 2892–2896.

BERENDT, A. R., SIMMONS, D. L., TANSEY, J., NEWBOLD, C. I. & MARSH, K. (1989). Intercellular adhesion molecule-1 is an endothelial cell adhesion receptor for *Plasmodium falciparum*. *Nature* **341**, 57–59.

BERENDT, A. R., TURNER, G. D. H. & NEWBOLD, C. I. (1994). Cerebral malaria: the sequestration hypothesis. *Parasitology Today* **10**, 412–414.

BRANNAN, L. R., TURNER, C. M. R. & PHILLIPS, R. S. (1994). Malaria parasites undergo antigenic variation at high rates *in vivo*. *Proceedings of the Royal Society London*, *Ser. B* **256**, 71–75.

BROWN, K. N. & BROWN, I. N. (1965). Antigenic variation and immunity to *Plasmodium knowlesi*: antibodies which induce antigenic variation and antibodies which destroy parasites. *Nature* **208**, 1286–1288.

BROWN, W. H. (1912). Malaria pigment (hematin) as a factor in the production of the malarial paroxysm. *Journal of Experimental Medicine* **15**, 579–597.

CHAIYAROJ, S. C., COPPEL, R. L., NOVAKOVIC, S. & BROWN, G. V. (1994). Multiple ligands for cytoadherence can be present simultaneously on the surface of *Plasmodium falciparum*-infected erythrocytes. *Proceedings of the National Academy of Sciences, USA* **91**, 10805–10808.

CRANDALL, I., LAND, K. M. & SHERMAN, I. W. (1994). *Plasmodium falciparum*–Pfalhesin and CD36 form an adhesin/receptor pair that is responsible for the pH-dependent portion of cytoadherence/sequestration. *Experimental Parasitology* **78**, 203–209.

CRANDALL, I., GUTHRIE, N. & SHERMAN, I. W. (1995). *Plasmodium falciparum*: sera of individuals living in a malaria-endemic region recognize peptide motifs of the human erythrocyte anion transport protein. *American Journal of Tropical Medicine and Hygiene* **52**, 450–455.

DAVID, P. H., HOMMEL, M., MILLER, L. H., UDEINYA, I. & OLIGINO, L. D. (1983). Parasite sequestration in *Plasmodium falciparum* malaria: spleen and antibody modulation of cytoadherence of infected erythrocytes. *Proceedings of the National Academy of Sciences, USA* **80**, 5075–5079.

DAVID, P. H., HOMMEL, M. & OLIGINO, L. D. (1981). Interactions of *Plasmodium falciparum*-infected erythrocytes with ligand coated agarose beads. *Molecular and Biochemical Parasitology* **4**, 195–204.

DENNISON, J. M. T. & HOMMEL, M. (1993). Cerebral sequestration of murine malaria parasites after receptor amplification in mixed infections. *Annals of Tropical Medicine and Parasitology* **87**, 665–666.

ELASED, K. & PLAYFAIR, J. H. L. (1994). Hypoglycaemia and hyperinsulinaemia in rodent models of severe malaria infection. *Infection and Immunity* **62**, 5157–5160.

FRIED, M. & DUFFY, P. E. (1996). Adherence of *Plasmodium falciparum* to chondroitin sulfate in the human placenta. *Science* **272**, 1502–1504.

GARDNER, J. P., PINCHES, R. A., ROBERTS, D. J. & NEWBOLD, C. I. (1996). Variant antigens and endothelial receptor adhesion in *Plasmodium falciparum*. *Proceedings of the National Academy of Sciences, USA* **93**, 3503–3508.

GINSBURG, H., KRUGLIAK, M., EIDELMAN, O. & CABANTCHIK, Z. I. (1985). New permeability pathways induced in membranes of *Plasmodium falciparum*-infected erythrocytes. *Molecular and Biochemical Parasitology* **8**, 177–190.

GOLDRING, J. D. & HOMMEL, M. (1992). Variation in the cytoadherence characteristics of malaria parasites: is this a true virulence factor? *Memórias do Instituto Oswaldo Cruz* **87**, Suppl. III, 313–322.

GOLDRING, J. D., MOLYNEUX, M. E., TAYLOR, T., WIRIMA, J. & HOMMEL, M. (1992). *Plasmodium falciparum*: diversity of isolates from Malawi in their cytoadherence to melanoma cells and monocytes *in vitro*. *British Journal of Haematology* **81**, 413–418.

GRAU, G. E. & BEHR, C. (1994). T cells and malaria: is Th1 cell activation a prerequisite for pathology? *Research in Immunology* **145**, 441–454.

GRAU, G. E., KINDLER, V., PIGUET, P. F., LAMBERT, P. H. & VASSALI, P. (1988). Prevention of experimental cerebral malaria by anticytokine antibodies. IL-3 and GM-CSF are intermediates in increased TNF production and macrophage accumulation. *Journal of Experimental Medicine* **168**, 1499–1504.

GRAU, G. E., TAYLOR, T. F., MOLYNEUX, M. E., WIRIMA, J. J., VASSALI, J., HOMMEL, M. & LAMBERT, P. H. (1989). Tumour necrosis factor and disease severity in children with falciparum malaria. *New England Journal of Medicine* **320**, 1586–1591.

GREENWOOD, B. M., EVANS-JONES, L. G. & STRATTON, D. (1975). Failure to detect endotoxin in serum of children with malaria. *Lancet* **ii**, 874–875.

GREENWOOD, B. M. & VICK, R. M. (1975). Evidence for a malaria mitogen in human malaria. *Nature* **257**, 592–594.

HALDAR, K. (1994). Ducts, channels and transporters in *Plasmodium*-infected erythrocytes. *Parasitology Today* **10**, 393–395.

HANDUNETTI, S. M., MENDIS, K. N. & DAVID, P. H. (1987). Antigenic variation of cloned *Plasmodium fragile* in its natural host *Macaca sinica*: sequential appearance of successive variant antigenic types. *Journal of Experimental Medicine* **165**, 1269–1283.

HELMBY, H., CAVELIER, L., PETERSSON, U. & WAHLGREN, M. (1993). Rosetting *Plasmodium falciparum*-infected erythrocytes express unique strain-specific antigens on their surface. *Infection and Immunity* **61**, 284–288.

HILL, A. V. S. (1992). Malaria resistance genes: a natural selection. *Transactions of the Royal Society of Tropical Medicine and Hygiene* **86**, 225–226.

HO, M., TONGSTAWE, P., KRIANGKUM, J., WIMONWATTRAWATEE, T., PATTANAPANYASAT, K., BRYANT, L., WEBSTER, L., WEBSTER, H. K. & ELLIOTT, J. F. (1994). Polyclonal expansion of peripheral $\gamma\delta$ T cells in human *Plasmodium falciparum* malaria. *Infection and Immunity* **62**, 855–862.

HOMMEL, M. (1985). The role of variant antigens in acquired immunity to *Plasmodium falciparum*. *Annales de la Société Belge de Médecine Tropicale* **65**, suppl. 2, 57–67.

HOMMEL, M. (1993). Amplification of cytoadherence in cerebral malaria: towards a more rational explanation of disease pathophysiology. *Annals of Tropical Medicine and Parasitology* **87**, 627–635.

HOMMEL, M., DAVID, P. H. & OLIGINO, L. D. (1983). Surface alterations of erythrocytes in *Plasmodium falciparum* malaria. Antigenic variation, antigenic diversity and role of the spleen. *Journal of Experimental Medicine* **157**, 1137–1148.

HOMMEL, M. & SEMOFF, S. (1988). Expression and function of erythrocyte-associated surface antigens in malaria. *Biology of the Cell* **64**, 183–203.

ISRAELI, A., SHAPIRO, M. & EPHROS, M. (1987). *Plasmodium falciparum* infection in an asplenic man. *Transactions of the Royal Society of Tropical Medicine and Hygiene* **81**, 233–234.

JAKOBSEN, P. H., BAECK, L. & JEPSEN, S. (1988). Demonstration of soluble *Plasmodium falciparum* antigens reactive with *Limulus* amoebocyte lysate and polymyxin B. *Parasite Immunology* **10**, 593–606.

JAKOBSEN, P. H., BATE, C. W., TAVERNE, J. & PLAYFAIR, J. H. L. (1995). Malaria: toxins, cytokines and disease. *Parasitology Immunology* **17**, 223–231.

JAKOBSEN, P. H., MOON, R. G., RIDLEY, R. G., BATE, C. A. W., TAVERNE, J., TAKACS, B., PLAYFAIR, J. H. L. & MCBRIDE, J. S. (1993). Tumour necrosis factor and interleukin-6 production induced by components associated with merozoite proteins of *Plasmodium falciparum*. *Parasite Immunology* **15**, 229–237.

KUMARATILAKE, L. M., & FERRANTE, A. (1994). T-cell cytokines in malaria: their role in the regulation of neutrophil- and macrophage-mediated killing of *Plasmodium falciparum* asexual blood forms. *Research in Immunology* **145**, 423–429.

KWIATKOWSKI, D. (1995). The biology of malarial fever. In *Baillière's Clinical Infectious Diseases – Malaria* (Ed. G. Pasvol), pp. 371–388. London, Baillière Tindall.

LEECH, J. H., BARNWELL, J. W., MILLER, L. H. & HOWARD, R. J. (1984). *Plasmodium falciparum* malaria: associations of knobs on the surface of infected erythrocytes with a histidine-rich protein and the erythrocyte skeleton. *Journal of Cell Biology* **98**, 1256–1264.

LE SCANF, C., FANDEUR, T., MORALES-BETOULE, M. E. & MERCEREAU-PUIJALON, O. (1997). *Plasmodium falciparum*: altered expression of erythrocyte membrane-associated antigens during antigenic variation. *Experimental Parasitology* **85**, 135–148.

LUCAS, R., LOU, J., MOREL, D. R., RICOU, B., SUTER, P. M. & GRAU, G. E. (1997). TNF receptors in the microvascular pathology of acute respiratory distress syndrome and cerebral malaria. *Journal of Leukocyte Biology* **61**, 551–557.

MCGUIRE, W., HILL, A. V. S., ALLSOPP, C. E., GREENWOOD, B. M. & KWIATKOWSKI, D. (1994). Variation in the TNF-alpha promoter region associated with susceptibility to cerebral malaria. *Nature* **371**, 508–510.

MCLEAN, S. A., PEARSON, C. D. & PHILLIPS, R. S. (1982). *Plasmodium chabaudi*: antigenic variation during recrudescent parasitaemias in mice. *Experimental Parasitology* **54**, 286–302.

MARSH, K. & HOWARD, R. J. (1986). Antigens induced on erythrocytes by *P. falciparum*: expression of diverse and conserved determinants. *Science* 231, 150–152.

MATTEELI, A., CALIGARIS, F., CASTELLI, F. & CAROSI, G. (1997). The placenta and malaria. *Annals of Tropical Medicine and Parasitology* 91, 803–810.

MAUBERT, B., GUILBERT, L. J. & DELORON, P. (1997). Cytoadherence of *Plasmodium falciparum* to intercellular adhesion molecule 1 and chondroitin-4-sulphate expressed by the syncytiotrophoblast in the human placenta. *Infection and Immunity* 65, 1251–1257.

MSHANA, R. N., BOULANDI, J. MSHANA, M. N., MAYOMBO, J. & MENDOME, G. (1994). Cytokines in the pathogenesis of malaria: levels of IL-1β, IL-4, IL-6, TNF-α, and IFN-γ in plasma of healthy individuals and malaria patients in a holoendemic area. *Journal of Clinical and Laboratory Immunology* 34, 131–139.

NEWBOLD, C. I., CRAIG, A. G., KYES, S., BERENDT, A. R., SNOW, R. W., PESHU, N. & MARSH, K. (1997). PfEMP1, polymorphism and pathogenesis. *Annals of Tropical Medicine and Parasitology* 91, 551–557.

OCKENHOUSE, C. F., KLOTZ, F. W., TANDON, N. N. & JAMIESON, G. A. (1991). Sequestrin, a CD36 recognition protein on *Plasmodium falciparum* malaria-infected erythrocytes identified by anti-idiotype antibodies. *Proceedings of the National Academy of Sciences, USA* 88, 3175–3179.

OQUENDO, P., HUNDT, E., LAWLER, J. & SEED, B. (1989). CD36 directly mediates cytoadherence of *Plasmodium falciparum* parasitized erythrocytes. *Cell* 58, 95–101.

PASLOSKE, B. L., BARUCH, D. I., VAN SCHRAVENDIJK, M. R., HANDUNNETTI, S. M., AIKAWA, M., FUJIOKA, H., TARASCHI, T. F., GORMLEY, J. A. & HOWARD, R. J. (1993). Cloning and characterization of a *Plasmodium falciparum* gene encoding a novel high-molecular weight host membrane-associated protein, PfEMP3. *Molecular and Biochemical Parasitology* 59, 59–72.

PAUL, R. E. L., PACKER, M. J., WALMSLEY, M., LAGOG, M., RANDFORD-CARTWRIGHT, L. C., PARU, R. & DAY, K. P. (1995). Mating patterns in malaria parasite populations of Papua New Guinea. *Science* 269, 1709–1711.

PICHYANGKUL, S., SAENGKRAI, P. & WEBSTER, H. K. (1994). *Plasmodium falciparum* pigment induces monocytes to release high levels of tumor necrosis factor alpha and interleukin-1 beta. *American Journal of Tropical Medicine and Hygiene* 51, 430–435.

PICOT, S., PEYRON, F., DELORON, P., BOUDIN, C., VHUMPITAZI, B., BARBE, G., VUILLEZ, J. P., DONADILLE, A. & AMBROISE-THOMAS, P. (1993). Ring-infected erythrocytes surface antigen (Pf155/RESA) induces tumour necrosis-alpha production. *Clinical and Experimental Immunology* 93, 184–188.

PLAYFAIR, J. H. L., TAVERNE, J., BATE, C. A. W. & DE SOUZA, J. B. (1990). The malaria vaccine: anti-parasite or anti-disease? *Immunology Today* 11, 25–27.

PORTA, J., CAROTA, A., PIZZOLATA, G. P., WILDI, E., WIDMER, E., MARGAIRAZ, C. & GRAU, G. E. (1993). Immunopathological changes in human cerebral malaria. *Clinical Neuropathology* 12, 142–146.

RASHEED, F. N., BULMER, J. N., DUNN, D. T., MENENDEZ, C., JAWLA, M. F., JEPSON, A., JAKOBSEN, P. H. & GREENWOOD, B. M. (1993). Suppressed peripheral and placental blood lymphoproliferative responses in first pregnancies: relevance to malaria. *American Journal of Tropical Medicine and Hygiene* 48, 154–160.

RILEY, E. M. (1996). The role of MHC- and non-MHC-associated genes in determining the human immune response to malaria antigens. *Parasitology* 112, S39–S51.

RINGWALD, P., LE BRAS, J. & SAVEL, J. (1992). Participation des cytokines et des sérums immuns à la cytoadhérence des hématies parasitées par *Plasmodium falciparum* sur des cellules endothéliales en culture. *Comptes Rendus de la Société de Biologie* 186, 215–225.

ROBERTS, D. D., SHERWOOD, J. A., SPITALNIK, S. L., PANTON, L. J., HOWARD, R. J., DIXIT, V. M., FRAZIER, W. A., MILLER, L. H. & GINSBURG, V. (1985). Thrombospondin binds falciparum malaria parasitized erythrocytes and may mediate cytoadherence. *Nature* 318, 64–66.

ROBERTS, D. J., CRAIG, A. G., BERENDT, A. R., PINCHES, R., NASH, G., MARSH, K. & NEWBOLD, C. I. (1992). Rapid switching to multiple antigenic and adhesive phenotypes in malaria. *Nature* 357, 689–691.

ROCK, E. P., MARSH, K., SAUL, A. J., WELLEMS, T. E., TAYLOR, D. W., MALOY, W. L. & HOWARD, R. J. (1988). Comparative analysis of the *Plasmodium falciparum* histidine-rich proteins HRP-1, HRP-2 and HRP-3 in malaria parasites of diverse origin. *Parasitology* 95, 209–227.

ROGERSON, S. J. & BROWN, G. V. (1997). Chondroitin sulphate A as an adherence receptor for *Plasmodium falciparum*-infected erythrocytes. *Parasitology Today* 13, 70–75.

ROSENBERG, Y. J. (1978). Autoimmune and polyclonal B responses during murine malaria. *Nature* 274, 170–172.

ROWE, J. A., MOULDS, J. M., NEWBOLD, C. I. & MILLER, L. H. (1997). *P. falciparum* rosetting mediated by a parasite variant erythrocyte membrane protein and complement receptor-1. *Nature* 388, 292–295.

RUDIN, W., FAVRE, N., BORDMANN, G. & RYFFEL, B. (1997). Interferon-γ is essential for the development of cerebral malaria. *European Journal of Immunology* 27, 810–815.

SCHOFIELD, L. & HACKETT, F. (1993). Signal transduction in host cells by a glycosylphosphatidylinositol toxin of malarial parasites. *Journal of Experimental Medicine* 177, 145–153.

SCHOFIELD, L., NOVAKOVIC, S., GEROLD, P., SCHWARZ, R. T., McCONVILLE, M. J. & TACHADO, S. D. (1996). Glycosylphosphatidylinositol toxin of *Plasmodium* upregulates intercellular adhesion molecule-1, vascular adhesion molecule-1 and E-selectin expression in vascular endothelial cells and increases leukocyte and parasite cytoadherence via a tyrosine kinase-dependent signal transduction. *Journal of Immunology* 156, 1886–1896.

SCHWARZER, E., TURRINI, F., ULLIERS, D., GIRIBALDI, G., GINSBURG, H. & ARESE, P. (1992). Impairment of macrophage functions after ingestion of *Plasmodium falciparum*-infected erythrocytes or isolated malarial pigment. *Journal of Experimental Medicine* 176, 1033–1041.

SMITH, J. D., CHITNIS, C. E., CRAIG, A. G., ROBERTS, D. J., HUDSON-TAYLOR, D. E., PETERSON, D. S., PINCHES, R.,

NEWBOLD, C. I. & MILLER, L. H. (1995). Switches in expression of *Plasmodium falciparum var* genes correlate with changes in antigenic and cytoadherent phenotypes of infected erythrocytes. *Cell* **82**, 101–110.

SU, X. Z., HEATWOLE, V. M., WERTHEIMER, S. P., GUINET, F., HERRFELDT, J. A., PETERSON, D. S., RAVETCH, J. A. & WELLEMS, T. E. (1995). The large diverse gene family *var* encodes proteins involved in cytoadherence and antigenic variation of *Plasmodium falciparum*-infected erythrocytes. *Cell* **82**, 89–100.

TAVERNE, J., SHEIKH, N., ELASED, K. & PLAYFAIR, J. H. L. (1995). Malaria toxins: hypoglycaemia and TNF production are induced by different components. *Parasitology Today* **11**, 462–463.

TAYLOR, K., CARR, R. E., PLAYFAIR, J. H. L. & SAGGERSON, E. D. (1992). Malarial toxic antigens synergistically enhance insulin signalling. *FEBS Letters* **311**, 231–234.

TAYLOR, T. E., MOLYNEUX, M. E., WIRIMA, J. J., BORGSTEIN, A., GOLDRING, J. D. & HOMMEL, M. (1992). Intravenous immunoglobulin in the treatment of pediatric cerebral malaria. *Clinical and Experimental Immunology* **90**, 357–362.

THAITHONG, S., BEALE, G. H., FENTON, B., MCBRIDE, J., ROSARIO, V., WALKER, A. & WALLIKER, D. (1984). Clonal diversity in a single isolate of the malaria parasite *Plasmodium falciparum*. *Transactions of the Royal Society of Tropical Medicine and Hygiene* **78**, 242–245.

THOMPSON, J. K., RUBIO, J. P., CARUANA, S., BROCKMAN, A., WICKHAM, M. E. & COWMAN, A. F. (1997). The chromosomal organization of *Plasmodium falciparum var* gene family is conserved. *Molecular and Biochemical Parasitology* **87**, 49–60.

TOURÉ-BALDE, A., SARTHOU, J. L., ARIBOT, G., MICHEL, P., TRAPE, J. F., ROGIER, C. & ROUSSILHON, C. (1996). *Plasmodium falciparum* induces apoptosis in human mononuclear cells. *Infection and Immunity* **64**, 744–750.

TRAGER, W., RUDZINSKA, M. A. & BRADBURY, P. C. (1966). The fine structure of *Plasmodium falciparum* and its host erythrocyte in natural malarial infections in man. *Bulletin of the World Health Organization* **35**, 883–885.

TREUTIGER, C. J., HEDLUND, I., HELMBY, H., CARLSON, J., JEPSON, A., TWUMASI, P., KWIATKOWSKI, D., GREENWOOD, B. M. & WAHLGREN, M. (1992). Rosette formation in *Plasmodium falciparum* isolates an anti-rosette activity of sera from Gambians with cerebral or uncomplicated malaria. *American Journal of Tropical Medicine and Hygiene* **46**, 503–510.

TUBBS, H. (1980). Endotoxin in human and murine malaria. *Transactions of the Royal Society of Tropical Medicine and Hygiene* **74**, 121–123.

TURNER, G. D., MORRISON, G., JONES, M., LOOAREESUWAN, S., DAVIS, T., BULEY, I., GATTER, K., NEWBOLD, C. I., WHITE, N. & BERENDT, A. (1994). An immunohistochemical study of the pathology of fatal malaria. Evidence for widespread endothelial activation and a potential role for intercellular adhesion molecule-1 in cerebral sequestration. *American Journal of Pathology* **145**, 1057–1069.

TURRINI, F., GIRIBALDI, G., VALENTE, E. & ARESE, P. (1997). *Mycoplasma* contamination of *Plasmodium* cultures – A case of parasite parasitism. *Parasitology Today* **13**, 367–368.

WAHLGREN, M., CARLSON, J. & UDOMSANGPETCH, R. (1987). Why do *Plasmodium falciparum*-infected erythrocytes form spontaneous rosettes. *Parasitology Today* **5**, 183–185.

WENISCH, C., PARSCHALK, B., BURGMANN, H., LOOAREESUWAN, S. & GRANINGER, W. (1995). Decreased serum levels of TGF-β in patients with acute *Plasmodium falciparum* malaria. *Journal of Clinical Immunology* **15**, 69–73.

WINOGRAD, E. & SHERMAN, I. W. (1989). Characterization of a modified red cell membrane protein expressed on erythrocytes infected with the human malarial parasite *Plasmodium falciparum*: possible role as a cytoadherence mediating protein. *Journal of Cell Biology* **108**, 23–30.

YESNER, L. M., HUH, N. Y., PEARCE, S. F. & SILVERSTEIN, R. L. (1996). Regulation of monocyte CD36 and thrombospondin-1 expression by soluble mediators. *Arteriosclerosis, Thrombosis and Vascular Biology* **16**, 1019–1025.

Immune evasion in malaria: altered peptide ligands of the circumsporozoite protein

M. PLEBANSKI, E. A. M. LEE *and* A. V. S. HILL

Nuffield Department of Medicine, Institute of Molecular Medicine, University of Oxford, John Radcliffe Hospital, Oxford OX3, UK.

SUMMARY

T cells are central to immunity in malaria. CD4[+] helper T cells favour the generation of high-affinity antibodies that are effective against blood stages and they are necessary to establish immunological memory. The intrahepatic stage of infection can be eliminated by specific CD8[+] cytotoxic T cells (CTL). Cytokines secreted by CD4[+] T cells may also contribute to liver stage immunity. Evolution has selected varied mechanisms in pathogens to avoid recognition by T cells. T cells recognize foreign epitopes as complexes with host major histocompatibility (MHC) molecules. Thus, a simple form of evasion is to mutate amino acid residues which allow binding to an MHC allele. Recently, more sophisticated forms of polymorphic evasion have been described. In *altered peptide ligand (APL) antagonism*, the concurrent presentation of particular closely related epitope variants can prevent memory T cell effector functions such as cytotoxicity, lymphokine production and proliferation. In *immune interference*, the effect of the concurrent presentation of such related epitope variants can go a step further and prevent the induction of memory T cells from naive precursors. The analysis of immune responses to a protein of *P. falciparum*, the circumsporozoite protein (CSP), indicates that the malaria parasite may utilize these evasion strategies.

Key words: evasion, malaria, T cells, antagonism, circumsporozoite protein.

INTRODUCTION

Malaria is a major health problem in the developing world affecting and killing millions each year. Drug-resistant parasite strains are appearing regularly and there is no generally effective vaccine in use (Charoenvit *et al.* 1991; Alonso *et al.* 1994; Beck *et al.* 1994; Ballou *et al.* 1995; D'Alessandro *et al.* 1995; Nosten *et al.* 1996). The complicated life cycle of this parasite may still provide the opportunities for intervention at various stages of its development (Good, Berzofsky & Miller, 1988; Nardin & Nussenzweig, 1993; Gilbert & Hill, 1997). Within minutes from an infectious mosquito bite, sporozoites specifically infect host liver cells. Parasites will differentiate and replicate inside liver cells for approximately a week before lysing them and proceeding onto the blood stage of infection. The infection of host red blood cells proceeds exponentially, with cycles of proliferation and release, the latter contributing to the periodic nature of the characteristic malaria fever. After a variable period of infection, sexual parasite stages appear in the blood. If these are taken up by a feeding mosquito, sexual recombination of gametocytes may occur in the mosquito midgut to generate new variants of sporozoites which then migrate to the mosquito salivary gland, ready to infect a human host.

Naturally exposed individuals have antibodies and CD4[+] T cells against both pre-erythrocytic and erythrocytic stage antigens (Good, 1988; Good *et al.* 1988; Nardin & Nussenzweig, 1993; Zevering *et al.* 1994; al Yaman *et al.* 1995), as well as cytotoxic T lymphocytes (CTL) which recognize *Plasmodium* antigens present during the liver stage of infection (Doolan, Houghten & Good, 1991; Malik *et al.* 1991; Hill *et al.* 1992; Aidoo *et al.* 1995). Although pre-erythrocytic immunity generated by infection with irradiated sporozoites can confer complete protection in animal and human models (Hoffman *et al.* 1989; Herrington *et al.* 1991; Moreno *et al.* 1993), the use of irradiated sporozoites is not a feasible option for large-scale human immunization. Thus there has been a search for candidate immunogens which can be produced as subunit vaccines. The last decade has seen the identification of B cell, CD4[+] T helper and CD8[+] cytotoxic T cell *Plasmodium* epitopes which are protective in animal models (Hoffman *et al.* 1989; Romero *et al.* 1989, 1992; Windmann *et al.* 1992; Nardin & Nussenzweig, 1993; Wang *et al.* 1995; Doolan *et al.* 1996). In many cases, the homologous target B and T cell epitopes for *P. falciparum* malaria in humans have been identified. Antibodies against the immunodominant B cell epitope of circumsporozoite protein (CS) are protective in mice (Hoffman *et al.* 1989; Romero *et al.* 1989; Nardin & Nussenzweig, 1993). However, immunization of humans with vaccines designed primarily to induce humoral responses have in general failed to demonstrate protection in field trials (Charoenvit *et al.* 1991; Alonso *et al.* 1994; Beck *et al.* 1994; Ballou *et al.* 1995; D'Alessandro *et al.* 1995; Nosten *et al.* 1996). It could, however, be argued that the full range of protective responses, including T cell activation, may be necessary for an effective vaccine (Nardin *et*

Variability in T cell epitopes of the circumsporozite protein of *P. falciparum*

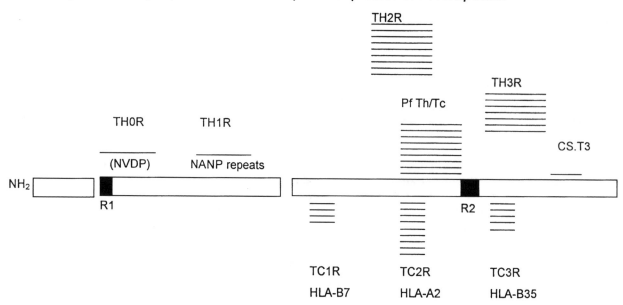

Fig. 1. Schematic representation of epitopes in the circumsporozoite protein of *P. falciparum*. It is not drawn to scale. The relative location of T helper epitopes is shown by lines on top of the CSP protein, and that of CTL epitopes on the bottom. Each line in the same location represents a variant of this epitope (Dame *et al.* 1984; Good, Berzofsky & Miller, 1988; Good *et al.* 1988; Lockyer, Marsh & Newbold, 1989; Good, 1991; Nardin *et al.* 1991; Doolan, Saul & Good, 1992; McCutchan *et al.* 1992; Shi *et al.* 1992; Qari *et al.* 1994; Udhayakumar *et al.* 1994; Zevering, Khamboonruang & Good, 1994; Aidoo *et al.* 1995; Blum Tirouvanziam *et al.* 1995). It is thus a minimal estimate of the possible variants that might exist. In addition, it does not exclude the notion that other variants may have existed at different times in this geographical location. The R2 conserved region contains the liver cell binding motif of the CSP protein (Dame *et al.* 1984).

al. 1992; Windmann *et al.* 1992; Nardin & Nussenzweig, 1993; Rodrigues, Nussenzweig & Zavala, 1993; Fell, Currier & Good, 1994).

The negative association of severe malaria and the HLA-B53 allele in West Africa indicated that class I restricted T cell responses are probably important in protection against human malaria (Hill *et al.* 1991). A conserved HLA-B53 binding peptide Ls6 from the liver stage protein 1 (LSA-1) was found to be the target of CTL responses in Gambian donors, suggesting a molecular basis for the HLA-B53 mediated protection (Hill *et al.* 1992; Aidoo *et al.* 1995). An effective liver stage malarial vaccine which could confer protection to a sizable proportion of the population would need to contain CTL epitopes that bind to a variety of common HLA class I alleles. CTL epitopes have now been identified in a range of proteins present during the liver stage of malaria, CSP, TRAP, STARP and LSA-1, which bind to some of the most common class I alleles found both in Europe and in West Africa, namely, HLA-B7, -B8, -B35, -B53, and A2 (Malik *et al.* 1991; Aidoo *et al.* 1995; Blum Tirouvanziam *et al.* 1995). Class II restricted responses are also predicted to be important, not only because CD4⁺ T cells promote the development of CTL and antibody responses, but because lymphokines which can be secreted by CD4 T cells such as TNFα and IFNγ can directly affect

parasite viability at both the liver (Mellouk *et al.* 1991; Nussler *et al.* 1991) and blood (Kwiatkowski, 1990; Kwiatkowski *et al.* 1990; McGuire *et al.* 1994; Jacobs, Radzich & Stevenson, 1996) stages of infection. CD4⁺ T cell epitopes have also been mapped in pre-erythrocytic and blood-stage proteins of *P. falciparum* (Good *et al.* 1988; Kumar *et al.* 1992; Nardin *et al.* 1992; Bilsborough, Carlisle & Good, 1993; Moreno *et al.* 1993; Udhayakumar *et al.* 1995).

Many research groups are working on the identification of further T cell epitopes. Moreover, novel vaccine delivery systems such as DNA (Hedstrom, Sedegah & Hoffman, 1994; Sedegah *et al.* 1994; Doolan *et al.* 1996), hepatitis B (Gilbert & Hill, 1997; Stoute *et al.* 1977) and yeast TY recombinant particles (Allsopp *et al.* 1996; Gilbert & Hill, 1997), modified vaccinia (Tine *et al.* 1996) and Salmonella (Schödel *et al.* 1994) live carriers, as well as new adjuvants (Allsopp *et al.* 1996; Stoute *et al.* 1997) are being investigated to optimize the induction of cellular T cell responses to malarial epitopes. Malaria has been prevented in animals vaccinated with some of these novel carriers and adjuvants (Schödel *et al.* 1994; Doolan *et al.* 1996) and recent human volunteer studies have shown high levels of protection to homologous challenge after vaccination with recombinant hepatitis B surface antigen particles and the C-terminal 189 aa from CSP (Stoute *et al.* 1997). It is expected that we will soon see some of these new T cell epitope-focused vaccines tested in the field. However, optimism must be tempered by the fact that in the field we are dealing with a heterogeneous population of parasites, which has had thousands of years to adapt to a dangerous lifestyle in their genetically heterogeneous human hosts. The most commonly used animal laboratory models of malaria not only fail to mimic this heterogeneity, but also fail to reproduce many clinical aspects of the human disease. Moreover, human laboratory-based protection studies usually utilize immunogens derived from a single antigenic strain and assay homologous protection in volunteers (Stoute *et al.* 1997). These volunteers are not normally part of the same population as the one to which the target parasite population has adapted its evasion strategies. In the field, a particular *P. falciparum* T cell epitope can have as many as 14 different polymorphic variants (Lockyer, Marsh & Newbold, 1989; Doolan, Saul & Good, 1992), and frequently CD8+ T cell responses to polymorphic variants are not crossreactive (Hill *et al.* 1992; Udhayakumar *et al.* 1994; Aidoo *et al.* 1995). In fact, most T cell epitopes in the CS protein are polymorphic (Fig. 1).

This present review will discuss how polymorphism of *P. falciparum* proteins, with CS as an example, can contribute to immune evasion. Novel immune evasion strategies based on the existence of altered peptide ligands (APLs) such as T cell antagonism and immune interference, will be described in this context. The more conventional strategies based on simple prevention of the presentation of complexes of the host's major histocompatibility molecules (MHC) and parasite-derived peptides, will also be reviewed. Moreover, the human host may have in turn adapted to protect itself against the parasite. The consideration of this ongoing molecular arms race should be useful in the rational design of a vaccine to work in genetically variable human populations in malaria endemic areas of the world.

THE CIRCUMSPOROZOITE PROTEIN

In many ways the circumsporozoite protein (CS) of *P. falciparum* appears to be an ideal immunogen for use in a pre-erythrocytic vaccine. It is the most abundant protein on the sporozoite coat and participates in its binding to liver cells (Dame *et al.* 1984). Antibodies against CS can block liver cell infection *in vivo* (Hoffman *et al.* 1989; Nardin *et al.* 1991) and *in vitro* (Romero *et al.* 1989). Cytotoxic T lymphocytes (CTL) against CS alone can confer complete protection in mice (Hoffman *et al.* 1989; Romero *et al.* 1989) suggesting it is an important target for liver stage immunity. As well as B cell and CTL epitopes,

the CS protein also contains CD4+ T cell epitopes (Fig. 1), and thus theoretically could induce a broad range of effector mechanisms.

The B cell epitope recurs in tandem within its large central repeat region (Fig. 1). The majority of individuals living in malaria-endemic areas have variable levels of anti-CS antibodies which confer partial immunity (Good *et al.* 1988; Marsh *et al.* 1988; Graves *et al.* 1989; Quakyi *et al.* 1989). Antibodies against this epitope can block liver cell infection *in vitro* (Ballou *et al.* 1985; Young *et al.* 1985; Zavala *et al.* 1985) and transfer of anti-CS antibodies in murine models can confer protection (Egan *et al.* 1987). As an added bonus for vaccine design, the repeat region of *P. falciparum* CSP is highly conserved in the parasite population (Dame *et al.* 1984; de la Cruz, Lal & T. F. 1987; Lockyer *et al.* 1989). Nevertheless, vaccines aimed at inducing humoral immunity against CS were only partly successful in human laboratory-based trials and have largely failed to protect against malaria in human field trials (Good *et al.* 1988; Herrington *et al.* 1991; Nardin & Nussenzweig, 1993). Moreover, high-affinity antibodies were not found frequently in these human trials (Good *et al.* 1988; Herrington *et al.* 1991). Hence, it could be argued that the failure to induce appropriate CD4+ helper T cells at the time of immunizing with the B cell epitope may have hindered the generation of high-affinity IgG antibodies (Scherle & Gerhard, 1986; Good *et al.* 1987). It could be speculated that the multi-repeat structure of the B cell epitope in CS could act as a standard T cell-independent polymeric antigen and preferentially induce low-affinity IgM antibodies. Whatever the mechanism for its poor immunogenicity, it may also be interesting to consider why a B cell epitope which is so obviously a target for protection is so remarkably conserved and multi-meric.

CD4+ T cell responsiveness against CS epitopes is found in $> 40\%$ of naturally exposed individuals (Good *et al.* 1988). CD4+ T cell epitopes are concentrated at the carboxy terminus (Fig. 1) (Good *et al.* 1988). Although no direct evidence exists for a protective role for CD4+ T cells *in vivo* in humans, in murine studies depletion of CD4 T cells prevents the induction of protection by immunization with irradiated sporozoites (Rodrigues *et al.* 1993). Hence, in mice CD4+ T cells could either play a direct role or be necessary to promote protective CTL responses. Human volunteers immunized with the carboxy terminal CS segment were found to have proliferating and IFNγ-secreting T cells but lacked class I restricted CTL (Stoute *et al.* 1997). The high levels of protection observed in this pilot trial did not correlate with titres of anti-CS antibodies and may suggest direct CD4+ T cell-mediated protection. The observation that IFNγ, IL-6 and TNF can promote parasite death at both the liver (Mellouk *et al.* 1991; Nussler *et al.* 1991) and blood stages of infection (Jacobs *et al.* 1996), suggests that CD4$^+$ T cells secreting these lymphokines could contribute to protection. It is of interest to note that the immunodominant CD4$^+$ T cell epitope of CS (TH2R, aa 326-345) contains the most known sequence variability of the whole of CS (Fig. 1) (Good *et al.* 1988). Fourteen variants have been observed (Dame *et al.* 1984; de la Cruz, Lal & McCutchan 1987; Lockyer *et al.* 1989), with 9 co-existing in The Gambia (Lockyer *et al.* 1989; Conway, Greenwood & McBride, 1991). Interestingly, the same variants can be found in widely different geographical regions, with some evidence for convergent evolution (de la Cruz *et al.* 1987; Lockyer *et al.* 1989; Yoshida *et al.* 1990; McCutchan *et al.* 1992; Qari *et al.* 1994; Gupta & Hill, 1995). This area of CS contains a higher than average number of non-synonymous substitutions from analysis of sequence variation (Hughes, 1991), further arguing for strong selective pressure at this locus (Lockyer *et al.* 1989; Yoshida *et al.* 1990; McCutchan *et al.* 1992; Gupta & Hill, 1995). The adjacent conserved RII region, shown to participate in liver cell binding (Dame *et al.* 1984), can also be partly recognized by T cells, but far less frequently than TH2R or TH3R, another highly polymorphic epitope, localized between aa 367–385 (Good *et al.* 1988). Thus, within the small spectrum of CS epitopes, polymorphism and immunodominance can be inversely related. It may be, however, premature to ascribe the polymorphism detected to immune selection pressure by CD4$^+$ T cells alone, since the immunodominant CS CTL epitopes are also contained within TH2R and TH3R (Good *et al.* 1988; Doolan *et al.* 1991; Malik *et al.* 1991; Shi *et al.* 1992; Udhayakumar *et al.* 1994; Aidoo *et al.* 1995; Blum Tirouvanziam *et al.* 1995).

Elegant early work in animal models demonstrated that CS specific CTL could be recovered from the lymphocytes that infiltrate foci of infection with irradiated sporozoites, and that these cells can transfer complete protection against malaria to naive recipients (Romero *et al.* 1989). Unfortunately, CS-specific CTL were found only infrequently and at low levels in naturally exposed humans (Doolan *et al.* 1993). Moreover, 3 out of 4 known human CTL CS epitopes are polymorphic (Fig. 1) (Aidoo *et al.* 1995). This observation stands in contrast to the CTL-identified epitopes of other proteins present during the liver stage which are generally conserved (Aidoo *et al.* 1995). The two most polymorphic CS CTL epitopes are the ones found within TH2R and TH3R, and which we will call TC2R and TC3R (Fig. 1). Most TC2R variants can bind HLA-A2 (M. Plebanski, unpublished observations), and some TC3R variants bind HLA-B35 (Aidoo *et al.* 1995). It is of note that both HLA-B35 and HLA-A2 are very common in West Africa, where *P. falciparum* malaria is endemic (Hill *et al.* 1991; Krausa *et al.* 1995).

HOW CAN A PATHOGEN EVADE RECOGNITION BY EPITOPE-SPECIFIC T CELLS?

Recognition of infected cells by T cells

A primitive but effective form of evasion is the escape of pathogens from specific antibodies by lodging inside host cells. Cells infected by intracellular pathogens can, however, be killed by specific CTL which either puncture the infected cell or transduce to it a suicide signal (Lowin *et al.* 1994). In addition, some CTL-derived lymphokines can stimulate the target cell to generate products toxic for the intracellular pathogen such as reactive NO-species (Mellouk *et al.* 1991; Doolan *et al.* 1996). Cytotoxic T cells are mostly CD8$^+$, but some CD4$^+$ CTL which recognize MHC class II/foreign peptide complexes have also been described (Nardin *et al.* 1992). CD8$^+$ CTL recognize on the surface of an infected cell a complex of the host's MHC class I molecule and a specific peptide derived from the intracellular pathogen. These peptides and/or slightly longer peptide precursors (8–25 amino acids (aa) in length) are generated from cytoplasmic proteins by the host's proteosome complex and then preferentially transported from the cytoplasm into the endoplasmic reticulum (ER) by TAP proteins where additional proteolytic trimming might occur and empty class I molecules are loaded (Goldberg & Rock 1992; Townsend & Trowsdale 1993). Subsequently, class I molecules loaded with peptides 8–11 aa long are transported to the cell surface. MHC class II loading with peptides normally follows a different route in which vacuoles containing either newly synthesized or surface recycled class II are fused with vacuoles in which proteins taken up by endocytosis have been partly degraded by specialized lysozomal proteases. Thus, normally class I is loaded with peptides derived from intracellular, and class II extracellular, proteins. However, it has been recently demonstrated that extracellularly delivered protein particles are a good source of peptides for class I loading (Kovacsovics Bankowski *et al.* 1993; Falo *et al.* 1995; Reis e Sousa & Germain, 1995). Interference by pathogens with class I and II presentation and T cell recognition is summarized in the following sections (see Fig. 2).

Immune evasion mechanisms

Preventing the presentation of all epitopes by MHC. Pathogens use an array of strategies to prevent the generation of target MHC/peptide complexes. The best examples have thus far been found in viral infections, probably both because studies of CTL-mediated protection to viruses have been studied the longest and because viruses are structurally relatively simple pathogens. For example, adenovirus specifically downregulates class I expression on the cells it infects by trapping class I molecules in the ER

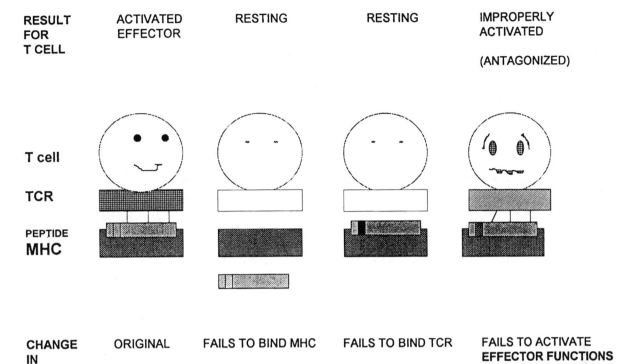

Fig. 2. Altered peptide ligands of cytotoxic T cell epitopes prevent correct T cell activation by three main mechanisms: prevent MHC binding, prevent T cell recognition through its T cell receptor (TCR) or prevent T cell activation.

through association with its E19 protein (Andersson *et al.* 1985) as well as decreasing the generation of newly formed class I through the inhibition of the transport of cellular mRNAs from the nucleus and the translation of host mRNA (Vaessen, Houweling & van der Eb, 1987). Human cytomegalovirus (HCMV) can increase the degradation of class I molecules before they can be expressed on the infected cell surface (Beersma, Bijlmakers & Ploegh, 1993). Intracellular bacteria such as *Mycobacterium tuberculosis* prevent digestion by host acid proteases when they lodge in the macrophage's phagosome by virtue of enzymes which prevent acidification (Chicurel, Garcia & Goodsaid, 1988). No molecular studies have addressed the general mechanism of MHC/peptide complex formation in malaria, to determine whether *Plasmodium* employ evasion strategies at this level.

Preventing binding to MHC of specific epitopes. Whether it is to avoid recognition by CD4+ or by CD8+ T cells specific for a given epitope, a simple strategy for the parasite would be to mutate the residues which are involved in the binding of this epitope to the MHC molecule. Generally MHC class I binding has strict requirements as to the nature of the aa involved at 2 peptide anchor positions (Rammensee, 1995). Mutation of either anchor position to a non-permissive aa would normally abrogate binding. Class II binding peptides may require more extensive changes to abrogate binding, since binding energy is also contributed by the

peptide backbone, there is a wider permissiveness in terms of peptide length, and for many alleles at least 3 anchor positions have been identified (Rammensee, 1995). No comprehensive studies exist on the ability of the *P. falciparum* TH2R and TH3R variants to bind different MHC class II alleles, or on the prevalence of certain helper epitope variant-bearing strains in individuals of a particular MHC class II type. In fact, there is a great need for studies addressing the role of MHC class II binding and escape variants for any pathogen population. The situation is clearer for MHC class I. Thus, high mutation rates in regions with class I binding potential have been demonstrated in viruses, particularly HIV (Phillips *et al.* 1991). Such mutations can alter the ability of a particular peptide epitope to bind the host's restricting class I allele, as clearly shown for a dominant epitope of Epstein Barr virus (de Campos Lima *et al.* 1993, 1994). Escape variants in HIV have been characterized which are able to overwhelm the host by virtue of a single mutation in the anchor position of a single CTL epitope which prevents its binding to the host's MHC class I (Phillips *et al.* 1991). A potential HLA-B35-restricted CTL epitope site has been localized within the TH3R region of *P. falciparum* CS protein (Áidoo *et al.* 1995). For convenience we will call this region TC3R. Of the 4 common variants of this epitope found in The Gambia (cp26, cp27, cp28, cp29), only two bound to HLA-B35 (cp26 and cp29) and were able to induce CTL responses in naturally exposed Gambians (Hill *et al.* 1992; Aidoo *et al.* 1995). Thus, it may be that the other strains (bearing cp27 and cp28) could have an advantage in HLA-B35 hosts, by evading the observed HLA-B35 restricted CTL responses.

Preventing T cell recognition. New variant peptide sequences may be generated by a pathogen which retains the ability to bind the same class I molecule. However, CTL specific for the original peptide/ MHC complex may not recognize them. This phenomenon has been described for viral epitopes, such as those in the NEF protein of HIV (Couillin *et al.* 1994). Studies on selected variants of the TH2R and TH3R regions of CPS in both human (Hill *et al.* 1992; Aidoo *et al.* 1995) and mice (Udhayakumar *et al.* 1994) indicated that such non-cross-reactive responses may exist in malaria. These non-cross-reactive responses could possibly confer an advantage to the selected variant. Our previous studies on the HLA-B35-restricted CTL response in naturally exposed Gambians also suggests that they are generally non-cross-reactive in standard CTL assays (Aidoo *et al.* 1995). Precursor frequency analysis studies by limiting dilution analysis (LDA) further suggests that reactivity to the two CTL epitopes of the TH3R region, cp26 and cp29, tends to occur in different donors (Plebanski *et al.* 1997). Therefore, limited cross-reactivity could favour the presence of CTL epitopes in this CS polymorphic region.

Preventing the generation of T cells. Surprisingly, we have observed that CTL reactivity specific to either cp26 or cp29 is maintained over 4 malaria seasons in a single individual ($n = 3$ individuals tested) (Plebanski *et al.* 1997). The prevalence of cp26- and cp29-bearing variants in this population is 25 and 30% respectively, which suggests that these donors have been expressed to both types of variants. Therefore it is possible that responses to 1 variant can be re-stimulated by the other. However, we do not see reactivity to both, thus it would also have to be postulated there was no induction of memory CTL from naive precursors to the second variant. The T cell responses would thus be 'deviated' away from the currently invading strain. Immune deviation has been observed in HIV, HBV and HCV infections (Bertoletti *et al.* 1994*a*; Klenerman *et al.* 1995; Kaneko *et al.* 1997). An HIV epitope has been described which stimulates *in vitro* CTL which are able to recognize a variant of the epitope, but not the epitope itself in CTL assays (Klenerman *et al.* 1995). In addition, CTL reactivity to an HBV epitope exists long after this epitope is substituted by a polymorphic variant to which there is no detectable reactivity (Bertoletti *et al.* 1994*a*). In malaria, the study of the induction of CTL from naive precursors (Plebanski *et al.* 1995) to the cp26 and cp29 variants supported the prediction that cp26 can induce cp29-specific CTL and vice versa (M. Plebanski *et al.*

unpublished). Moreover, when these variants were presented simultaneously to the T cells, they could completely prevent the generation of CTL from naive precursors, a phenomenon of repertoire selection which we propose to call 'immune interference'.

Preventing effector T cell functions. Once epitope-specific T cells are generated, the pathogen may still modulate their effector functions by altered peptide ligand (APL) antagonism. In this phenomenon, the concomitant presentation of closely related variant peptides provides an altered activation signal to the T cell, resulting in the inactivation of some of its effector functions (Sette *et al.* 1994; Jameson & Bevan, 1995). Thus, for class I restricted CTL, it has been observed that the simultaneous presentation of 2 natural variants of an HIV (Klenerman *et al.* 1994) or an HBV (Bertoletti *et al.* 1994*b*) epitope can prevent killing of the target cells. The production of IFNγ was also downregulated (Bertoletti *et al.* 1994*b*). Minimal changes may be necessary to convert a class I binding peptide into an antagonist (Jameson, Carbone & Bevan 1993; Bertoletti *et al.* 1994*b*; Klenerman *et al.* 1994; Dong *et al.* 1996). Thus, single conservative amino acid changes in residues predicted to contact the T cell receptor (TCR) frequently yield antagonistic peptides (Jameson *et al.* 1993; Bertoletti *et al.* 1994*b*; Klenerman *et al.* 1994). Interestingly, natural variants of TC2R region frequently have conservative substitutions at such positions. However, a single change in the anchor position of an Influenza HLA-B35 binding epitope, serine to proline at position 2 of the peptide, may also result in an antagonist ligand (Dong *et al.* 1996). Of note is that the 2 HLA-B35-binding variant CS epitopes cp26 and cp29 of TC3R only differ by a serine to proline change at this anchor position (Aidoo *et al.* 1995). This opens the exciting possibility that cp26 and cp29 may be such APL antagonists, and promote the survival of these variants within the parasite population. This hypothesis would require that cp26 and cp29 co-habit in the host. Co-infections are indeed very common, even in an area of medium and perennial transmission such as The Gambia. In fact, in The Gambia, over 25% of all co-infections are with the cp26 and cp29 variants (S. Gilbert *et al.* unpublished). A further advantage to co-habitation by parasites bearing these particular variants would be the ability to effect the 'immune interference' described in the previous section, and thus go a step back to prevent the effective generation of cp26 or cp29 reactive CTL from naive precursors.

Effector functions other than killing may fail to be activated by APLs of CTL epitopes. In fact, a variant peptide form of the Tax protein of HTLV-1 failed to induce IL2 production or proliferation from CTL specific for the index peptide, but was strongly recognized in CTL killing assays (Hollsberg *et al.*

1995). In another study, this time with cytotoxic T cells recognizing variants of an ovalbumin-peptide (Jameson, *et al.* 1993), serine esterase release, calcium flux, production of GM-CSF and IL-3 were all shown to have varying degrees of susceptibility to antagonism. The selective failure to activate secretion of particular types of lymphokines by APL has also been described for CD4$^+$ T cells (Evavold & Allen, 1991; Evavold *et al.* 1993). Thus, a class II restricted clone which normally would respond to its index epitope ligand by proliferation as well as IL-2 and IL-4 secretion, selectively downregulated proliferation and IL-2 secretion, but not IL-4 production, if stimulated by an APL (Evavold & Allen, 1991). The use of such mechanisms by a pathogen *in vivo*, could lead to modulation of the immune response towards the T-helper type 2 (Th2) pathway, and thus avoid T-helper type 1 (Th1) responses, where these normally play a role in protection. This type of *in vivo* modulation by peptide analogues has in fact been successfully utilised artificially to alter the pathogenic Th1 profile of autoimmune cells in experimental autoimmune encephalomyelitis (EAE) to the non-harmful Th2 type (Karin *et al.* 1994). It is not known whether *P. falciparum* variant peptide epitopes also act in this manner.

COMPARISON OF IMMUNE EVASION STRATEGIES

Strains which have generated an epitope which does not bind to the MHC class I may have an advantage if the cytotoxic T cell response is restricted to that particular epitope. However, lack of binding of one epitope may allow others which were competitively excluded to bind. In addition, subdominant responses to other epitopes may become dominant. A similar argument can be utilized for the generation of non-cross-reactive variants. It is thus difficult to justify any permanent dominant advantage to strains which employ lack of MHC binding or cross-reactivity as an evasion strategy. The exceptions will be found in cases where other epitopes cannot compensate for the dominant epitope that has been lost (de Campos Lima *et al.* 1993, 1994). Allowing an epitope to continue binding but modifying it slightly so that it can modulate the immune response may be a more successful strategy. Thus, by interfering with the generation of primary T cell responses ('immune interference'), and by diverting the immune response to another variant, a pathogen effectively perpetuates a population of 'naive' hosts it can repeatedly infect. In addition, by modulating the type of T cell effector functions by APL antagonism, it can perpetuate a population of 'immune' hosts which nevertheless continue being susceptible to infection. The ability to modulate lymphokine production, in particular, may make APL-bearing

variants able to divert the quality of the response against other epitopes of the pathogen into a non-protective pattern (Evavold & Allen, 1991). The ability to modulate lymphokine responses can also have systemic effects. At the population level these strategies should be devastatingly effective. The study of antagonism in viral infections has raised the question of whether newly generated antagonist strains will still have an advantage when they eventually induce specific CTL from naive precursors. The antagonistic strain could become an 'altruist', turning off responses against the parent strain, but not benefiting in the process (Davenport, 1995). The combination of antagonism with 'immune interference', however, would ensure that such naive T cells would not be efficiently stimulated (see Fig. 3). Studies of successful pathogens such as HIV and HBV as well as our recent work on malaria suggest that 'immune interference' and antagonism may indeed be caused by the same variant APLs (Bertoletti *et al.* 1994*a*; Klenerman *et al.* 1994; M. Plebanski *et al.* unpublished).

IMMUNE EVASION AND MHC POLYMORPHISM

Conserved amino acids, usually at 2 key positions, bind in deep pockets within the MHC class I groove (Bjorkman *et al.* 1987) and largely determine the ability of any given peptide to bind to a particular class I molecule allele (Rammensee, Friede & Stevanoviic, 1995). These positions provide a characteristic 'peptide-binding motif'. Genetic analysis of HLA molecules has shown a high proportion of nucleotide substitutions leading to amino acid changes in the 3 dimensional region involved in peptide/MHC interactions, which suggests diversifying selection pressures rather than genetic drift (Hughes & Nei, 1988). Thus, the natural polymorphism of HLA molecules may reflect the continuous need of human populations to recognize and deal with a varying challenge of environmental pathogens. The selection of a particular HLA allelic type by a specific variant pathogen strain has been difficult to prove, although different MHC alleles have been demonstrated to be associated with protection or susceptibility to disease. Thus, HIV susceptibility is associated with the HLA haplotype-A1-B8-DR3 (Steel *et al.* 1988; Kaslow *et al.* 1990) and of HLA-B35 with disease progression (Itescu *et al.* 1992), protection against severe malaria with HLA-B53 (Hill *et al.* 1992) and leprosy is associated with HLA-DR2 (Rani *et al.* 1993), to mention but a few. These associations are usually found locally; thus, HLA-B53 is associated with protection in West but not East Africa (Hill *et al.* 1992). One interpretation of this finding is that different strains of the same pathogen are indeed the main cause of disease in different geographical locations, but no direct evidence exists to support this hypothesis.

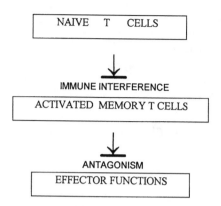

Fig. 3. Altered peptide ligands can affect repertoire selection as well as effector functions.

IMPLICATIONS FOR VACCINE DESIGN

The CS protein is recognized by the immune system; antibodies and T cells against it can confer protection (Good *et al.* 1988; Marsh *et al.* 1988; Graves *et al.* 1989; Quakyi *et al.* 1989). The stumbling block, however, has been the extensive polymorphism in its T cell epitopes (Dame *et al.* 1984; de la Cruz *et al.* 1987; Lockyer *et al.* 1989). This may have been overcome by identifying variants which generated cross-reactive immunity or by including the majority of variants in a single vaccine. However, many of the T cell epitopes generate non-cross-reactive responses (Hill *et al.* 1992; Aidoo *et al.* 1995). Moreover, with the potential for antagonism and 'immune interference' for some of its CTL epitopes, inclusion of many variants in a single vaccine could even be detrimental to the induction of a protective immune response. In general, it may be safer to concentrate on immunogens, and epitopes, which are not obviously immunodominant and do not have the ability to generate multiple variant epitopes with antagonistic potential. In this context, it may also be speculated that if we concentrate on epitope based vaccines rather than whole proteins as immunogens we are less likely to include the some of the 'booby-traps' that *P. falciparum* has successfully evolved during its elusive game of hide-and-seek with the human immune system.

ACKNOWLEDGEMENTS

The authors are funded by the Wellcome Trust. Thanks to Dr Margaret Pinder and Dr Katie Flanagan for comments on the manuscript. Special thanks are due to members of the MRC Laboratory in The Gambia, particularly Dr Margaret Pinder, Dr Hilton Whittle, Professor Brian Greenwood and Professor Keith McAdam for their help and support during our malaria studies in The Gambia.

REFERENCES

AIDOO, M., LALVANI, A., ALLSOPP, C. E. M., PLEBANSKI, M., MEISNER, S. J., KRAUSA, P., BROWNING, M., MORRIS JONES, S., GOTCH, F., FIDOCK, D. A., TAKIGUCHI, M.,

ROBSON, K. J. H., GREENWOOD, B. M., DRUILHE, P., WHITTLE, H. C. & HILL, A. V. S. (1995). Identification of conserved antigenic components for a cytotoxic T lymphocyte-inducing vaccine against malaria. *Lancet* **345**, 1003–1007.

AL YAMAN, F., GENTON, B., ANDERS, R., TARAIKA, J., GINNY, M., MELLOR, S. & ALPERS, M. P. (1995). Assessment of the role of the humoral response to *Plasmodium falciparum* MSP2 compared to RESA and SPf66 in protecting Papua New Guinean children from clinical malaria. *Parasite Immunology* **17**, 493–501.

ALLSOPP, C. E. M., PLEBANSKI, M., GILBERT, S., SINDEN, R. E., HARRIS, S., FRANKEL, G., DOUGAN, G., HIOE, C., NIXON, D., PAOLETTI, E., LAYTON, G. & HILL, A. V. S. (1996). Comparison of numerous delivery systems for the induction of cytotoxic T lymphocytes by immunization. *European Journal of Immunology* **26**, 1951–1959.

ALONSO, P. L., SMITH, T., SCHELLENBERG, J. R., MASANJA, H., MWANKUSYE, S., URASSA, H., BASTOS DE AZEVEDO, I., CHONGELA, J., KOBERO, S., MENENDEZ, C., HURT, N., THOMAS, M. C., LYIMO, E., WEISS, N. A., HAYES, R., KITUA, A. Y., LOPEZ, M. C., KILAMA, W. L., TEUSCHER, T. & TANNER, M. (1994). Randomised trial of efficacy of SPf66 vaccine against *Plasmodium falciparum* malaria in children in southern Tanzania. *Lancet* **344**, 1175–1181.

ANDERSSON, M., PAABO, S., NILSSON, T. & PETERSON, P. A. (1985). Impaired intracellular transport of class I MHC antigens as a possible means for adenoviruses to evade immune surveillance. *Cell* **43**, 215–222.

BALLOU, W. R., BLOOD, J., CHONGSUPHAJAISSIDHI, T., GORDON, D. M., HEPPNER, D. G., KYLE, D. E., LUXEMBURGER, C., NOSTEN, F., SADOFF, J. C., SINGHASIVANON, P., WHITE, N. J., WEBSTER, K. H., WITTES, J. & WONGSRICHANALAI, C. (1995). Field trials of an asexual blood stage malaria vaccine: studies of the synthetic peptide polymer SPf66 in Thailand and the analytic plan for a phase IIb efficacy study. *Parasitology* **110**, S25–S36.

BALLOU, W. R., ROTHBARD, J., WIRTZ, R. A., GORDON, D. M., WILLIAMS, J. S., GORE, R. W., SCHNEIDER I., HOLLINGDALE, M. R., BEAUDOIN, R. L., MALOY, W. L., MILLER, L. H. & HOCKMEYER, W. T. (1985). Immunogenicity of synthetic peptides from the circumsporozoite protein of *Plasmodium falciparum*. *Science* **228**, 996–999.

BECK, H. P., FELGER, I., KABINTIK, S., TAVUL, L., GENTON, B., ALEXANDER, N., BHATIA, K. K., AL YAMAN, F., HII, J. & ALPERS, M. (1994). Assessment of the humoral and cell-mediated immunity against the *Plasmodium falciparum* vaccine candidates circumsporozoite protein and SPf66 in adults living in highly endemic malarious areas of Papua New Guinea. *American Journal of Tropical Medicine and Hygiene* **51**, 356–364.

BEERSMA, M. F., BIJLMAKERS, M. J. & PLOEGH, H. L. (1993). Human cytomegalovirus down-regulates HLA class I expression by reducing the stability of class I H chains. *Journal of Immunology* **151**, 4455–4464.

BERTOLETTI, A., COSTANZO, A., CHISARI, F. V., LEVRERO, M., ARTINI, M., SETTE, A., PENNA, A., GIUBERTI, T., FIACCADORI, F. & FERRARI, C. (1994a). Cytotoxic T lymphocyte response to a wild-type hepatitis B virus epitope in patients chronically infected by variant viruses carrying substitutions within the epitope. *Journal of Experimental Medicine* **180**, 933–943.

BERTOLETTI, A., SETTE, A., CHISARI, F. V., PENNA, A., LEVRERO, M., DE CARLI, M., FIACCADORI, F. & FERRARI, C. (1994*b*). Natural variants of cytotoxic epitopes are T-cell receptor antagonists for antiviral cytotoxic T cells. *Nature* **369**, 407–410.

BILSBOROUGH, J., CARLISLE, M. & GOOD, M. F. (1993). Identification of caucasian CD4 T cell epitopes on the circumsporozoite protein of *Plasmodium vivax*: T cell memory. *Journal of Immunology* **151**, 890–899.

BJORKMAN, P. J., SAPER, M. A., SAMRAOUI, B., BENNETT, W. S., STROMINGER, J. L. & WILEY, D. C. (1987). Structure of the human class I histocompatibility antigen, HLA-A2. *Nature* **329**, 506–512.

BLUM TIROUVANZIAM, U., SERVIS, C., HABLUETZEL, A., VALMORI, D., MEN, Y., ESPOSITO, F., DEL NERO, L., HOLMES, N., FASEL, N. & CORRADIN, G. (1995). Localization of HLA-A2.1-restricted T cell epitopes in the circumsporozoite protein of *Plasmodium falciparum*. *Journal of Immunology* **154**, 3922–3931.

CHAROENVIT, Y., COLLINS, W. E., JONES, T. R., MILLET, P., YUAN, L., CAMPBELL, G. H., BEAUDOIN, R. L., BRODERSON, J. R. & HOFFMAN, S. L. (1991). Inability of malaria vaccine to induce antibodies to a protective epitope within its sequence. *Science* **251**, 668–671.

CHICUREL, M., GARCIA, E. & GOODSAID, F. (1988). Modulation of macrophage lysosomal pH by *Mycobacterium tuberculosis*-derived proteins. *Infection and Immunity* **56**, 479–483.

CONWAY, D. J., GREENWOOD, B. M. & McBRIDE, J. S. (1991). The epidemiology of multiple-clone *Plasmodium falciparum* infections in Gambian patients. *Parasitology* **103**, 1–6.

COUILLIN, I., CULMANN PENCIOLELLI, B., GOMARD, E., CHOPIN, J., LEVY, J. P., GUILLET, J. G. & SARAGOSTI, S. (1994). Impaired cytotoxic T lymphocyte recognition due to genetic variations in the main immunogenic region of the human immunodeficiency virus 1 NEF protein. *Journal of Experimental Medicine* **180**, 1129–1134.

D'ALESSANDRO, U., LEACH, A., DRAKELEY, C. J., BENNETT, S., OLALEYE, B. O., FEGAN, G. W., JAWARA, M., LANGEROCK, P., GEORGE, M. O., TARGETT, G. A. & GREENWOOD, B. M. (1995). Efficacy trial of malaria vaccine SPf66 in Gambian infants. *Lancet* **346**, 462–467.

DAME, J. B., WILLIAMS, J. L., McCUTCHAN, T. F., WEBER, J. L., WIRTZ, R. A., HOCKMEYER, W. T., MALOY, W. L., HAYNES, D., SCHNEIDER, I., ROBERTS, D., SANDERS, G. S., REDDY, E. P., DIGGS, C. L. & MILLER, L. H. (1984). Structure of the gene encoding the immunodominant surface antigen on the sporozoite of the human malaria parasite *Plasmodium falciparum*. *Science* **255**, 593–599.

DAVENPORT, M. P. (1995). Antagonists or altruists: do viral mutants modulate T-cell responses? *Immunology Today* **16**, 432–436.

DE CAMPOS LIMA, P. O., GAVIOLI, R., ZHANG, Q. J., WALLACE, L. E., DOLCETTI, R., ROWE, M., RICKINSON, A. B. & MASUCCI, M. G. (1993). HLA-A11 epitope loss isolates of Epstein–Barr virus from a highly A11+ population. *Science* **260**, 98–100.

DE CAMPOS LIMA, P. O., LEVITSKY, V., BROOKS, J., LEE, S. P., HU, L. F., RICKINSON, A. B. & MASUCCI, M. G. (1994). T cell responses and virus evolution: loss of HLA A11-restricted CTL epitopes in Epstein–Barr virus isolates from highly A11-positive populations by selective mutation of anchor residues. *Journal of Experimental Medicine* **179**, 1297–1305.

DE LA CRUZ, V. F., LAL, A. A. & McCUTCHAN, T. F. (1987). Sequence variation in putative functional domains of the circumsporozoite protein of *Plasmodium falciparum*: implications for vaccine development. *Journal of Biological Chemistry* **262**, 11935–11939.

DONG, T., BOYD, D., ROSENBERG, W., ALP, A., TAKIGUCHI, M., McMICHAEL, A. & ROWLAND-JONES, S. (1996). An HLA-B35-restricted epitope modified at an anchor residue results in an antagonist peptide. *European Journal of Immunology* **26**, 335–339.

DOOLAN, D. L., HOUGHTEN, R. A. & GOOD, M. F. (1991). Location of human cytotoxic T cell epitopes within a polymorphic domain of the *Plasmodium falciparum* circumsporozoite protein. *International Immunology* **3**, 511–516.

DOOLAN, D. L., KHAMBOONRUANG, C., BECK, H. P., HOUGHTEN, R. A. & GOOD, M. F. (1993). Cytotoxic T lymphocyte (CTL) low-responsiveness to the *Plasmodium falciparum* circumsporozoite protein in naturally exposed endemic populations: analysis of human CTL response to most known variants. *International Immunology* **5**, 37–46.

DOOLAN, D. L., SAUL, A. J. & GOOD, M. F. (1992). Geographically restricted heterogeneity of the *Plasmodium falciparum* circumsporozoite protein: relevance for vaccine development. *Infection and Immunity* **60**, 675–682.

DOOLAN, D. L., SEDEGAH, M., HEDSTROM, R. C., HOBART, P., CHAROENVIT, Y. & HOFFMAN, S. L. (1996). Circumventing genetic restriction of protection against Malaria with Multigene DNA Immunization: CD8+ T cell-, Interferon g-, and Nitric Oxide-dependent immunity. *Journal of Experimental Medicine* **183**, 1739–1746.

EGAN, J. E., WEBER, J. L., BALLOU, W. R., HOLLINGDALE, M. R., MAJARIAN, W. R., GORDON, D. M., MALOY, W. L., HOFFMAN, S. L., WIRTZ, R. A., SCHNEIDER, I., WOOLLETT, G. R., YOUNG, J. F. & HOCKMEYER, W. T. (1987). Efficacy of murine malaria sporozoite vaccines: implications for human vaccine development. *Science* **236**, 453–456.

EVAVOLD, B. D. & ALLEN, P. M. (1991). Separation of IL-4 production from Th cell proliferation by an altered T cell receptor ligand. *Science* **252**, 1308–1310.

EVAVOLD, B. D., SLOAN LANCASTER, J., HSU, B. L. & ALLEN, P. M. (1993). Separation of T helper 1 clone cytolysis from proliferation and lymphokine production using analog peptides. *Journal of Immunology* **150**, 3131–3140.

FALO, L. D., KOVACSOVICS-BANKOWSKI, M., THOMPSON, K. & ROCK, K. L. (1995). Targetting antigen into the phagocytic pathway *in vivo* induces protective immunity. *Nature Medicine* **1**, 649–653.

FELL, A. H., CURRIER, J. & GOOD, M. F. (1994). Inhibition of *Plasmodium falciparum* growth *in vitro* by CD4+ and CD8+ T cells from non-exposed donors. *Parasite Immunology* **16**, 579–586.

GILBERT, S. C. & HILL, A. V. S. (1997). Protein particle vaccines for malaria. *Parasitology Today* **13**, 302–306.

GOLDBERG, A. L. & ROCK, K. L. (1992). Proteolysis, proteosomes and antigen presentation. *Nature* **357**, 375–379.

GOOD, M. F., BERZOFSKY, J. A. & MILLER, L. H. (1988). The T cell response to the malaria circumsporozoite protein: an immunological approach to vaccine development. *Annual Review of Immunology* **6**, 663–668.

GOOD, M. F., MALOY, W. L., LUNDE, M. N., MARGALIT, H., CORNETTE, J. L., SMITH, G. L., MOSS, B., MILLER, L. H. & BERZOFSKY, J. A. (1987). Construction of synthetic immunogen: use of new T helper epitope on malaria circumsporozoite protein. *Science* **235**, 1059–1062.

GOOD, M. F., POMBO, D., QUAKYI, I., RILEY, E. M., HOUGHTEN, R. A., MENON, A., ALLING, D. W., BERZOFSKY, J. A. & MILLER, L. H. (1988). Human T-cell recognition of the circumsporozoite protein of *Plasmodium falciparum*: immunodominant T-cell domains map to the polymorphic regions of the molecule. *Proceedings of the National Academy of Sciences, USA* **85**, 1199–1203.

GRAVES, P. M., BHATIA, K., BURKOT, T. R., PRASAD, M., WIRTZ, R. A. & BECKERS, P. (1989). Association between HLA type and antibody response to malaria sporozoite and gametocyte epitopes is not evident in immune Papua New Guineans. *Clinical and Experimental Immunology* **78**, 418–423.

GUPTA, S. & HILL A. V. S. (1995). Dynamic interactions in malaria: host heterogeneity meets parasite polymorphism. *Proceedings of the Royal Society of London Series B (Biological Sciences)* **261**, 271–277.

HEDSTROM, R. C., SEDEGAH, M. & HOFFMAN, S. L. (1994). Prospects and strategies for development of DNA vaccines against malaria. *Research in Immunology* **145**, 476–483.

HERRINGTON, D., DAVIS, J., NARDIN, E., BEIER, M., CORTESE, J., EDDY, H., LOSONSKY, G., HOLLINGDALE, M., SZTEIN, M., LENNE, M., NUSSENZWEIG, R. S., CLYDE, D. & EDELMAN, R. (1991). Successful immunization of humans with irradiated malaria sporozoites: humoral and cellular responses of the protected individuals. *American Journal of Tropical Medicine and Hygiene* **45**, 539–547.

HILL, A. V. S., ALLSOPP, C. E. M., KWIATKOWSKI, D., ANSTEY, N. M., TWUMASI, P., ROWE, P. A., BENNETT, S., BREWSTER, D., McMICHAEL, A. J. & GREENWOOD, B. M. (1991). Common West African HLA antigens are associated with protection from severe malaria. *Nature* **352**, 595–600.

HILL, A. V. S., ELVIN, J., WILLIS, A. C., AIDOO, M., ALLSOPP, C. E. M., GOTCH, F. M., GAO, X. M., TAKIGUCHI, M., GREENWOOD, B. M., TOWNSEND, A. R. M., McMICHAEL, A. J. & WHITTLE, H. C. (1992). Molecular analysis of the association of HLA-B53 and resistance to severe malaria. *Nature* **360**, 434–439.

HOFFMAN, S. L., ISENBARGER, D., LONG, G. W., SEDEGAH, M., SZARFMAN, A., WATERS, L., HOLLINGDALE, M. R., VAN DER MEIDE, P. H., FINBLOOM, D. S. & BALLOU, W. R. (1989). Sporozoite vaccine induces genetically restricted T cell elimination of malaria from hepatocytes. *Science* **244**, 1078–1081.

HOLLSBERG, P., WEBER, W. E., DANGOND, F., BATRA, V., SETTE, A. & HAFFLER, D. A. (1995). Differential activation of proliferation and cytotoxicity in human T cell lymphotropic virus type 1 Tax-specific CD8 T cells by an altered peptide ligand. *Proceedings of the National Academy of Sciences, USA* **92**, 4036–4040.

HUGHES, A. L. (1991). Circumsporozoite protein genes of malaria parasites (*Plasmodium* spp.): evidence for positive selection on immunogenic regions. *Genetics* **127**, 345–354.

HUGHES, A. L. & NEI, M. (1988). Pattern of nucleotide substitution at major histocompatibility complex class I loci reveals overdominant selection. *Nature* **335**, 167–170.

ITESCU, S., MATHUR WAGH, U., SKOVRON, M. L., BRANCATO, L. J., MARMOR, M., ZELENIUCH JACQUOTTE, A. & WINCHESTER, R. (1992). HLA-B35 is associated with accelerated progression to AIDS. *Journal of the Acquired Immunodeficiency Syndrome* **5**, 37–45.

JACOBS, P., RADZIOCH, D. & STEVENSON, M. M. (1996). A Th1-associated increase in tumor necrosis factor alpha expression in the spleen correlates with resistance to blood-stage malaria in mice. *Infection and Immunity* **64**, 535–541.

JAMESON, S. C. & BEVAN, M. J. (1995). T cell receptor antagonists and partial agonists. *Immunity* **2**, 1–11.

JAMESON, S. C., CARBONE, F. R. & BEVAN, M. J. (1993). Clone-specific T cell receptor antagonists of major histocompatibility complex class I-restricted cytotoxic T cells. *Journal of Experimental Medicine* **177**, 1541–1550.

KANEKO, T., MORIYAMA, T., UDAKA, K., HIROISHI, K., KITA, H., OKAMOTO, H., YAGITA, H., OKUMURA, K. & IMAWARI, M. (1997). Impaired induction of cytotoxic T lymphocytes by antagonism of a weak agonist borne by a variant hepatitis C virus epitope. *European Journal of Immunology* **27**, 1782–1787.

KARIN, N., MITCHELL, D. J., BROCKE, S., LING, N. & STEINMAN, L. (1994). Reversal of experimental autoimmune encephalomyelitis by a soluble peptide variant of a myelin basic protein epitope: T cell receptor antagonism and reduction of interferon gamma and tumor necrosis factor alpha production. *Journal of Experimental Medicine* **180**, 2227–2237.

KASLOW, R. A., DUQUESNOY, R., VANRADEN, M., KINGSLEY, L., MARRARI, M., FRIEDMAN, H., SU, S., SAAH, A. J., DETELS, R., PHAIR, J. & RINALDO, C. (1990). A1, Cw7, B8, DR3 HLA antigen combination associated with rapid decline of T-helper lymphocytes in HIV-1 infection. A report from the Multicentre AIDS Cohort Study. *Lancet* **335**, 927–930.

KLENERMAN, P., MEIER, U. C., PHILLIPS, R. E. & McMICHAEL, A. J. (1995). The effects of natural altered peptide ligands on the whole blood cytotoxic T lymphocyte response to human immunodeficiency virus. *European Journal of Immunology* **25**, 1927–1931.

KLENERMAN, P., ROWLAND JONES, S., McADAM, S., EDWARDS, J., DAENKE, S., LALLOO, D., KOPPE, B., ROSENBERG, W., BOYD, D., EDWARDS, A., GIANGRANDE, P., PHILLIPS, R. & McMICHAEL, A. (1994). Cytotoxic T-cell activity antagonized by naturally occurring HIV-1 Gag variants. *Nature* **369**, 403–407.

KOVACSOVICS BANKOWSKI, M., CLARK, K., BENACERRAF, B. & ROCK, K. L. (1993). Efficient major histocompatibility complex class I presentation of exogenous antigen upon phagocytosis by macrophages. *Proceedings of the National Academy of Sciences, USA* **90**, 4942–4946.

KRAUSA, P., BRYWKA, M., SAVAGE, D., HUI, K. M., BUNCE, M., NGAI, J. L. F., TEO, D. L. T., ONG, Y. W., BAROUCH, D., ALLSOP, C. E. M., HILL, A. V. S., McMICHAEL, A. J., BODMER, J. G. & BROWNING, M. J. (1995). Genetic polymorphism within HLA-A*02: Significant allelic variation revealed in different populations. *Tissue Antigens* **45**, 223–231.

KUMAR, A., ARORA, R., KAUR, P., CHAUHAN, V. S. & SHARMA, P. (1992). 'Universal' T helper cell determinants enhance immunogenicity of a *Plasmodium falciparum* merozoite surface antigen peptide. *Journal of Immunology* **148**, 1499–1505.

KWIATKOWSKI, D. (1990). Tumour necrosis factor, fever and fatality in falciparum malaria. *Immunological Letters* **25**, 213–216.

KWIATKOWSKI, D., HILL, A. V. S., SAMBOU, I., TWUMASI, P., CASTRACANE, J., MANOGUE, K. R., CERAMI, A., BREWSTER, D. R. & GREENWOOD, B. M. (1990). TNF concentration in fatal cerebral, non-fatal cerebral, and uncomplicated *Plasmodium falciparum* malaria. *Lancet* **336**, 1201–1204.

LOCKYER, M. J., MARSH, K. & NEWBOLD, C. I. (1989). Wild isolates of *Plasmodium falciparum* show extensive polymorphism in T cell epitopes of the circumsporozoite protein. *Molecular and Biochemical Parasitology* **37**, 275–280.

LOWIN, B., HAHNE, M., MATTMANN, C. & TSCHOPP, J. (1994). Cytolytic T cell cytotoxicity is mediated through perforin and Fas lytic pathways. *Nature* **370**, 650–652.

MALIK, A., EGAN, J. E., HOUGHTEN, R. A., SADOFF, J. C. & HOFFMAN, S. L. (1991). Human cytotoxic T lymphocytes against the *Plasmodium falciparum* circumsporozoite protein. *Proceedings of the National Academy of Sciences, USA* **88**, 3300–3304.

MARSH, K., HAYES, R. H., CARSON, D. C., OTOO, L. N., SHENTON, F., BYASS, P., ZAVALA, F. & GREENWOOD, B. M. (1988). Anti-sporozoite antibodies and immunity to malaria in a rural Gambian population. *Transactions of the Royal Society of Tropical Medicine and Hygiene* **82**, 532–537.

McCUTCHAN, T. F., LAL, A. A., DO ROSARIO, V. & WATERS, A. P. (1992). Two types of sequence polymorphism in the circumsporozoite gene of *Plasmodium falciparum*. *Molecular and Biochemical Parasitology* **50**, 37–46.

McGUIRE, W., HILL, A. V. S., ALLSOPP, C. E. M., GREENWOOD, B. M. & KWIATKOWSKI, D. (1994). Variation in the TNF-alpha promoter region associated with susceptibility to cerebral malaria. *Nature* **371**, 508–511.

MELLOUK, S., GREEN, S. J., NACY, C. A. & HOFFMAN, S. L. (1991). IFN-gamma inhibits development of *Plasmodium berghei* exoerythrocytic stages in hepatocytes by an L-arginine-dependent effector mechanism. *Journal of Immunology* **146**, 3971–3976.

MORENO, A., CLAVIJO, P., EDELMAN, R., DAVID, J., SZTEIN, M., SINIGAGLIA, F. & NARDIN, E. (1993). CD4 positive T cell clones obtained from *Plasmodium falciparum* sporozoite-immunized volunteers recognize polymorphic sequences of the circumsporozoite protein. *Journal of Immunology* **151**, 489–499.

NARDIN, E., CLAVIJO, P., MONS, B., VAN BELKUM, A., PONNUDURAI, T. & NUSSENZWEIG, R. S. (1991). T cell epitopes of the circumsporozoite protein of *Plasmodium vivax*: recognition by lymphocytes of a sporozoite-immunized chimpanzee. *Journal of Immunology* **146**, 1674–1678.

NARDIN, E., MUNESINGHE, Y. D., MORENO, A., CLAVIJO, P., CALLE, M. C., EDELMAN, R., DAVIS, J., HERRINGTON, D. & NUSSENZWEIG, R. S. (1992). T cell responses to repeat and non-repeat regions of the circumsporozoite protein detected in volunteers immunized with *Plasmodium falciparum* sporozoites. *Memorias do Institutio Oswaldo Cruz* **3**, 223–227.

NARDIN, E. H. & NUSSENZWEIG, R. S. (1993). T cell responses to pre-erythrocytic stages of malaria: role in protection and vaccine development against pre-erythrocytic stages. *Annual Review of Immunology* **11**, 687–727.

NOSTEN, F., LUXEMBURGER, C., KYLE, D. E., BALLOU, W. R., WITTES, J., WAH, E., CHONGSUPHAJAISIDDHI, T., GORDON, D. M., WHITE, N. J., SADOFF, J. C. & HEPPNER, D. G. (1996). Randomised double-blind placebo-controlled trial of SPf66 malaria vaccine in children in northwestern Thailand. Shoklo SPf66 Malaria Vaccine Trial Group. *Lancet* **348**, 701–707.

NUSSLER, A., PIED, S., GOMA, J., RENIA, L., MILTGEN, F., GRAU, G. E. & MAZIER, D. (1991). TNF inhibits malaria hepatic stages *in vitro* via synthesis of IL-6. *International Immunology* **3**, 317–321.

PHILLIPS, R. E., ROWLAND JONES, S., NIXON, D. F., GOTCH, F. M., EDWARDS, J. P., OGUNLESI, A. O., ELVIN, J. G., ROTHBARD, J. A., BANGHAM, C. R., RIZZA, C. R. & McMICHAEL, A. J. (1991). Human immunodeficiency virus genetic variation that can escape cytotoxic T cell recognition. *Nature* **354**, 453–459.

PLEBANSKI, M., AIDOO, M., WHITTLE, H. C. & HILL, A. V. S. (1997). Precursor frequency analysis of cytotoxic T lymphocytes to pre-erythrocytic antigens of *Plasmodium falciparum* in West Africa. *Journal of Immunology* **158**, 2849–2855.

PLEBANSKI, M., ALLSOPP, C. E. M., AIDOO, M., REYBURN, H. & HILL, A. V. S. (1995). Induction of peptide-specific primary cytotoxic T lymphocyte responses from human peripheral blood. *European Journal of Immunology* **25**, 1783–1787.

QARI, S. H., COLLINS, W. E., LOBEL, H. O., TAYLOR, F. & LAL, A. A. (1994). A study of polymorphism in the circumsporozoite protein of human malaria parasites. *American Journal of Tropical Medicine and Hygiene* **50**, 45–51.

QUAKYI, I. A., OTOO, L. N., POMBO, D., SUGARS, L. Y., MENON, A., DEGROOT, A. S., JOHNSON, A., ALLING, D., MILLER, L. H. & GOOD, M. F. (1989). Differential non-responsiveness in humans of candidates *Plasmodium falciparum* vaccine antigens. *American Journal of Tropical Medicine and Hygiene* **41**, 125–134.

RAMMENSEE, H. G. (1995). Chemistry of peptides associated with MHC class I and class II molecules. *Current Opinions in Immunology* **7**, 85–96.

RAMMENSEE, H. G., FRIEDE, T. & STEVANOVIIC, S. (1995). MHC ligands and peptide motifs: first listing. *Immunogenetics* **41**, 178–228.

RANI, R., FERNANDEZ-VINA, M. A., ZAHEER, S. A., BEENA, K. R. & STASTNY, P. (1993). Study of HLA class II alleles by PCR oligotyping in leprosy patients from North India. *Tissue Antigens* **42**, 133–137.

REIS E SOUSA, C. & GERMAIN, R. N. (1995). Major Histocompatibility Complex class I presentation of peptides derived from soluble exogenous antigen by a subset of cells engaged in phagocytosis. *Journal of Experimental Medicine* **182**, 841–851.

RODRIGUES, M., NUSSENZWEIG, R. S. & ZAVALA, F. (1993). The relative contribution of antibodies, CD4+ and CD8+ T cells to sporozoite-induced protection against malaria. *Immunology* **80**, 1–5.

ROMERO, P., EBERL, G., CASANOVA, J.-L., CORDEY, A.-S., WINDMANN, C., LUESCHER, I. F., CORRADIN, G. & MARYANSKI, J. L. (1992). Immunization with synthetic peptides containing a defined malaria epitope induces a highly diverse cytotoxic T lymphocyte response. *Journal of Immunology* **148**, 1871–1878.

ROMERO, P., MARYANSKI, J. L., CORRADIN, G., NUSSENZWEIG, R. S., NUSSENZWEIG, V. & ZAVALA, F. (1989). Cloned cytotoxic T cells recognize an epitope in the circumsporozoite protein and protect against malaria. *Nature* **341**, 323–326.

SCHERLE, P. A. & GERHARD, W. (1986). Functional analysis of influenza-specific helper T cell clones *in vivo*. T cells specific for internal viral proteins provide cognate help for B cell responses to hemagglutinin. *Journal of Experimental Medicine* **164**, 1114–1128.

SCHÖDEL, F., KELLY, S. M., PETERSON, D. L., MILICH, D. R. & CURTISS, R. R. (1994). Hybrid hepatitis B virus core-pre-S proteins synthesized in avirulent *Salmonella typhimurium* and *Salmonella typhi* for oral vaccination. *Infection and Immunity* **62**, 1669–1676.

SEDEGAH, M., HEDSTROM, R., HOBART, P. & HOFFMAN, S. L. (1994). Protection against malaria by immunisation with plasmid DNA encoding circumsporozoite protein. *Proceedings of the National Academy of Sciences, USA* **91**, 9866–9870.

SETTE, A., ALEXANDER, J., RUPPERT, J., SNOKE, K., FRANCO, A., ISHIOKA, G. & GREY, H. M. (1994). Antigen analogs/MHC complexes as a specific T cell receptor antagonists. *Annual Review of Immunology* **12**, 413–431.

SHI, Y. P., ALPERS, M. P., POVOA, M. M. & LAL, A. A. (1992). Diversity in the immunodominant determinants of the circumsporozoite protein of *Plasmodium falciparum* parasites from malaria-endemic regions of Papua New Guinea and Brazil. *American Journal of Tropical Medicine and Hygiene* **47**, 844–851.

STEEL, C. M., LUDLAM, C. A., BEATSON, D., PEUTHERER, J. F., CUTHBERT, R. J., SIMMONDS, P., MORRISON, H. & JONES, M. (1988). HLA haplotype A1 B8 DR3 as a risk factor for HIV-related disease. *Lancet* **1**, 1185–1188.

STOUTE, J. A., SLAOUI, M., HEPPNER, G., MOMIN, P., KESTER, K. E., DESMONS, P., WELLDE, B. T., GARCON, N., KRZYCH, U., MARCHAND, M., BALLOU, R. & COHEN, J. (1997). A preliminary evaluation of a recombinant circumsporozoite protein vaccine against *Plasmodium Falciparum* malaria. *New England Journal of Medicine* **336**, 86–91.

TINE, J. A., LANAR, D. E., SMITH, D. M., WELLDE, B. T., SCHULTHEISS, P., WARE, L. A., KAUFFMAN, E. B., WIRTZ, R. A., DE TAISNE, C., HUI, G. S. N., CHANG, S. P., CHURCH, P., HOLLINGDALE, M. R., KASLOW, D. C., HOFFMAN, S., GUITO, K. P., BALLOU, W. R., SADOFF, J. C. & PAOLETTI, E. (1996). NYVAC-Pf7: a poxvirus-vectored, multiantigen, multistage vaccine candidate for *Plasmodium Falciparum* malaria. *Infection and Immunity* **64**, 3833–3844.

TOWNSEND, A. & TROWSDALE, J. (1993). The transporters associated with antigen presentation. *Seminars in Cell Biology* **257**, 927–934.

UDHAYAKUMAR, V., ANYONA, D., KARIUKI, S., SHI, Y. P., BLOLAND, P. B., BRANCH, O. H., WEISS, W., NAHLEN, B. L., KASLOW, D. C. & LAL, A. L. (1995). Identification of T and B cell epitopes recognized by humans in the C-terminal 42-kDa domain of the *Plasmodium Falciparum* merozoite surface protein (MSP). 1. *Journal of Immunology* **154**, 6022–6030.

UDHAYAKUMAR, V., SHI, Y. P., KUMAR, S., JUE, D. L., WOHLHUETER, R. M. & LAL, A. A. (1994). Antigenic diversity in the circumsporozoite protein of *Plasmodium falciparum* abrogates cytotoxic T-cell recognition. *Infection and Immunity* **62**, 1410–1413.

VAESSEN, R. T., HOUWELING, A. & VAN DER EB, A. J. (1987). Post-transcriptional control of class I MHC mRNA expression in adenovirus 12-transformed cells. *Science* **235**, 1486–1488.

WANG, R., CHAROENVIT, Y., CORRADIN, G., PORROZZI, R., HUNTER, R. L., GLENN, G., ALVING, C. R., CHURCH, P. & HOFFMAN, S. L. (1995). Induction of protective polyclonal antibodies by immunization with *Plasmodium yoelii* circumsporozoite protein multiple antigen peptide vaccine. *Journal of Immunology* **154**, 2784–2793.

WINDMANN, C., ROMERO, P., MARYANSKI, J. L., CORRADIN, G. & VALMORI, D. (1992). T helper epitopes enhance the cytotoxic response of mice immunized with MHC class I-restricted malaria peptides. *Journal of Immunological Methods* **155**, 95–99.

YOSHIDA, N., DI SANTI, S. M., DUTRA, A. P., NUSSENZWEIG, R. S., NUSSENZWEIG, V. & ENEA, V. (1990). *Plasmodium falciparum*: restricted polymorphism of T cell epitopes of the circumsporozoite protein in Brazil. *Experimental Parasitology* **71**, 386–392.

YOUNG, J. F., HOCKMEYER, W. T., GROSS, M., BALLOU, W. R., WIRTZ, R. A., TROSPER, J. H., BEAUDOIN, R. L., HOLLINGDALE, M. R., MILLER, L. H., DIGGS, C. L. & ROSENBERG, M. (1985). Expression of *Plasmodium falciparum* circumsporozoite proteins in *Escherichia coli* for potential use in a human malaria vaccine. *Science* **228**, 958–962.

ZAVALA, F., TAM, J. P., HOLLINGDALE, M. R., COCHRANE, A. H., QUAKYI, I., NUSSENZWEIG, R. S. & NUSSENZWEIG, V. (1985). Rationale for development of a synthetic vaccine against *Plasmodium falciparum* malaria. *Science* **228**, 1436–1440.

ZEVERING, Y., KHAMBOONRUANG, C., RUNGRUENGTHANAKIT, K., TUNGVIBOONCHAI, L., RUENGPIPATTANAPAN, J., BATHURST, I., BARR, P. & GOOD, M. F. (1994). Life-spans of human T-cell responses to determinants from the circumsporozoite proteins of *Plasmodium falciparum* and *Plasmodium vivax*. *Proceedings of the National Academy of Sciences, USA* **91**, 6118–6122.

Role of the immune response induced by superantigens in the pathogenesis of microbial infections

I. MAILLARD[1], F. LUTHI[1], H. ACHA-ORBEA[2] *and* H. DIGGELMANN[1]*

[1] *Institute of Microbiology, University of Lausanne, Rue du Bugnon 44, 1011 Lausanne, Switzerland*
[2] *Institute of Biochemistry and Ludwig Institute for Cancer Research, Lausanne Branch, Ch. des Boveresses 155, 1066 Epalinges, Switzerland*

SUMMARY

Superantigens (SAgs) are microbial proteins which have potent effects on the immune system. They are presented by major histocompatibility complex (MHC) class II molecules and interact with a large number of T cells expressing specific T cell receptor $V\beta$ domains. Encounter of a SAg leads initially to the stimulation and subsequently to the clonal deletion of reactive T cells. SAgs are expressed by a wide variety of microorganisms which use them to exploit the immune system to their own advantage. Bacterial SAgs are exotoxins which are linked to several diseases in humans and animals. A classical example is the toxic shock syndrome in which the massive release of cytokines by SAg-reactive cells is thought to play a major pathogenic role. The best characterized viral SAg is encoded by mouse mammary tumour virus (MMTV) and has proved to have a major influence on the viral life cycle by dramatically increasing the efficiency of viral infection. In this paper, we review the general properties of SAgs and discuss the different types of microorganisms which produce these molecules, with a particular emphasis on the role played by the SAg-induced immune response in the course of microbial infections.

Key words: superantigens, microbial infections, bacterial exotoxins, MMTV.

INTRODUCTION

Superantigens (SAgs) constitute a group of proteins with potent effects on the immune system. Although different SAgs are expressed by a wide variety of microorganisms, they share the ability to stimulate a large number of T cells through similar mechanisms. SAgs associate with MHC class II molecules and interact with T cells expressing specific variable domains in the T cell receptor β chain (TCR $V\beta$ domains) independently of their specificity for classical antigens. This property led to their designation as 'super'-antigens since it allows them to stimulate a much higher proportion of the T cell repertoire than conventional antigens do. The strong immune response induced by SAgs is thought to play a role in the pathogenesis of several disease processes in humans and animals. Furthermore, several microorganisms use SAgs to exploit the immune system to their own advantage. Such a phenomenon has been described in particular for the SAg expressed by a murine retrovirus, mouse mammary tumour virus (MMTV).

In this paper, we will first summarize the main properties of SAgs in comparison with conventional peptide antigens. We will then critically discuss the identification of the different microbial SAgs and present current evidence for their role in the pathogenesis of microbial infections *in vivo*.

* Corresponding author.

PROPERTIES OF SUPERANTIGENS

The main differences between SAgs and conventional antigens are summarized in Fig. 1 together with a model of the TCR-MHC-antigen and TCR-MHC-SAg interactions.

Processing and presentation

Conventional antigens are actively processed and presented to T cells as small peptides. In contrast, SAgs are usually presented as unprocessed molecules and can induce a T cell response even if they are added to fixed antigen-presenting cells. Some SAgs (e.g. MMTV SAgs) may require a limited number of processing steps (Park *et al.* 1995; Mix & Winslow, 1996).

SAgs interact exclusively with MHC class II and cannot be presented by MHC class I molecules. In contrast to conventional antigens, SAgs do not bind to the most polymorphic part of MHC, but associate with the more conserved regions on the lateral side of the molecule (Dellabona *et al.* 1990) (see Fig. 1). Therefore, the recognition of SAgs is not MHC-restricted. SAgs can thus be presented by a wide variety of MHC alleles and isotypes, although a hierarchy is observed in the affinity of binding and in the efficiency of presentation (White *et al.* 1989; Herman *et al.* 1990). Binding to MHC from different species is generally possible, even though the highest affinities are usually reported in the species which is

Parasitology (1997) **115**, S67–S78. © 1997 Cambridge University Press

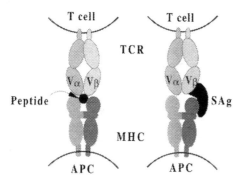

	Conventional antigen	Superantigen
Processing	Yes	No or limited
Presentation	MHC I or II	MHC II
TCR elements	Vα, Jα, Vβ, Dβ, Jβ	Vβ
Frequency of reacting T cells	0.01-0.0001%	1-40%

Fig. 1. Comparison between conventional antigens and superantigens. The main differences between conventional antigens and SAgs are summarized. The schematic model of trimolecular interactions between T cell receptor (TCR), major histocompatibility complex (MHC) on an antigen-presenting cell (APC) and peptide or SAg is based on available crystallographic data. A typical extracellular staphylococcal SAg is shown, as the structure of other SAgs has not been elucidated yet. A peptide was present in the peptide-binding groove of the SAg-MHC cocrystal (Jardetzky *et al.* 1994), but it is not drawn here for the sake of simplicity.

the natural host of the SAg-producing organism (Mollick *et al.* 1991). Such a lack of MHC restriction is one of the important criteria to distinguish SAgs from classical antigens.

Vβ specificity

The specificity of a T cell for a given SAg is conferred almost exclusively by the Vβ domain of the T cell receptor (TCR) (Kappler *et al.* 1988, 1989; MacDonald *et al.* 1988). The other parts of the TCR do not contribute to the specificity except for the Vα domain which has a minor influence (see for example Vacchio *et al.* 1992). This is in contrast to the recognition of classical antigens in which several additional polymorphic elements of the TCR are involved (Dβ, Jβ, Vα, Jα). In particular, the so-called CDR3 region of the TCR β chain, which is made by the junction of the V, D and J elements, forms together with the homologous region of the α chain the most critical region for the recognition of a specific peptide–MHC combination, but it has no influence on the interaction of SAgs with the TCR. Therefore, a population of SAg-reactive T cells expressing a given Vβ domain will contain a full range of pre-existing CDR3 sequences characteristic of a true polyclonal expansion. In contrast, some CDR3 patterns will be overrepresented in oligo-

clonal responses. This constitutes another important difference between SAgs and classical antigens, especially since several conventional antigens induce a markedly skewed Vβ usage in reactive T cells (see for example Boitel *et al.* 1992; Casanova *et al.* 1992; Pantaleo *et al.* 1994).

An important result of the peculiar interaction of SAgs with the TCR is the much higher precursor frequency of responding T cells than for classical antigens. The typical frequency of cells recognizing a specific peptide-MHC complex is in the order of 1 in 10^4–1 in 10^6 in a naïve population of T cells. As there are only about 20 Vβ elements in the mouse and 60 in humans, the proportion of cells reacting to a given SAg will be between 1 in 100 and 1 in 3 depending on the basal level of the reactive Vβ subset(s).

Immune response induced by superantigens

The effects of SAgs on mature T cells are biphasic. The encounter of a SAg first leads to the expansion of reactive T cells which then enter a phase of anergy (i.e. unresponsiveness) and are later eliminated from the T cell repertoire by a process of clonal deletion (White *et al.* 1989; Webb, Morris & Sprent, 1990; MacDonald, Baschieri & Lees, 1991). If immature T cells encounter the SAg during thymic development, they undergo intrathymic deletion (Kappler *et al.* 1988; MacDonald *et al.* 1988). The deletion of reactive T cells can be partial or complete and transient or permanent depending on the dose and type of SAg. Thus, SAgs are able to stimulate a huge number of T cells and can also shape the T cell repertoire by leading to the unresponsiveness and to the elimination of entire subsets of T cells.

Both CD4[+] and CD8[+] T cells have the ability to respond to SAgs. Surprisingly, the response of CD8[+] T cells to several SAgs was found to be nearly as strong as that of CD4[+] T cells (MacDonald *et al.* 1990, 1991; Webb *et al.* 1990). The effects of SAgs on CD8[+] T cells break the rules of classical antigen recognition, since they do not involve MHC class I but instead require the presence of the same MHC class II molecules as for CD4[+] T cells (Herrmann *et al.* 1990). CD8[+] T cells are not MHC class II-restricted, but the interaction with the SAg-MHC class II complex is thought not to require as intimate TCR-MHC contacts as in classical antigen recognition, due to a strong bridging effect of the SAg (Fields *et al.* 1996).

SAgs can induce the differentiation of responding T cells which are then able to perform most or all of the functions of classically activated T cells. For example, SAg-responsive T cells have been shown to secrete many cytokines (e.g. IL-2, IFN-γ) which by themselves can have powerful effects (Herrmann *et al.* 1992; Miethke *et al.* 1992). Activated CD4[+] T cells can provide classical T cell help to B cells and CD8$^+$ T cells can acquire a specific cytolytic activity (Herrmann *et al.* 1990; Tumang *et al.* 1991). In addition, as SAgs can stimulate T cells irrespective of their antigen specificity, they can potentially activate autoreactive cells. For this reason, SAgs have been supposed to play a role in autoimmune diseases, especially in those which might be linked to infectious agents (e.g. insulin-dependent diabetes mellitus, multiple sclerosis, reactive polyarthritis in Reiter's disease).

Pathogens have developed many different strategies to evade the immune system of their host. In the case of SAgs, the strategy is an active manipulation of the immune system. We will now discuss how this strategy is exploited by different microorganisms.

BACTERIAL SUPERANTIGENS

Bacterial superantigens are exotoxins produced by many strains of bacteria

Many bacterial SAgs were initially described as bacterial exotoxins. Their strong mitogenic potential for T cells has been known since then, but the mechanism of T cell stimulation which defines them as SAgs was elucidated much later (Kappler *et al.* 1989; White *et al.* 1989). A list of *bona fide* bacterial SAgs is given in Table I. It is important to point out that the identification and characterization of bacterial SAgs suffers from the extreme potency of these molecules. Fleischer *et al.* (1995) have demonstrated that many reported SAg activities resulted from minute contaminations of the toxin preparations which were undetectable by conventional biochemical methods but still gave rise to Vβ-specific T cell stimulation. In this respect, the use of recombinant proteins facilitates the classification of bacterial products as true SAgs. For example, fragments of the streptococcal M proteins were initially reported to have a SAg activity which was later traced to contaminants of the preparation. Similar observations were made for the staphylococcal exfoliative toxin A (ETA) and for the streptococcal pyrogenic exotoxin B (SPEB) (reviewed by Fleischer *et al.* 1995). In addition to the SAgs listed in Table I, candidate SAgs have been reported from *Yersinia enterocolitica*, *Clostridium perfringens*, *Pseudomonas aerunginosa* and *Mycobacterium tuberculosis*. These reports await confirmation using recombinant proteins.

The bacterial SAgs which have been most extensively studied are produced by Staphylococci, Streptococci and Mycoplasma (reviewed in Marrack & Kappler, 1990). The prototypic bacterial SAgs are the staphylococcal enterotoxins A to E (SEA to SEE). These toxins were first known for their involvement in food poisoning, but they are also produced by bacteria causing a severe clinical syndrome with high fever, hypotension and systemic organ failure, called toxic shock syndrome (TSS). Another staphylococcal SAg, TSST-1, was the first to be associated with TSS (see for example Schlievert *et al.* 1981) and is probably also involved in Kawasaki syndrome, a febrile illness of children with muco-cutaneous inflammation which can lead to severe vascular complications (Leung *et al.* 1993). Known streptococcal SAgs are SPEA and C (streptococcal pyrogenic exotoxins A and C) and the recently identified mitogenic factor (MF) (Toyosaki *et al.* 1996). SPEA/C are thought to contribute to the pathogenesis of scarlet fever and to the more recently described streptococcal TSS (Stevens *et al.* 1989). The SAg expressed by *Mycoplasma arthritidis* (MAM or mycoplasma arthritidis mitogen) is involved in the chronic relapsing arthritis induced by this natural pathogen of rodents (Cole & Atkin, 1991). A SAg is also encoded by *Yersinia pseudotuberculosis* (see for example Ito *et al.* 1995). This Gram-negative bacterium causes gastrointestinal infection and mesenteric lymphadenitis, but has also been isolated in Japan from patients with a systemic disease matching the definition of Kawasaki syndrome and from patients with reactive polyarthritis (Ito *et al.* 1995).

Structural and biochemical features of bacterial superantigens

Much information is available on the structure and biochemistry of bacterial SAgs. These molecules are small proteins of 15–30 kDa. Information obtained by classical biochemical methods and by mutagenesis has been corroborated recently by the crystal structure of several bacterial SAgs (reviewed by Schlievert *et al.* 1995). Furthermore, the structures of SAg-MHC and SAg-TCRβ cocrystals have now been reported (Jardetzky *et al.* 1994; Fields *et al.* 1996) and have shown what are the important binding regions of the three partners in the SAg-MHC-TCR interaction.

Role in the pathogenesis of bacterial infections

Bacterial SAgs have clearly been shown to have a causal relationship to several disease processes in humans and animals. However, there is controversy as to the specific effects of SAgs with respect to the different manifestations of these diseases.

In toxic shock syndrome, selective expansion of Vβ2-expressing cells has been seen in patients at the acute phase of the disease (Choi *et al.* 1990). The massive release of cytokines triggered by SEs or TSST-1 is thought to play a major pathogenic role in TSS. It is likely that cytokines are released mainly by the reactive T cell subsets, but some bacterial SAgs can also activate monocytes–macrophages directly, possibly by transducing signals through MHC class II molecules (Mourad *et al.* 1993). The T cell activation might be needed to potentiate this

Table 1. Well-characterized bacterial superantigens

(Additional candidate bacterial SAgs are mentioned in the text. Abbreviations:
SE, staphylococcal enterotoxin; TSS, toxic shock syndrome; TSST, toxic shock
syndrome toxin; SPE, streptococcal pyrogenic exotoxin; MF, mitogenic factor;
MAM, mycoplasma arthritidis mitogen; YPM, *Yersinia pseudotuberculosis*-
derived mitogen.)

Producing strain	Protein	Associated diseases
Staphylococcus aureus	SEA to SEE	Food poisoning, TSS
	TSST-1	TSS, Kawasaki syndrome
Streptococcus pyogenes	SPEA and SPEC	Streptococcal TSS, scarlet fever
	MF	Undefined
Mycoplasma arthritidis	MAM	Chronic arthritis in rodents
Yersinia pseudotuberculosis	YPM	Mesenteric lymphadenitis Kawasaki syndrome, reactive arthritis

phenomenon (Grossmann *et al.* 1992). At least in mice, the systemic toxicity of SEB has been shown to require the presence of T cells. Marrack *et al.* (1990) have demonstrated that the toxicity of SEB, as assessed by weight loss, was abolished in nude mice and diminished upon treatment with cyclosporin A. Furthermore, the degree of weight loss was proportional to the number of SAg-reactive cells in mice bred to have varying numbers of the responsive Vβ subsets (Marrack *et al.* 1990). Miethke *et al.* (1992) obtained similar results in a murine model in which injection of SEB after sensitization with D-galactosamine resulted in a severe shock syndrome with a high mortality. The release of tumour necrosis factor (TNF) was found to play a crucial role in this model (Miethke *et al.* 1992). In this respect, TSS has significant similarities to endotoxin-mediated shock, except that the latter is mostly mediated by macrophages rather than by T cells.

Kawasaki syndrome (KS) has been convincingly linked to the effects of bacterial SAgs. TSST-1-producing *S. aureus* isolates have been identified in many KS cases and YPM-producing *Y. pseudotuberculosis* in some of them in Japan (Leung *et al.* 1993; Ito *et al.* 1995). Polyclonal expansion of specific Vβ subsets has been demonstrated in the blood of some patients with acute KS (Abe *et al.* 1993). Furthermore, coronary vasculitis is the most severe complication of KS and a recent report from a fatal case of KS showed expansion of Vβ2-expressing T cells in the affected myocardium, supporting a crucial role for SAg-reactive T cells in this disease (Leung *et al.* 1995).

Potential advantages for the bacteria

Although SAgs are encoded by different strains of bacteria and show often little homology, they have conserved a very similar mechanism of action and comparable effects on the immune system. There-

fore, it seems likely that the effects of SAgs might confer advantages to the bacteria and that selective forces have contributed to the maintenance of functional SAgs in the bacterial populations. An interesting possibility is that SAgs might be able to down-regulate the immune response to convential antigens by 'focusing' on the SAg response, thus impeding immune recognition of the invading bacteria. Surprisingly, however, few studies so far have shown clearcut advantages of SAgs for the producing bacteria.

Rott & Fleischer (1994) have studied the course of experimental systemic staphylococcal infection in mice and compared strains of bacteria containing or not a SAg-encoding plasmid. This approach had the advantage of using strains of bacteria which were isogenic except for the plasmid. A significant delay in the clearance from several organs was observed for SAg-expressing bacteria, together with a reduced T cell responsiveness of the mice to *S. aureus* classical antigens. However, the overall effect of the SAg was small. Bremell & Tarkowski (1995) have compared the development of septic arthritis and the mortality induced in rats by various strains of SAg-producing staphylococci. SEB and SED-producing strains were associated with increased mortality and increased incidence of arthritis. In this experiment, however, the different strains were not isogenic, thus limiting the interpretation of the results.

In summary, much is known already about the identity, structure, mechanisms of action and involvement in disease of bacterial SAgs. However, it is still unclear what the relevant effects of these molecules are for the bacteria themselves.

VIRAL SUPERANTIGENS

A SAg activity has been reported for several viruses, but the protein which exerts this activity has not yet been identified in all cases. Furthermore, the evi-

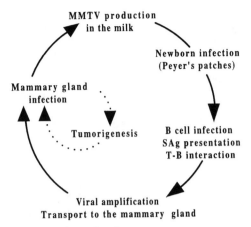

Fig. 2. Life-cycle of mouse mammary tumour virus.

dence supporting the existence of a true SAg is controversial for some of the viruses. The best characterized viral SAg is encoded by mouse mammary tumour virus (MMTV).

Mouse mammary tumour virus superantigens

Different forms of MMTV. Mouse mammary tumour virus (MMTV) is a retrovirus leading to the development of mammary carcinoma in susceptible mice. The infectious or exogenous form of MMTV is transmitted from mothers to pups via milk. Most mouse strains also bear numerous MMTV insertions in their genome, typically 2–8 different proviruses (reviewed by Luther & Acha-Orbea, 1997). These endogenous *Mtvs* (or *Mtv* loci) are thought to result from rare integration events of exogenous MMTVs into the germline which are then stably transmitted by Mendelian inheritance.

The life-cycle of exogenous MMTV is depicted schematically in Fig. 2. Viral particles are produced in large amounts by mammary epithelial cells and the pups get infected upon nursing in the first days of life. Peyer's patches are the site of entry of the virus in the gut (Karapetian *et al.* 1994). MMTV is first found in cells of the immune system, and initially in B cells (Held *et al.* 1993*a*). Expression of the viral SAg plays a crucial role at this stage of the infection (see below). The virus later spreads to the mammary gland of adult animals. Due to the presence of strong hormone response elements in the viral promoter, the expression of viral genes is highly up-regulated by steroid hormones (Buetti & Diggelmann, 1983). This allows the virus to be secreted in large amounts in the milk as a result of the high levels of steroid hormones during pregnancy and lactation. The active viral replication often leads to the development of mammary tumours by insertional mutagenesis.

Identification of MMTV superantigens. The proviral form of MMTV has a classical retroviral structure with two long terminal repeats (LTRs) flanking the

gag, pol and *env* genes (Fig. 3). The 5′ LTR contains the viral promoter, including the elements which are required for the hormone regulation and for the tissue-specific expression of the viral genes. In addition, the 3′ LTR contains a long open reading frame (*orf*) encoding a protein of 320 amino acids. Although the existence of this *orf* was known since the viral LTR was first sequenced (see for example Fasel *et al.* 1982), its function remained unclear. Thanks to a link to the previously unrelated field of minor lymphocyte stimulatory (Mls) antigens, the corresponding protein was found to behave as a SAg.

Indeed, immunologists had been working since 1970 on mysterious endogenous antigens which were able to induce strong mixed lymphocyte reactions between MHC-identical strains of mice. These antigens were named Mls for 'minor lymphocyte stimulatory' antigens. It was later discovered that Mls antigens interact only with T cells expressing specific TCR Vβ domains (Kappler *et al.* 1988; MacDonald *et al.* 1988). Studies of genetic linkage showed then that Mls loci could not be segregated from endogenous *Mtv* loci. Using different experimental strategies, two groups obtained independent evidence that Mls products were actually encoded by the 3′LTR open reading frame of MMTV proviruses (Acha-Orbea *et al.* 1991; Choi, Kappler & Marrack, 1991). Functional SAgs were also shown to be encoded by infectious MMTV (Marrack, Kushnir & Kappler, 1991; Held *et al.* 1992) and their presence proved to play a crucial role in the viral life cycle (Held *et al.* 1993*b*).

Structural and biochemical features of MMTV superantigens. Due to their low expression levels, MMTV SAgs have been more difficult to study biochemically than bacterial SAgs. MMTV SAgs are glycoproteins of about 320 amino acids with a molecular weight of 47 kDa. The putative structure of the molecule is depicted in Fig. 3. MMTV SAgs are type II transmembrane proteins with an extracellular C-terminus. This has been confirmed both biochemically (Choi, Marrack & Kappler, 1992; Korman *et al.* 1992) and functionally (see for example Acha-Orbea *et al.* 1992). Five in-frame initiation codons are found along the sequence (Fig. 3). Recombinant proteins starting at the second ATG were functional in transfected cells, whereas truncation at the 3rd, 4th or 5th ATG resulted in the loss of SAg activity (Choi *et al.* 1992). Similarly, constructs starting at the 1st but not the 5th ATG had a SAg activity in a recombinant vaccinia virus (Krummenacher, Diggelmann & Acha-Orbea, 1996) and in transgenic mice (Lambert *et al.* 1993). Presumably, the transmembrane segment of the protein is necessary for the correct transport of the SAg to the cell surface. The amino acid sequence contains five potential *N*-glycosylation sites and two or three putative protease cleavage sites in the membrane-proximal extracel-

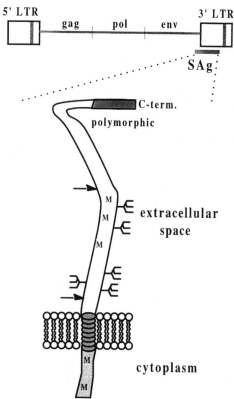

Fig. 3. Proviral genome of mouse mammary tumour virus and putative structure of MMTV superantigen. Abbreviations: LTR, long terminal repeat: M, methionine encoded by an in-frame ATG initiation codon. The putative protease cleavage sites are indicated by arrows and the potential glycosylation sites by schematic trees.

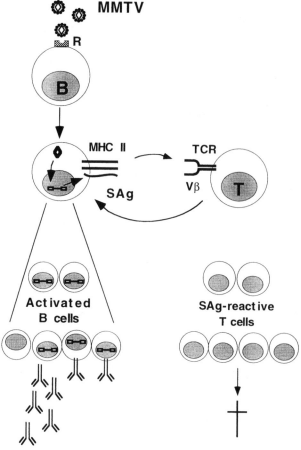

Fig. 4. Early immune response to mouse mammary tumour virus. MMTV first infects B cells through a still unknown receptor (R). The viral genome is reverse transcribed, integrates into the cellular genome and the viral superantigen (SAg) is expressed at the surface of the B cells in association with major histocompatibility complex class II molecules (MHC II). T cells bearing a T cell receptor (TCR) with the appropriate Vβ domain are stimulated and accumulate locally. SAg-reactive T cells provide potent T cells help to infected B cells which get activated and divide, leading to a dramatic increase in the amount of proviral DNA. During the course of this immune response, many B cells differentiate and produce large amounts of immunoglobulins. SAg-reactive T cells are later eliminated by a process of clonal deletion.

lular region. In contrast to bacterial SAgs, a limited processing of the molecule by proteases is probably required for an efficient presentation of the SAg (Park *et al.* 1995; Mix & Winslow, 1996).

The 20–30 C-terminal amino acids of the protein show a highly variable sequence between different strains of MMTV. This polymorphism is tightly correlated to the Vβ specificity of the SAg. Indeed, this region is involved in the specific interaction with the TCR (Yazdanbakhsh *et al.* 1993). MMTV SAgs can thus be grouped into several families which share a very similar SAg C-terminal amino acid sequence and a common Vβ specificity (reviewed by Luther & Acha-Orbea, 1997). It is interesting to note that the families of MMTV SAgs which have been described so far accommodate a large proportion of the Vβ elements present in the T cell repertoire of the mouse (see for example Marrack *et al.* 1991; Held *et al.* 1992; Luther *et al.* 1994; Maillard *et al.* 1996).

Immune response induced by MMTV superantigens. Peyer's patches are the site of entry of the virus in the neonatal gut (Karapetian *et al.* 1994). However, due to the small size of Peyer's patches in newborn animals, the early immune response to MMTV has mostly been characterized in adult mice after injection of the virus into the footpad. This route of

injection induces a SAg response in the draining popliteal lymph node and leads to a productive infection which is comparable to the natural infection (Held *et al.* 1992).

The early immune response to MMTV is depicted in Fig. 4. The main initial target cell of MMTV is the B cell. The cellular receptor which mediates binding and entry of the virus has not been identified yet. Semi-quantitative PCR and *in situ* hybridization on infected lymph node cells have shown that few B cells are initially infected (probably a few hundred cells only) (Held *et al.* 1993*b*). At this stage, the viral SAg gets expressed at the surface of B cells in association with MHC class II molecules. SAg-reactive T cells are activated and accumulate locally.

The increase in percentage of the reactive $V\beta$ subset(s) is first visible at day 2 and peaks between day 4 and 6 after infection. In contrast to other SAgs, this stimulation mainly involves $CD4^+$ and not $CD8^+$ T cells. The SAg-reactive T cells provide potent help to infected B cells through cognate T-B interaction which leads to a large increase in the number of B cells and therefore in the amount of proviral DNA in the infected lymph node (Held *et al.* 1993*a*). This dramatic increase in viral load has been shown to result from the division of infected B cells and not from new rounds of B cell infection (Held *et al.* 1994*a*). During the course of this immune response, many B cells differentiate into plasma cells and produce large amounts of immuno-globulins (Held *et al.* 1993*a*; Luther *et al.* 1994). In summary, MMTV uses its SAg as a tool to increase the initial viral load by recruiting a potent T cell help for infected B cells. SAg-reactive T cells are later eliminated from the T cell repertoire by clonal deletion (Held *et al.* 1992).

Proviral DNA is initially detectable in B but not in T cells. However, the virus can be transferred between lymphocyte subsets at later stages of the viral life cycle (Waanders *et al.* 1993). MMTV eventually gets access to its ultimate replication site in the mammary gland after being transported there by cells of the immune system, but the exact mechanisms of this transfer are not known.

Role of the superantigen in the viral life cycle. It has been known from the work of Tsubura *et al.* that a functional immune system is required to establish an efficient MMTV infection. These authors showed that nude mice had low levels of infection in the mammary gland after foster nursing to MMTV-infected mothers. Transfer of T cells from infected mice was able to establish efficient infection in nude recipients (Tsubura *et al.* 1988). In contrast to observations made in mice with a functional immune system (Waanders *et al.* 1993), transfer of infected B cells only did not result in efficient infection of nude mice, showing that both T and B cell populations are needed to propagate the viral infection.

The absolute requirement for the presence of B cells was formally demonstrated by Beutner *et al.* (1994) who studied MMTV infection in B cell-deficient mice. These mice fail to be infected efficiently by MMTV and do not transfer the virus to their offspring, showing that B cells are critical to sustain milk-borne infection.

More specifically, several groups have shown that the immune response induced by the SAg plays a critical role in the viral life cycle and is required for an efficient transmission of the virus to the next generation. Golovkina *et al.* (1992) have studied MMTV infection in mice expressing a MMTV SAg as a transgene, resulting in the neonatal deletion of T cells with the appropriate $V\beta$-specificity. These mice

were poorly infectable by an exogenous virus expressing a SAg with the same $V\beta$-specificity. Similarly, Held *et al.* (1993*b*) have studied the infection with MMTV (SW) in mice carrying *Mtv*-7, an endogenous *Mtv* which interacts with the same subsets of T cells. Expression of the *Mtv*-7 SAg results in the neonatal deletion of all the T cells that are able to respond to the SAg of MMTV(SW). These mice fail to produce MMTV in the milk and to transmit it to the next generation. In the 2 situations, neonatal expression of a SAg either as a transgene or as a natural endogenous SAg was able to protect the mice from exogenous MMTV infection by inducing a loss of SAg-reactive T cells and thus of the SAg-induced viral amplification step.

Furthermore, the efficiency of different MHC class II alleles to present the viral SAg was shown to correlate with the strength of the SAg response and with the overall efficiency of the infection (Held *et al.* 1994*b*). Thus any interference with the SAg response such as the lack of the appropriate MHC molecule for presentation or the lack of responding T cells strongly reduces the efficiency of viral infection.

In summary, MMTV uses the SAg to exploit the immune system to its own advantage. It induces an efficient T-B interaction which increases dramatically the amount of provirus in the B cell compartment, a process which then allows the virus to be transferred to its distant target organ via cells of the immune system.

Rabies virus superantigen

Superantigen activity of the rabies virus nucleoprotein. The rabies virus nucleocapsid (NC) has been shown to behave as a SAg both in humans and in mice (Lafon *et al.* 1992, 1994). Indeed, rabies NC specifically stimulates human $V\beta8^+$ and murine $V\beta6^+$ $CD4^+$ T cells in the presence of MHC class II-expressing antigen-presenting cells, but does not require active processing and is not strictly MHC-restricted. Similar results were obtained with purified or recombinant forms of the main component of the nucleocapsid, the viral N protein (Lafon *et al.* 1992). In mice, injection of rabies NC to adult animals leads to an expansion of $V\beta6^+ CD4^+$ T cells whereas repeated injections in the neonatal period induce a transient but nearly complete clonal deletion of these cells (Lafon *et al.* 1994). Similarly to MMTV SAgs but not to bacterial SAgs, the rabies NC stimulates predominantly $CD4^+$ rather than $CD8^+$ T cells. All these findings are reminiscent of the effects of other SAgs, even if the high doses of NC used in these experiments are in contradiction with the nanomolar range of activity observed for bacterial SAgs.

Potential role in the virus–host interaction. The role played by the NC SAg activity in the virus–host interaction has not yet been completely elucidated. In the mouse model of rabies infection, SAg-reactive cells are crucial for the appearance of limb paralysis, a symptom which results from an immunopathological injury to peripheral nerves (Lafon *et al.* 1994). An increased mortality was reported in some mouse strains in comparison to mice lacking SAg-reactive T cells, but it has not been clearly shown that the mice die of encephalitic rabies and not as a result of a severe form of peripheral nerve injury (Lafon & Galelli, 1996). The effects of the NC SAg activity on the production of antibodies are somewhat contradictory. Rabies NC has been reported to trigger the production of antibodies by human B lymphocytes in the presence of T cells and to behave as an adjuvant in the mouse when coinjected with an unrelated antigen, although the effects are rather weak (reviewed by Lafon & Galelli, 1996). Earlier studies had shown that rabies NC is able to potentiate the neutralizing antibody response upon immunization with the viral surface glycoprotein (Dietzschold *et al.* 1987). In contrast, the titres of virus-specific IgG2a antibodies during infection with rabies virus were slightly higher in mice lacking than in mice having SAg-reactive T cells, although the titres of all the isotypes were not reported (Lafon & Galelli, 1996).

Altogether, the NC SAg seems to contribute to rabies immunopathology at least in the peripheral nervous system and has the potential to modulate the antiviral antibody response, although the magnitude and the significance of this effect *in vivo* are unclear.

Other putative viral superantigens

Cytomegalovirus. The existence of a SAg encoded by human cytomegalovirus (CMV) has been suggested recently by a group which was working initially on a putative HIV-1 SAg. This group had first observed a preferential replication of HIV in $V\beta12$-expressing CD4$^+$ T cell lines (Laurence, Hodtsev & Posnett, 1992). Such an effect required the presence of MHC class II-expressing antigen-presenting cells (APCs) without being strictly MHC-restricted. It was suggested that a putative HIV SAg was activating $V\beta12^+$ CD4$^+$ T cells and thus driving HIV replication in these cells. The *in vivo* viral load of many patients was also higher in this $V\beta$ subset which was thus presented as a potential reservoir for HIV-1. Finally, a subsequent report showed that CMV infection in the APC fraction was required for the $V\beta12$-selective HIV replication (Dobrescue *et al.* 1995). Several experiments suggested the existence of a CMV-encoded SAg such as the stimulation of $V\beta12^+$ CD4$^+$ and CD8$^+$ T cells upon exposure to a CMV-infected monocytoid cell line. However, the polyclonal nature of the response was not analysed, the viral protein exerting the SAg activity has not been identified yet and no deletion of $V\beta12^+$ T cells

was observed in chronically infected patients. If the putative CMV SAg might influence the course of HIV infection *in vivo*, its role in CMV infection itself is unclear.

The existence of a SAg in the homologous murine CMV (MCMV) was investigated by footpad injection of purified virus in mice (I. Maillard *et al.*, unpublished). No detectable expansion of any $V\beta$ subset was seen during the acute phase of the infection, indicating that a strong SAg activity is not apparent during acute MCMV infection.

Epstein–Barr virus. After CMV, Epstein–Barr virus (EBV) is the second human herpesvirus which has been reported to encode a putative SAg (Sutkowski *et al.* 1996). T cells were shown to be activated upon stimulation with lytically infected EBV-transformed B cells. The $V\beta$ specificity of the stimulation was suggested by the early appearance of activation markers on $V\beta13^+$ T cells, but no preferential expansion of this subset was reported. The overall stimulation was MHC class II-dependent but not MHC-restricted. Here again, the identity of the putative SAg is not known. A role for this SAg has been suggested in the establishment and maintenance of persistent EBV infection, but it is not yet supported by experimental data.

Herpesvirus saimiri. The third herpesvirus which might encode a SAg is the Herpesvirus saimiri (HVS), a virus mainly infecting monkeys. The HVS genome contains an open reading frame (HVS14) which has 22% amino acid identity to the *Mtv-7* SAg. Recombinant HVS14 was recently shown to bind to HLA-DR and to induce proliferation of primary T cells in the presence of antigen-presenting cells (Yao *et al.* 1996). The $V\beta$ usage and polyclonality of the responding T cells remain to be shown. The role of HVS14 in the pathogenesis of HVS infection is unknown.

Human immunodeficiency virus. Many hypotheses have been made to explain the progressive depletion of CD4$^+$ T cells in HIV infection. In this context, a putative HIV SAg has been postulated to contribute to the loss of CD4$^+$ lymphocytes and to the immunopathogenesis of AIDS. Several groups have studied the $V\beta$ repertoire of HIV-1-infected patients using either semi-quantitative RT–PCR or flow cytometry. Several $V\beta$ subsets were first reported to be underrepresented in a group of HIV-infected individuals at late stages of the disease (Imberti *et al.* 1991). However, this study did not distinguish CD4$^+$ from CD8$^+$ T cells and was made in patients with very low CD4 counts. Furthermore, the T cell repertoire is influenced by genetic polymorphisms, in particular by the HLA genes. A later study was devoted to the study of monozygotic twins discordant for HIV-1 infection (Rebai *et al.* 1994), which had the advantage of ensuring a perfect HLA matching between the infected and the non-infected twins. Alterations in the percentage of cells expressing certain Vβ subsets were seen, but without massive deletions.

Pantaleo *et al.* have studied in detail the Vβ repertoire of patients during primary HIV-1 infection when a putative HIV SAg would be expected to stimulate reactive Vβ subsets (Pantaleo *et al.* 1994). Major expansions of specific Vβ subsets were observed, but appeared to predominate in the CD8 population. Furthermore, analysis of the TCR CDR3 region from the reactive cells showed a small number of recurrent sequences, which is characteristic of oligoclonal responses to classical antigens and has later been shown to result from stimulation by immunodominant HIV antigens. Altogether, the existence of a SAg encoded by HIV has not been formally ruled out, but seems unlikely.

Murine leukaemia virus. A mixture of replication-competent and -defective murine leukaemia virus (MuLV) induces the murine acquired immuno-deficiency syndrome (MAIDS) in certain mouse strains. A SAg-like activity was first identified in a B cell line expressing proteins of the defective virus and was proposed to contribute to the early T cell activation which is characteristic of MAIDS (Hügin, Vacchio & Morse, 1991). Later studies by other groups did not confirm the existence of an MuLV-encoded SAg and did not show any change in the course of MAIDS in mice lacking the reactive Vβ subset (Doyon *et al.* 1996). It is possible that the SAg activity identified by Hügin *et al.* was in fact due to the overexpression of SAgs from endogenous *Mtvs* in the transformed B cell line. Further studies will be needed to answer this question definitively.

Human foamy virus. The Bel 3 protein of human foamy virus (a retrovirus belonging to the family of lentiviridae) has been reported to have a SAg activity (Weissenberger *et al.* 1994). Recombinant Bel 3 protein induced stimulation of Vβ18$^+$ human T cells *in vitro*, but the requirement for MHC class II has not been studied. The existence and the possible role of this SAg activity during viral infection have not been examined. Unlike rabies NC, the SAg activity of Bel 3 is not conserved in mice, since injection of recombinant Bel 3 into the footpad of mice did not result in detectable expansion of any Vβ subset (H. Acha-Orbea *et al.*, unpublished).

OTHER SUPERANTIGENS

Toxoplasma gondii

The existence of a SAg encoded by *Toxoplasma gondii* has been proposed by the group of Alan Sher (Denkers *et al.* 1994, 1996). The capacity of nonimmune splenocytes to respond *in vitro* to intact and soluble tachyzoite antigen was examined. Both types of stimuli induced high levels of proliferation as well as IFN-γ secretion. The response was inhibited by anti-MHC class II antibodies, but was not strictly MHC-restricted. A preferential expansion of Vβ5$^+$CD8$^+$ T cells was observed. Furthermore, mice had moderate increases in the percentage of Vβ5$^+$ T cells during acute *T. gondii* infection, as well as specific unresponsiveness of these cells in the chronic phase of the infection.

Altogether, these results suggest the existence of a SAg in *T. gondii*. The protozoal protein which is responsible for this effect has not been identified. Moreover, it has not been excluded that *T. gondii* infection upregulates the expression of an endogenous SAg. In addition, the exclusive expansion of CD8$^+$ cells is in contradiction with data from other SAgs which always induce a response of CD4$^+$ T cells. Interestingly CD4$^+$ T cells might still play an important role, since depletion of CD4$^+$ T cells prior to exposure to the parasite antigens abolishes the stimulation of CD8$^+$ T cells, implying a helper activity provided by the CD4$^+$ T cells.

The potential advantage of a SAg for *T. gondii* remains to be demonstrated. The SAg-induced early production of IFN-γ has been suggested to contribute to a strong cell-mediated immunity and thus to the control of the infection by the host immune system. However, this is in contradiction with the finding that mouse strains expressing high levels of Vβ5$^+$ cells tend to be more susceptible to *T. gondii*, suggesting that the presence of these cells might also be detrimental for the host (Denkers *et al.* 1994, 1996).

CONCLUDING REMARKS

In summary, proteins with SAg activity are expressed by many different microorganisms. Although evolutionarily distant, these proteins have conserved a very similar mechanism of action and have comparable effects on the immune system. Much is known already about the identity, structure, activity and involvement in disease of bacterial SAgs: they can induce a very potent systemic immune response which has clearly been shown to be causally associated with several diseases in humans and animals. However, it is still unclear what are the advantages conferred by these molecules to the bacteria themselves. This is in contrast to MMTV SAgs whose structural and biochemical features are difficult to study, but whose utility for the virus is much better understood. An important future task will be to strengthen the evidence supporting the existence of all the currently described putative SAgs and possibly to find new members for this family of proteins. In addition, further studies on the role of these molecules in microbial infections might allow a better understanding of host–pathogen interactions.

ACKNOWLEDGEMENTS

I.M. is supported by the Foundation Max Cloëtta. F.L., H.A.O. and H.D. are recipients of grants from the Swiss National Science Foundation. We would like to thank Daniela Finke and Sanjiv Luther for useful discussions and critical reading of the manuscript.

REFERENCES

ABE, J., KOTZIN, B. L., MEISSNER, C., MELISH, M. E., TAKAHASHI, M., FULTON, D., ROMAGNE, F., MALISSEN, B. & LEUNG, D.Y. (1993). Characterization of T cell repertoire changes in acute Kawasaki disease. *Journal of Experimental Medicine* **177**, 791–796.

ACHA-ORBEA, H., SCARPELLINO, L., SHAKHOV, A. N., HELD, W. & MACDONALD, R. (1992). Inhibition of mouse mammary tumor virus-induced T cell responses *in vivo* by antibodies to an open reading frame protein. *Journal of Experimental Medicine* **176**, 1769–1772.

ACHA-ORBEA, H., SHAKHOV, A. N., SCARPELLINO, L., KOLB, E., MULLER, V., VESSAZ-SHAW, A., FUCHS, R., BLOCHLINGER, K., ROLLINI, P., BILLOTTE, J., SARAFIDOU, M., MACDONALD, H. R. & DIGGELMANN, H. (1991). Clonal deletion of Vβ14-bearing T cells in mice transgenic for mammary tumour virus. *Nature* **350**, 207–211.

BEUTNER, U., KRAUS, E., KITAMURA, D., RAJEWSKY, K. & HUBER, B. T. (1994). B cells are essential for murine mammary tumor virus transmission, but not for presentation of endogenous superantigens. *Journal of Experimental Medicine* **179**, 1457–1466.

BOITEL, B., ERMONVAL, M., PANINA-BORDIGNON, P., MARIUZZA, R.A., LANZAVECCHIA, A. & ACUTO, O. (1992). Preferential Vβ gene usage and lack of junctional sequence conservation among human T cell receptors specific for a tetanus toxin-derived peptide: evidence for a dominant role of a germline-encoded V region in antigen/major histocompatibility complex recognition. *Journal of Experimental Medicine* **175**, 765–777.

BREMELL, T. & TARKOWSKI, A. (1995). Preferential induction of septic arthritis and mortality by superantigen-producing staphylococci. *Infection and Immunity* **63**, 4185–4187.

BUETTI, E. & DIGGELMANN, H. (1983). Glucocorticoid regulation of mouse mammary tumor virus: identification of a short essential DNA region. *EMBO Journal* **2**, 1423–1429.

CASANOVA, J. L., CEROTTINI, J. C., MATTHEWS, M., NECKER, A., GOURNIER, H., BARRA, C., WIDMANN, C., MACDONALD, H. R., LEMONNIER, F. & MALISSEN, B. (1992). H-2-restricted cytolytic T lymphocytes specific for HLA display T cell receptors of limited diversity. *Journal of Experimental Medicine* **176**, 439–447.

CHOI, Y., KAPPLER, J. W. & MARRACK, P. (1991). A superantigen encoded in the open reading frame of the 3′ long terminal repeat of mouse mammary tumour virus. *Nature* **350**, 203–207.

CHOI, Y., LAFFERTY, J. A., CLEMENTS, J. R., TODD, J. K., GELFAND, E. W., KAPPLER, J., MARRACK, P. & KOTZIN, B. L. (1990). Selective expansion of T cells expressing Vβ2 in toxic shock syndrome. *Journal of Experimental Medicine* **172**, 981–984.

CHOI, Y., MARRACK, P. & KAPPLER, J. W. (1992). Structural analysis of a mouse mammary tumor virus superantigen. *Journal of Experimental Medicine* **175**, 847–852.

COLE, B. C. & ATKIN, C. L. (1991). The Mycoplasma arthritidis T-cell mitogen, MAM: a model superantigen. *Immunology Today* **12**, 271–276.

DELLABONA, P., PECCOUD, J., KAPPLER, J., MARRACK, P., BENOIST, C. & MATHIS, D. (1990). Superantigens interact with MHC class II molecules outside of the antigen groove. *Cell* **62**, 1115–1121.

DENKERS, E. Y., CASPAR, P., HIENY, S. & SHER, A. (1996). *Toxoplasma gondii* infection induces specific nonresponsiveness in lymphocytes bearing the Vβ5 chain of the mouse T cell receptor. *Journal of Immunology* **156**, 1089–1094.

DENKERS, E. Y., CASPAR, P. & SHER, A. (1994). *Toxoplasma gondii* possesses a superantigen activity that selectively expands murine T cell receptor Vβ5-bearing CD8$^+$ lymphocytes. *Journal of Experimental Medicine* **180**, 985–994.

DIETZSCHOLD, B., WANG, H. H., RUPPRECHT, C. E., CELIS, E., TOLLIS, M., ERTL, H., HEBER-KATZ, E. & KOPROWSKI, H. (1987). Induction of protective immunity against rabies by immunization with rabies virus ribonucleoprotein. *Proceedings of the National Academy of Sciences, USA* **84**, 9165–9169.

DOBRESCU, D., URSEA, B., POPE, M., ASCH, A. S. & POSNETT, D. N. (1995). Enhanced HIV-1 replication in Vβ12 T cells due to human cytomegalovirus in monocytes: evidence for a putative herpesvirus superantigen. *Cell* **82**, 753–763.

DOYON, L., SIMARD, C., SEKALY, R. P. & JOLICOEUR, P. (1996). Evidence that the murine AIDS defective virus does not encode a superantigen. *Journal of Virology* **70**, 1–9.

FASEL, N., PEARSON, K., BUETTI, E. & DIGGELMANN, H. (1982). The region of mouse mammary tumor virus DNA containing the long terminal repeat includes a long coding sequence and signals for hormonally regulated transcription. *EMBO Journal* **1**, 3–7.

FIELDS, B. A., MALCHIODI, E. L., LI, H., YSERN, X., STAUFFACHER, C. V., SCHLIEVERT, P. M., KARJALAINEN, K. & MARIUZZA, R. A. (1996). Crystal structure of a T-cell receptor β-chain complexed with a superantigen. *Nature* **384**, 188–192.

FLEISCHER, B., GERLACH, D., FUHRMANN, A. & SCHMIDT, K. H. (1995). Superantigens and pseudosuperantigens of gram-positive cocci. *Medical Microbiology and Immunology* **184**, 1–8.

GOLOVKINA, T. V., CHERVONSKY, A., DUDLEY, J. P. & ROSS, S. R. (1992). Transgenic mouse mammary tumor virus superantigen expression prevents viral infection. *Cell* **69**, 637–645.

GROSSMAN, D., LAMPHEAR, J. G., MOLLICK, J. A., BETLEY, M. J. & RICH, R. R. (1992). Dual roles for class II major histocompatibility complex molecules in staphylococcal enterotoxin-induced cytokine production and *in vivo* toxicity. *Infection & Immunity* **60**, 5190–5196.

HELD, W., SHAKHOV, A. N., IZUI, S., WAANDERS, G. A., SCARPELLINO, L., MACDONALD, H. R. & ACHA-ORBEA, H. (1993*a*). Superantigen-reactive CD4$^+$ T cells are required to stimulate B cells after infection with mouse mammary tumor virus. *Journal of Experimental Medicine* **177**, 359–366.

HELD, W., SHAKHOV, A. N., WAANDERS, G., SCARPELLINO, L., LUETHY, R., KRAEHENBUHL, J.-P., MACDONALD, H. R. & ACHA-ORBEA, H. (1992). An exogenous mouse mammary tumor virus with properties of Mls-1a (*Mtv*7). *Journal of Experimental Medicine* **175**, 1623–1633.

HELD, W., WAANDERS, G. A., ACHA-ORBEA, H. & MACDONALD, H. R. (1994a). Reverse transcriptase-dependent and -independent phases of infection with mouse mammary tumor virus: implications for superantigen function. *Journal of Experimental Medicine* **180**, 2347–2351.

HELD, W., WAANDERS, G. A., MACDONALD, H. R. & ACHA-ORBEA, H. (1994b). MHC class II hierarchy of superantigen presentation predicts efficiency of infection with mouse mammary tumor virus. *International Immunology* **6**, 1403–1407.

HELD, W., WAANDERS, G. A., SHAKHOV, A. N., SCARPELLINO, L., ACHA-ORBEA, H. & MACDONALD, H. R. (1993b). Superantigen-induced immune stimulation amplifies mouse mammary tumor virus infection and allows virus transmission. *Cell* **74**, 529–540.

HERMAN, A., CROTEAU, G., SEKALY, R. P., KAPPLER, J. & MARRACK, P. (1990). HLA-DR alleles differ in their ability to present staphylococcal enterotoxins to T cells. *Journal of Experimental Medicine* **172**, 709–717.

HERRMANN, T., BASCHIERI, S., LEES, R. K. & MACDONALD, H. R. (1992). *In vivo* responses of CD4[+] and CD8[+] cells to bacterial superantigens. *European Journal of Immunology* **22**, 1935–1938.

HERRMANN, T., MARYANSKI, J. L., ROMERO, P., FLEISCHER, B. & MACDONALD, H. R. (1990). Activation of MHC class I-restricted CD8[+] CTL by microbial T cell mitogens. Dependence upon MHC class II expression of the target cells and Vβ usage of the responder T cells. *Journal of Immunology* **144**, 1181–1186.

HUGIN, A. W., VACCHIO, M. S. & MORSE, H. C. (1991). A virus-encoded 'superantigen' in a retrovirus-induced immunodeficiency syndrome of mice. *Science* **252**, 424–427.

IMBERTI, L., SOTTINI, A., BETTINARDI, A., PUOTI, M. & PRIMI, D. (1991). Selective depletion in HIV infection of T cells that bear specific T cell receptor Vβ sequences. *Science* **254**, 860–862.

ITO, Y., ABE, J., YOSHINO, K., TAKEDA, T. & KOHSAKA, T. (1995). Sequence analysis of the gene for a novel superantigen produced by *Yersinia pseudotuberculosis* and expression of the recombinant protein. *Journal of Immunology* **154**, 5896–5906.

JARDETZKY, T., BROWN, J. H., GORGA, J. C., STERN, L. J., URBAN, R. G., CHI, Y. I, STAUFFACHER, C., STROMINGER, J. L. & WILEY, D. C. (1994). Three-dimensional structure of a human class II histocompatibility molecule complexed with superantigen. *Nature* **368**, 711–718.

KAPPLER, J., KOTZIN, B., HERRON, L., GELFAND, E. W., BIGLER, R. D., BOYLSTON, A., CARREL, S., POSNETT, D. N., CHOI, Y. & MARRACK, P. (1989). Vβ-specific stimulation of human T cells by staphylococcal toxins. *Science* **244**, 811–813.

KAPPLER, J. W., STAERZ, U., WHITE, J. & MARRACK, P. C. (1988). Self-tolerance eliminates T cells specific for

Mls-modified products of the major histocompatibility complex. *Nature* **332**, 35–40.

KARAPETIAN, O., SHAKHOV, A. N., KRAEHENBUHL, J.-P. & ACHA-ORBEA, H. (1994). Retroviral infection of neonatal Peyer's patch lymphocytes through an intact epithelium. *Journal of Experimental Medicine* **180**, 1511–1516.

KORMAN, A. J., BOURGAREL, P., MEO, T. & RIECKHOF, G. E. (1992). The mouse mammary tumor virus long terminal repeat encodes a type II transmembrane glycoprotein. *EMBO Journal* **11**, 1901–1905.

KRUMMENACHER, C., DIGGELMANN, H. & ACHA-ORBEA, H. (1996). *In vivo* effects of a recombinant vaccinia virus expressing a mouse mammary tumor virus superantigen. *Journal of Virology* **70**, 3026–3031.

LAFON, M. & GALELLI, A. (1996). Superantigen related to rabies. *Springer Seminars in Immunopathology*, **17**, 307–318.

LAFON, M., LAFAGE, M., MARTINEZ-ARENDS, A., RAMIREZ, R., VUILLIER, F., CHARRON, D., LOTTEAU, V. & SCOTT-ALGARA, D. (1992). Evidence for a viral superantigen in humans. *Nature* **358**, 507–510.

LAFON, M., SCOTT-ALGARA, D., MARCHE, P. N., CAZENAVE, P. A. & JOUVIN-MARCHE, E. (1994). Neonatal deletion and selective expansion of mouse T cells by exposure to rabies virus nucleocapsid superantigen. *Journal of Experimental Medicine* **180**, 1207–1215.

LAMBERT, J.-F., ACHA-ORBEA, H., KOLB, E. & DIGGELMANN, H. (1993). The 3′ half of the mouse mammary tumor virus *orf* gene is not sufficient for its superantigen function in transgenic mice. *Molecular Immunology* **30**, 1399–1404.

LAURENCE, J., HODTSEV, A. S. & POSNETT, D. N. (1992). Superantigen implicated in dependence of HIV-1 replication in T cells on TCR Vβ expression. *Nature* **358**, 255–259.

LEUNG, D. Y., GIORNO, R. C., KAZEMI, L. V., FLYNN, P. A. & BUSSE, J. B. (1995). Evidence for superantigen involvement in cardiovascular injury due to Kawasaki syndrome. *Journal of Immunology* **155**, 5018–5021.

LEUNG, D. Y., MEISSNER, H. C., FULTON, D. R., MURRAY, D. L., KOTZIN, B. L. & SCHLIEVERT, P. M. (1993). Toxic shock syndrome toxin secreting *Staphylococcus aureus* in Kawasaki syndrome. *Lancet* **342**, 1385–1388.

LUTHER, S. & ACHA-ORBEA, H. (1997). Mouse mammary tumor virus: immunological interplays between virus and host. *Advances in Immunology* **65**, in press.

LUTHER, S., SHAKHOV, A. N., XENARIOS, I., HAGA, S., IMAI, S. & ACHA-ORBEA, H. (1994). New infectious mammary tumor virus superantigen with Vβ-specificity identical to staphylococcal enterotoxin B (SEB). *European Journal of Immunology* **24**, 1757–1764.

MACDONALD, H. R., BASCHIERI, S. & LEES, R. K. (1991). Clonal expansion precedes anergy and death of Vβ8[+] peripheral T cells responding to staphylococcal enterotoxin B *in vivo*. *European Journal of Immunology* **21**, 1963–1966.

MACDONALD, H. R., LEES, R. K. & CHVATCHKO, Y. (1990). CD8[+] T cells respond clonally to Mls-1ᵃ-encoded determinants. *Journal of Experimental Medicine* **171**, 1381–1386.

MACDONALD, H. R., SCHNEIDER, R., LEES, R. K., HOWE, R. C., ACHA-ORBEA, H., FESTENSTEIN, H., ZINKERNAGEL, R. M. & HENGARTNER, H. (1988). T cell receptor Vβ use predicts reactivity and tolerance of Mls^a-encoded antigens. *Nature* **332**, 40–45.

MAILLARD, I., ERNY, K., ACHA-ORBEA, H. & DIGGELMANN, H. (1996). A Vβ4-specific superantigen encoded by a new exogenous mouse mammary tumor virus. *European Journal of Immunology* **26**, 1000–1006.

MARRACK, P., BLACKMAN, M., KUSHNIR, E. & KAPPLER, J. (1990). The toxicity of staphylococcal enterotoxin B in mice is mediated by T cells. *Journal of Experimental Medicine* **171**, 455–464.

MARRACK, P. & KAPPLER, J. W. (1990). The staphylococcal enterotoxins and their relatives. *Science* **248**, 705–711.

MARRACK, P., KUSHNIR, E. & KAPPLER, J. (1991). A maternally inherited superantigen encoded by a mammary tumour virus. *Nature* **349**, 524–526.

MIETHKE, T., WAHL, C., HEEG, K., ECHTENACHER, B., KRAMMER, P. H. & WAGNER, H. (1992). T cell-mediated lethal shock triggered in mice by the superantigen staphylococcal enterotoxin B: critical role of tumor necrosis factor. *Journal of Experimental Medicine* **175**, 91–98.

MIX, D. & WINSLOW, G. M. (1996). Proteolytic processing activates a viral superantigen. *Journal of Experimental Medicine* **184**, 1549–1554.

MOLLICK, J. A., CHINTAGUMPALA, M., COOK, R. G. & RICH, R. R. (1991). Staphylococcal exotoxin activation of T cells. Role of exotoxin–MHC class II binding affinity and class II isotype. *Journal of Immunology* **146**, 463–468.

MOURAD, W., AL-DACCAK, R., CHATILA, T. & GEHA, R. S. (1993). Staphylococcal superantigens as inducers of signal transduction in MHC class II-positive cells. *Seminars in Immunology* **5**, 47–55.

PANTALEO, G., DEMAREST, J. F., SOUDEYNS, H., GRAZIOSI, C., DENIS, F., ADELSBERGER, J. W., BORROW, P., SAAG, M. S., SHAW, G. M. & SEKALY, R. P. (1994). Major expansion of CD8⁺ T cells with a predominant V beta usage during the primary immune response to HIV. *Nature* **370**, 463–467.

PARK, C. G., JUNG, M. Y., CHOI, Y. & WINSLOW, G. M. (1995). Proteolytic processing is required for viral superantigen activity. *Journal of Experimental Medicine* **181**, 1899–1904.

REBAI, N., PANTALEO, G., DEMAREST, J. F., CIURLI, C., SOUDEYNS, H., ADELSBERGER, J. W., VACCAREZZA, M., WALKER, R. E., SEKALY, R. P. & FAUCI, A. S. (1994). Analysis of the T-cell receptor beta-chain variable-region (Vβ) repertoire in monozygotic twins discordant for human immunodeficiency virus: evidence for perturbations of specific Vβ segments in CD4⁺ T cells of the virus-positive twins. *Proceedings of the National Academy of Sciences, USA* **91**, 1529–1533.

ROTT, O. & FLEISCHER, B. (1994). A superantigen as virulence factor in an acute bacterial infection. *Journal of Infectious Diseases* **169**, 1142–1146.

SCHLIEVERT, P. M., BOHACH, G. A., OHLENDORF, D. H., STAUFFACHER, C. V., LEUNG, D. Y., MURRAY, D. L., EARHART, C. A., JABLONSKI, L. M., HOFFMANN, M. L. & CHI, Y. I. (1995). Molecular structure of staphylococcus and streptococcus superantigens. *Journal of Clinical Immunology* **15**, 4S–10S.

SCHLIEVERT, P. M., SHANDS, K. N., DAN, B. B., SCHMID, G. P. & NISHIMURA, R. D. (1981). Identification and characterization of an exotoxin from *Staphylococcus aureus* associated with toxic-shock syndrome. *Journal of Infectious Diseases* **143**, 509–516.

STEVENS, D. L., TANNER, M. H., WINSHIP, J., SWARTS, R., RIES, K. M., SCHLIEVERT, P. M. & KAPLAN, E. (1989). Severe group A streptococcal infections associated with a toxic shock-like syndrome and scarlet fever toxin A. *New England Journal of Medicine* **321**, 1–7.

SUTKOWSKI, N., PALKAMA, T., CIURLI, C., SEKALY, R.-P., THORLEY-LAWSON, D. A. & HUBER, B. T. (1996). An Epstein–Barr virus-associated superantigen. *Journal of Experimental Medicine* **184**, 971–980.

TOYOSAKI, T., YOSHIOKA, T., TSURUTA, Y., YUTSUDO, T., IWASAKI, M. & SUZUKI, R. (1996). Definition of the mitogenic factor (MF) as a novel streptococcal superantigen that is different from streptococcal pyrogenic exotoxins A, B, and C. *European Journal of Immunology* **26**, 2693–2701.

TSUBURA, A., MUNEO, I., IMAI, S., MURAKAMI, A., OYAIZU, N., YASUMIZU, R., OHNISHI, Y., TANAKA, H., MORII, S. & IKEHARA, S. (1988). Intervention of T-cells in transportation of mouse mammary tumor virus (milk factor) to mammary gland cells *in vivo*. *Cancer Research* **48**, 6555–6559.

TUMANG, J. R., CHERNIACK, E. P., GIETL, D. M., COLE, B. C., RUSSO, C., CROW, M. K. & FRIEDMAN, S. M. (1991). T helper cell-dependent, microbial superantigen-induced murine B cell activation: polyclonal and antigen-specific antibody responses. *Journal of Immunology* **147**, 432–438.

VACCHIO, M. S., KANAGAWA, O., TOMONARI, K. & HODES, R. J. (1992). Influence of T cell receptor Vα expression on Mls^a superantigen-specific T cell responses. *Journal of Experimental Medicine* **175**, 1405–1408.

WAANDERS, G. A., SHAKHOV, A. N., HELD, W., KARAPETIAN, O., ACHA-ORBEA, H. & MACDONALD, H. R. (1993). Peripheral T cell activation and deletion induced by transfer of lymphocyte subsets expressing endogenous or exogenous mouse mammary tumor virus. *Journal of Experimental Medicine* **177**, 1359–1366.

WEBB, S., MORRIS, C. & SPRENT, J. (1990). Extrathymic tolerance of mature T cells: clonal elimination as a consequence of immunity. *Cell* **63**, 1249–1256.

WEISSENBERGER, J., ALTMANN, A., MEUER, S. & FLUGEL, R. M. (1994). Evidence for superantigen activity of the Bel 3 protein of the human foamy virus. *Journal of Medical Virology* **44**, 59–66.

WHITE, J., HERMAN, A., PULLEN, A. M., KUBO, R., KAPPLER, J. W. & MARRACK, P. (1989). The Vβ-specific superantigen staphylococcal enterotoxin B: stimulation of mature T cells and clonal deletion in neonatal mice. *Cell* **56**, 27–35.

YAO, Z., MARASKOVSKY, E., SPRIGGS, M. K., COHEN, J. I., ARMITAGE, R. J. & ALDERSON, M. R. (1996). Herpesvirus saimiri open reading frame 14, a protein encoded by a T lymphotropic herpesvirus, binds to MHC class II molecules and stimulates T cell proliferation. *Journal of Immunology* **156**, 3260–3266.

YAZDANBAKHSH, K., PARK, C. G., WINSLOW, G. M. & CHOI, Y. (1993). Direct evidence for the role of COOH terminus of mouse mammary tumor virus superantigen in determining T cell receptor Vβ specificity. *Journal of Experimental Medicine* **178**, 737–741.

Macrophage apoptosis in microbial infections

H. HILBI[1], A. ZYCHLINSKY[1] *and* P. J. SANSONETTI[2]*

[1] *The Skirball Institute, Department of Microbiology and Kaplan Cancer Center, New York University School of Medicine, 540 First Avenue, New York, NY 10016 USA*
[2] *Unité de Pathogenie Microbienne Moleculaire, Unité 389 Institut National de la Santé et de la Recherche Médicale, Institut Pasteur, 28 rue du Dr. Roux, F-75724 Paris Cédex 15, France*

SUMMARY

Upon infection with a pathogen, eukaryotic cells can undergo programmed cell death as an ultimate response. Therefore, modulation of apoptosis is often a prerequisite to establish a host-pathogen relationship. Some pathogens kill macrophages by inducing apoptosis and thus overcome the microbicidal arsenal of the phagocyte. Apoptotic macrophages, on the other hand, can elicit an inflammation by secretion of proinflammatory cytokines. *Shigella flexneri*, the aetiological agent of bacillary dysentery, induces apoptosis in macrophages which, in agony, specifically release mature interleukin-1β (IL-1β). This cytokine attracts neutrophils (PMN) to the site of infection resulting in the massive colonic inflammation characteristic of bacillary dysentery. Shigellosis represents a paradigm of a proinflammatory apoptosis in a bacterial infection. The molecular link between apoptosis and inflammation is interleukin-1β converting enzyme (ICE) which is activated during macrophage apoptosis and binds to IpaB, a secreted *Shigella* protein.

Key words: Bacteria, pathogenicity, macrophages, apoptosis, inflammation, parasites.

PROGRAMMED CELL DEATH IN DISEASES

Normal development of multicellular organisms requires death and disposal of certain individual cells for the benefit of the whole organism. Surplus, damaged or undesirable cells commit suicide by enacting an evolutionarily conserved, intrinsic programme referred to as apoptosis or programmed cell death (Ameisen, 1996; Arends & Wyllie, 1991; Steller, 1995). Apoptotic cells are characterized by a distinct morphology and, on a biochemical level, by the cleavage of nuclear DNA at the internucleosomal boundaries to fragments of about 200 bp. Apoptosis not only takes place in structural processes like morphogenesis, metamorphosis or tissue homeostasis (Raff, 1992), but it is also involved in eliminating potentially harmful cells such as autoreactive thymocytes and senescent neutrophils (PMN) (Ellis, Yuan & Horvitz, 1991). Macrophages recognize and phagocytose apoptotic cells and thus clear the debris of the dying cells from the tissue. In summary, the very purpose of canonical programmed cell death is the noninflammatory removal of bothersome cells.

Many diseases are associated with an imbalance of apoptosis resulting either in a detrimental increase or in depletion of certain cells. In malignancies, autoimmune disorders and viral infections apoptosis is blocked, whereas in neurodegenerative disorders or AIDS increased apoptosis occurs (Thompson, 1995). During viral infections, programmed cell death is employed as a defense strategy. Virally infected cells are prone to die by apoptosis triggered

either by effector cells of the immune system or endogenously. It is, therefore, not surprising that a number of viruses are capable of suppressing apoptosis of their host cells, thus sustaining an intracellular infection. Interestingly, some viral anti-apoptotic proteins directly interfere with the components of the cell death machinery. CrmA of cowpox virus and p35 of baculovirus inhibit ICE-like proteases which are crucial components of the apoptotic programme. Inhibition of ICE not only prevents apoptosis but also the production of mature IL-1β, a potent proinflammatory cytokine which, upon release by infected cells, can alert the immune system (Vaux, Haecker & Strasser, 1994). Other viral proteins like BHRF1 of Epstein–Barr virus are homologous to Bcl-2, an inhibitor of apoptosis, or they inactivate p53 which mediates the apoptotic response to DNA damage (Thompson, 1995).

Intracellular pathogens can also cause apoptosis of their host cells. For example, human immunodeficiency virus (HIV) infects and kills CD4$^+$ lymphocytes. Specific depletion of these cells accounts for immunodeficiency and might also aid in the establishment of a chronic infection (Barr & Tomei, 1994). Another example is *Shigella flexneri*, an intracellular bacterium that kills macrophages by inducing apoptosis and thereby overcomes the bactericidal potential of this host cell. The dying macrophages secrete IL-1β which triggers an acute inflammation of the colonic mucosa. Shigellosis is a self-limiting disease, probably because the macrophage's apoptosis is not an immunologically silent, altruistic suicide, but rather a proinflammatory process eventually leading to clearance of the infection (Zychlinsky & Sansonetti, 1997).

* Corresponding author.

Parasitology (1997) **115**, S79–S87. © 1997 Cambridge University Press

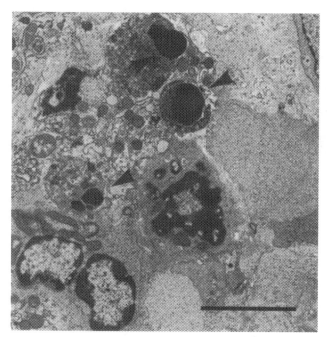

Fig. 1. Electronmicrograph of apoptotic cells from a lymphoid follicle infected with *Shigella flexneri*. Arrowheads point to apoptotic nuclei. Bar represents 5 μm.

After an overview of the events involved in Shigella-directed phagocytosis we will focus in this review on macrophage apoptosis induced by *Shigella* and other pathogens. We will summarize the evidence that in shigellosis macrophage apoptosis is a proinflammatory event and, in light of the concept that pathogen-induced apoptosis might provoke an inflammation, we will also review recent findings on the pathogenesis of other infections.

PATHOGENESIS OF SHIGELLOSIS

The enterobacterium *Shigella* is the aetiological agent of bacillary dysentery, a severe form of diarrhoea characterized by blood and mucus in the stool. Typical symptoms are painful abdominal cramps, nausea and fever. The pathology of the disease is due to colonic epithelial cell destruction and massive inflammation (reviewed by Siebers & Finlay, 1995; Zychlinsky *et al.* 1994*c*). Dysentery is self-limiting in healthy individuals; however, because of dehydration, shigellosis can be a life-threatening disease and is frequently fatal in children.

The pathogenicity of *Shigella* is associated with its ability to invade colonic cells which it can enter only from the basolateral side but not from the apical side (Mounier *et al.* 1992). As a consequence, *Shigella* relies on M-cells, specialized antigen sampling cells of the mucosa-associated lymphoid tissue (MALT), to penetrate the epithelial barrier (Perdomo *et al.* 1994*b*; Wassef, Keren & Mailloux, 1989). After passage through M cells, associated resident tissue macrophages readily phagocytose the pathogen.

Inside the macrophages, *Shigella* escapes the phagolysosome, and kills its host cells by inducing apoptosis (Zychlinsky, Prévost & Sansonetti, 1992). Fig. 1 shows apoptotic cells in a lymphoid follicle infected with *S. flexneri*.

The dying macrophages secrete IL-1 which initiates inflammation (Zychlinsky *et al.* 1996). PMN are chemoattracted to the site of infection and destroy the integrity of the epithelium as they traverse into the lumen of the colon (Perdomo *et al.* 1994*b*). Transmigration of PMN facilitates mucosal access for the bacteria by opening intercellular junctions. The current model of inflammation-enhanced shigellosis is shown in Fig. 2. This model is substantiated by the following findings: (1) a monoclonal antibody against $\alpha_M\beta_2$ integrin (CD18) which neutralizes PMN adhesion to epithelial cells, largely reduced infection in the rabbit ligated ileal loop model (Perdomo *et al.* 1994*b*) and blocks PMN transmigration *in vitro* (Perdomo *et al.* 1994*b*); (2) intravenous infusion of IL-1 receptor antagonist reduces inflammation and decreases the number of bacteria in mucosal tissue of rabbit ileal loops (Sansonetti *et al.* 1995); and (3) disruption of tight cell junctions by treatment with EGTA greatly enhances cell infection of a polarized colonic cell monolayer with *Shigella* via the basolateral route (Mounier *et al.* 1992).

BOTH SHIGELLA-INDUCED PHAGOCYTOSIS AND INTRACELLULAR MOVEMENT EMPLOY THE ACTIN CYTOSKELETON OF EPITHELIAL CELLS

Epithelial cells of the colonic mucosa internalize *Shigella* by a process resembling macropinocytosis. This pathogen-directed phagocytosis is characterized by membrane ruffling, polymerization of actin filaments and actin bundling by plastin (reviewed by Ménard, Dehio & Sansonetti, 1996*a*). Concomitant with bacterial entry, a number of host cell signalling events take place, including cortactin tyrosine phosphorylation and recruitment of small Rho GTPases (Ménard *et al.* 1996*a*).

Bacteria-induced uptake of *Shigella* is a complex process involving more than 30 gene products (Ménard *et al.* 1996*a*). The genes conferring entry into epithelial cells are located on a 31 kb fragment of the 220 kb virulence plasmid characteristic of all virulent *Shigella* strains. The gene products include IpaA, IpaB, IpaC and IpaD (*i*nvasion *p*lasmid *a*ntigen) and the chaperone IpgC. The Ipa proteins are indispensable for entry into epithelial cells and lysis of the phagocytic vacuole (High *et al.* 1992; Ménard, Sansonetti & Parsot, 1993). A type III secretion apparatus, encoded by the *Mxi* and *Spa* operons, secretes the Ipa proteins (Ménard *et al.* 1996*a*) and is controlled by IpaB and IpaD (Ménard, Sansonetti & Parsot, 1994*a*).

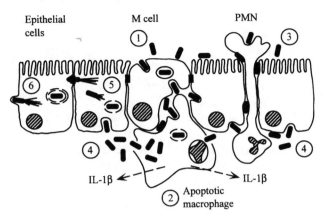

Fig. 2. Passage of *Shigella* from the colonic lumen to the mucosa. (1) Shigellae enter the intestinal epithelium via M cells; (2) resident macrophages phagocytose the bacteria and are killed by apoptosis. The dying macrophages release mature IL-1β which initiates inflammation; (3) chemoattracted PMN destroy the integrity of the epithelial barrier; (4) *Shigella* invades enterocytes basolaterally through bacteria-induced phagocytosis; (5) endocytosed bacteria escape the vacuole, move in the cytoplasm by actin polymerization, spread to neighbouring cells *via* cell junctions; and (6) lyse the double membrane of the protrusions to continue the intraepithelial cell-to-cell spread.

Presynthesized IpaB and IpaC are separately stored in the bacterial cytoplasm, each in transient association with the chaperone IpgC (*invasion plasmid gene*) which prevents the degradation of IpaB (Ménard *et al.* 1994*b*). Upon contact with a target cell or components of the extracellular matrix, a soluble IpaB–IpaC complex is rapidly released (Ménard *et al.* 1994*a*; Watarai *et al.* 1995). Latex beads coupled to immunopurified IpaB-IpaC complexes are efficiently internalized into HeLa cells thus providing that this complex is sufficient to promote uptake. Interestingly, these beads elicit membrane projections and actin polymerization as seen during wild type *Shigella* invasion (Ménard *et al.* 1996*b*).

After entry and lysis of the phagosome, *Shigella* moves intracellularly and spreads to adjacent epithelial cells by exploiting components of the host cell cytoskeleton. IcsA (*intra/intercellular spread*) mediates the directional polymerization of actin filaments (Bernardini *et al.* 1989). Actin-based motility contributes largely to the virulence of *Shigella*, since strains lacking *icsA* only form small localized inflammatory lesions in lymphoid follicles but do not cause dysentery (Sansonetti *et al.* 1991). IcsA contains a consensus sequence for cAMP-dependent kinases which upon phosphorylation by host cell kinases limits the intracellular spread of *Shigella* (d'Hauteville & Sansonetti, 1992). This phosphorylation consensus site is also the cleavage site for proteolytic processing of IcsA which is a prerequisite for its unipolar localization in the outer membrane (d'Hauteville *et al.* 1996).

The fibres formed by IcsA-dependent actin polymerization propel the bacteria through the cell and eventually deliver it to adjacent cells (Goldberg & Theriot, 1995). Cell to cell spread is dependent on cadherins, a family of calcium-dependent cell adhesion molecules that are components of intermediate junctions (Sansonetti *et al.* 1994). Finally, the endocytosed protrusions, consisting of two membranes, are lysed by IcsB (Allaoui *et al.* 1992), and the invasion may continue.

SHIGELLA KILLS MACROPHAGES BY APOPTOSIS

Shigella flexneri does not sustain an infection within macrophages but rapidly after infection escapes the phagolysosome and kills its host cells by apoptosis. Dying macrophages infected either with a virulent laboratory strain of *S. flexneri* (Zychlinsky *et al.* 1992) or with clinical *Shigella* isolates (Guichon & Zychlinsky, 1997) exhibit the distinct characteristics of apoptotic morphology: blebbing of the cytoplasmic membrane, vacuolization of the cytoplasm yet overall maintenance of organelle structure and chromatin condensation at the nuclear boundary. Conversly, HeLa cells infected with *S. flexneri* stay viable and intact for at least 12 h and show only signs of metabolic stress but not the ultrastructural hallmarks of programmed cell death (Mantis, Prévost & Sansonetti, 1996).

Induction of apoptosis by virulent *Shigella* is significant *in vivo*. After infection of rabbit Peyer's patches, apoptotic cells were labelled by the TUNEL technique (*terminal d-transferase-mediated dUTP-biotin nick end labelling*) and quantified. In this assay, wild type *Shigella* caused apoptosis in cells of the lympoid tissue, in contrast to a plasmidless, nonvirulent strain and an avirulent strain capable of colonizing Peyer's patches. Apoptotic cells were identified by immunofluorescence as macrophages, B and T cells (Zychlinsky *et al.* 1996).

APOPTOTIC MACROPHAGES RELEASE MATURE IL-1β, A PROINFLAMMATORY CYTOKINE

Macrophages are among the most effective cells in signalling inflammation by secreting proinflammatory cytokines such as IL-1, IL-6 and tumour necrosis factor α (TNFα) (Dinarello, 1992). IL-1, in contrast to IL-6 and TNFα, is not secreted by activated macrophages but rather stored in the cytoplasm. IL-1α and IL-1β, the two forms of IL-1, bind to the same receptor and have very similar biological activity including chemotactic attraction of PMN and macrophages. IL-1α and IL-1β lack classical signal sequences and are cleaved by either calpain, a calcium-dependent protease, or by ICE, the interleukin-1β converting enzyme, respectively (Dinarello & Margolis, 1995). Whereas IL-1α is biologically active as precursor and mature protein,

IL-1β is active only after cleavage of the 31 kDa pro-protein to the 17·5 kDa mature form by ICE. Hence, ICE is a pivotal immunomodulatory enzyme.

Processing and secretion of IL-1 by macrophages depends on the nature of injury imposed (Hogquist *et al.* 1991). Only during apoptotic death, but not as a consequence of necrosis, is pro-IL-1β processed and mature IL-1β secreted. Mature IL-1β, there-fore, might be a marker cytokine for apoptotic macrophages.

Upon *in vitro* infection of stimulated peritoneal macrophages, *S. flexneri* was shown to elicit specifi-cally the rapid secretion of both preformed precursor IL-1α and mature IL-1β. Stimulation of IL-1β secretion was restricted to the wild type *S. flexneri* strain and neither a noninvasive derivative nor a noncytotoxic yet invasive control strain provoked the release of mature cytokine. Conversely, IL-6 and TNFα were not detected, since the macrophages were killed by *Shigella* before *de novo* biosynthesis could take place (Zychlinsky *et al.* 1994*a*; 1996).

To substantiate the role of IL-1 in the patho-genesis of *Shigella*, rabbit ileal loops were infected with the pathogen after intravenous administration of IL-1 receptor antagonist (IL-1ra). IL-1ra is a macrophage product which inhibits IL-1 competi-tively binding to the IL-1 receptor. Continuous infusion of the animals with recombinant IL-1ra reduced inflammation, destruction of mucosal tissue and bacterial invasion thus proving the crucial role of IL-1 in the pathogenesis of shigellosis (Sansonetti *et al.* 1995).

In summary, there is convincing evidence that *Shigella*-infected apoptotic macrophages secrete ma-ture IL-1β which initiates inflammation. IL-1, in turn, boosts the secretion of other proinflammatory cytokines leading to a full-fledged inflammation. The orchestrated upregulation of proinflammatory cytokines in the colon chemoattracts PMN which destroy the epithelium and in shigellosis finally clear the infection. Thus, in *Shigella* pathogenesis the apoptotic release of IL-1β is a proinflammatory event.

IPAB BINDS ICE, THE ENZYME LINKING APOPTOSIS AND INFLAMMATION

Only recently the molecular link of *Shigella*-induced macrophage apoptosis and the onset of inflammation was unravelled. *Shigella* has to reach the cytoplasm of macrophages to induce apoptosis. To show which protein is necessary for apoptosis, the requirement of Ipa proteins for lysis of the phagosome had to be bypassed. This was accomplished by constructing non-polar mutant *Shigella* strains that were deficient in single Ipa proteins but secreted low amounts of *E. coli* hemolysin. With this system IpaB, a 62 kDa protein, but not IpaC or IpaD, was shown to be necessary to induce apoptosis (Zychlinsky *et al.*

1994*b*). Microinjecting purified IpaB into peritoneal macrophages caused these cells to undergo apoptosis thus proving that IpaB is sufficient to elicit pro-grammed cell death (Chen *et al.* 1996*b*).

Macrophage target proteins of IpaB were identi-fied by using purified IpaB as a ligand in affinity chromatography. Four proteins of 45, 33, 20 and 10 kDa, respectively, were specifically retained by an IpaB affinity column and were identified as precur-sors and mature fragments of ICE or a closely related protein. Furthermore, inhibition of ICE with the peptide inhibitor acetyl-YVAD-CHO abrogated *Shigella*-induced macrophage apoptosis as well as release of mature IL-1β (Chen *et al.* 1996*b*). Hence, in macrophage apoptosis induced by *Shigella* the proapoptotic and proinflammatory potential of ICE converges (Fig. 3).

ICE originally was identified as a novel cysteine protease that processes pro-IL-1β to its mature, proinflammatory form (Cerretti *et al.* 1992; Thorn-berry *et al.* 1992). Subsequently, it was found that the cell death gene *ced-3* of the nematode *Caenor-habditis elegans* is homologous to ICE (Yuan *et al.* 1993) and that overexpression of either cysteine protease in fibroblasts results in programmed cell death (Miura *et al.* 1993). These findings suggest that mediation of programmed cell death by cysteine proteases is evolutionarily conserved. ICE is synthe-sized as a 45 kDa proenzyme, which after heterolo-gous cleavage or autoprocessing at aspartate sites releases a 11 kDa N-terminal prodomain and an internal 2 kDa fragment to form the active (p20)$_2$/(p10)$_2$ heterotetramer (Whyte, 1996).

To date, 10 homologues of the still growing family of ICE/CED-3 cysteine proteases have been identi-fied and subdivided into three groups based on their degree of similarity (ICE-, CPP32- and ICH-1-like proteases). For these related proteases the term 'caspase' ('*c*ysteine protease cleaving after *asp*artic acid') has been suggested (Alnemri *et al.* 1996). Interestingly, ICE is largely redundant for apoptosis but indispensable for processing of IL-1β since mutant mice deficient in ICE appear healthy but require a larger dose of endotoxin to induce toxic shock (Kuida *et al.* 1995; Li *et al.* 1995). CPP32-deficient mice, on the other hand, suffer from decreased apoptosis in the brain and die prematurely (Kuida *et al.* 1996).

MACROPHAGE APOPTOSIS IN OTHER BACTERIAL INFECTIONS

Induction of apoptosis by bacterial pathogens is widely used as a cytotoxic strategy among both extra- and intracellular bacteria (reviewed by Chen & Zychlinsky, 1994). Apart from IpaB mentioned above, the apoptotic arsenal comprises (1) A-B toxins which inhibit protein synthesis, (2) enzymes that generate second messengers like cAMP, (3) strepto-

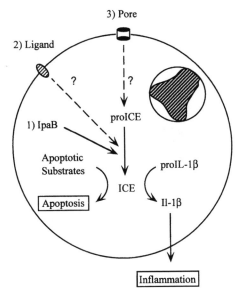

Fig. 3. Activation of the bi-functional enzyme ICE in macrophages results in convergence of apoptosis and inflammation. Macrophage apoptosis can be induced by (1) factors directly delivered to the cytoplasm like IpaB; (2) transmembrane signalling events involving, e.g., extracellular ATP, or (3) poreforming proteins.

and staphylococcal superantigens that induce apoptosis by crosslinking certain T-cell receptors and (4) pore-forming proteins like *Staphylococcus aureus* α-toxin or hemolysin of enteropathogenic *Escherichia coli*. The variety of mechanistically different apoptogenic toxins reflects the plethora of pathways leading to programmed cell death.

A subset of bacteria or microbial products cause apoptosis in macrophages thus killing the phagocyte by exploiting its intrinsic cell death programme (Table 1). These pathogens very efficiently overcome the microbicidal activities of the phagocyte by expression of comparably small amounts of toxin. However, since macrophage apoptosis also bears the potential of a proinflammatory event, the initial success of the pathogen finally might contribute to its restriction and elimination. Host–pathogen relationships involving proinflammatory apoptosis may not be restricted to shigellosis but rather may apply to other diseases such as anthrax, diphtheria, whooping cough, periodontitis, listeriosis, salmonellosis and parasitic infections (Zychlinsky & Sansonetti, 1997).

Only recently it was found that both *Salmonella typhi* and *S. typhimurium* induce apoptosis in macrophages, although with slower kinetics than *Shigella* (Chen, Kaniga & Galan, 1996 a; Lindgren, Stojilkovic & Heffron, 1996; Monack *et al.* 1996). Hence, *Salmonella* not only evades the host immune system by sustaining an intramacrophage infection but also by depletion of phagocytes. The ability to cause macrophage programmed cell death is restricted to invasive *S. typhimurium*; noninvasive mutant strains do not induce apoptosis (Monack *et al.* 1996). Interestingly, *Salmonella* spp. synthesize

SipB, a protein homologous to IpaB from *Shigella*. This protein, like IpaB, is required for entry into epithelial cells (Kaniga *et al.* 1995) and is necessary to induce apoptosis (Chen *et al.* 1996 a). This is rather surprising considering the different ways *Shigella* and *Salmonella* interact with phagocytic and nonphagocytic cells and the differences in the course of shigellosis and salmonellosis. Whereas the former disease manifests as an acute, self-limiting diarrhoea, localized to the colon, the latter can cause a severe chronic, systemic illness (typhoid fever). However, the phenotype of a particular infection depends on the complete pattern of the pathogen's virulence factors which likely are not identical for *Shigella* and *Salmonella*.

Additionally, IpaB and SipB are related to YopB (*Yersinia outer protein*), a *Yersinia*-secreted protein involved in translocation of Yop effector proteins across the target cell plasma membrane (Hakansson *et al.* 1993, 1996). YopB is also implicated in suppression of TNFα expression (Beuscher *et al.* 1995). IpaB, as outlined above, causes apoptosis in macrophages which is accompanied by the secretion of proinflammatory IL-1β. Hence, closely related bacterial virulence factors can act either as suppressor or as effector of inflammation depending on the pathogenic strategy of the infectious agent.

Listeria monocytogenes does not induce apoptosis in a macrophage cell line (Zychlinsky *et al.* 1992) but in hepatocytes (Rogers *et al.* 1996). The latter cells were shown to chemoattract PMN to the site of infection by an as yet unidentified factor. Consequently, the concept of apoptosis as a proinflammatory event might not only apply to macrophages releasing mature IL-1β but also to other apoptotic cells releasing different proinflammatory cytokines.

Lethal anthrax toxin is a putative zinc protease synthesized by *Bacillus anthracis*. At high concentrations, this toxin lyses macrophages *in vitro*. The systemic toxic shock, characteristic of anthrax, is specifically mediated by macrophages and can be counteracted *in vivo* by administering antiserum against IL-1 or IL-1 receptor antagonist (Hanna, Acosta & Collier, 1993). Since sublytic stimulation of macrophages with lethal anthrax toxin also provokes the secretion of IL-1, this toxin might be a mediator of apoptosis possibly by proteolytically activating ICE.

Inhibition of protein synthesis efficiently induces apoptosis in eukaryotic cells. Diphtheria toxin of *Corynebacterium diphtheriae* and exotoxin A of *Pseudomonas aeruginosa* mediate apoptosis by ADP-ribosylation of elongation factor-2 (Morimoto & Bonavida, 1992). These toxins and other inhibitors of protein synthesis induce apoptosis in a human monoblastoid cell line (Kochi & Collier, 1993) and, therefore, might also be cytotoxic for macrophages *in vivo*. Proinflammatory macrophage apoptosis, in turn, might contribute to the formation of a

Table 1. Bacterial toxins inducing apoptosis in macrophages (for references see text)

Bacterium	Toxin	Mode of action
Shigella flexneri	IpaB	Binding of ICE
Salmonella typhi(murium)	SipB	?
Bacillus anthracis	Lethal antrax toxin	Proteolysis?
Corynebacterium diphtheriae	Diphteria toxin	Inhibition of protein synthesis
Pseudomonas aeruginosa	Exotoxin A	Inhibition of protein synthesis
Bordetella pertussis	Adenylate cyclase Hemolysin	Generation of cAMP
Actinobacillus actinomycetemcomitans	Leukotoxin	Formation of pores Changes in Ca^{2+} concentration
Escherichia coli	Hemolysin	Formation of pores
Staphylococcus aureus	α-toxin	Formation of pores

diagnostically characteristic feature of diphtheria: the pharyngal pseudomembranes, consisting of fibrin, bacteria and inflammatory cells.

Bordetella pertussis, the causative agent of whooping cough, can enter and survive in epithelial cells and macrophages. In a monocyte/macrophage cell line *B. pertussis* causes apoptosis which indispensably involves the calmodulin-dependent adenylate cyclase hemolysin, a RTX toxin, but not pertussis toxin, an ADP-ribosylating A-B toxin (Khelef, Zychlinsky & Guiso, 1993). Murine alveolar macrophages also undergo apoptosis upon infection with *B. pertussis* indicating that promotion of cell death might contribute to bacterial survival and escape from the host immune response *in vivo*.

Pore-forming toxins can provoke either apoptosis or necrosis depending on their concentration. *Actinobacillus actinomycetemcomitans*, a causative agent of periodontitis, synthesizes a pore-forming leukotoxin which is another member of the bacterial RTX toxin family (reviewed by Czuprynski & Welch, 1995). This toxin is selectively cytotoxic for monocytes and PMN and kills T-lymphocytes by apoptosis and necrosis (Mangan *et al.* 1991). Leukotoxicity might be critical in establishing a chronic periodontal infection by *A. actinomycetemcomitans*. In sublytic concentrations, hemolysin of pathogenic *Escherichia coli* stimulates IL-1 release from monocytes *in vitro* (Bhakdi *et al.*, 1990), thereby likely causing apoptosis in these cells. Both hemolysin and staphylococcal α-toxin (Bhakdi *et al.*, 1990) induce internucleosomal DNA fragmentation and apoptosis in T-lymphocytes (reviewed by Chen & Zychlinsky, 1994) and thus probably contribute to immunosuppression by pathogenic *E. coli* and *S. aureus*.

APOPTOSIS IN PARASITIC DISEASES

Depending on their pathogenic strategy, protozoan parasites either downregulate or induce apoptosis of immune cells. Macrophages are the exclusive host cells of *Leishmania*. Visceral leishmaniasis (kala-azar) is caused by *L. donovani* which was shown to suppress macrophage apoptosis by induction of

TNFα expression thereby sustaining an intracellular infection (Moore & Matlashewski, 1994). In contrast, *Entamoeba histolytica* causes contact-dependent apoptosis and necrosis in PMN and macrophages (Ragland *et al.* 1994). Upon infection with *Trypanosoma cruzi*, $CD4^+$ lymphocytes are killed by apoptosis and, therefore, $CD4^+$ T cell depletion could contribute to parasite persistence in Chagas' disease (Lopes *et al.* 1995). Similarly, in malaria caused by *Plasmodium falciparum* increased programmed cell death occurs in mononuclear cells (Toure-Blade *et al.* 1996). Taken together, in parasite infections apoptosis might play an important role in immunosuppression and the development of the diseases.

CONCLUDING REMARKS

Pathogens overcome the microbicidal activities of macrophages by either surviving intracellularly or by killing the immune cell. A strategy increasingly found to be employed by bacteria and parasites is to provoke apoptosis in macrophages. Thus, the phagocyte is efficiently killed since its pre-established cell death arsenal can be exploited. The best studied example of bacteria-induced apoptosis is *Shigella flexneri* which has evolved a peculiar host–pathogen relationship in that it does not establish a homeostatic condition within macrophages but rapidly kills this cell through apoptosis. However, although initially successful in overcoming the host's innate immune response, shigellosis is self-limiting in healthy individuals presumably due to an acute inflammation which is the direct aftermath of IL-1β secretion by the dying macrophage. Apoptosis as a trigger of inflammation might represent a novel mechanism in the establishment of complex host-pathogen relationships. Apart from surviving the macrophages, antibacterial activities, *Shigella*'s benefit of evolving IpaB as a potent macrophage toxin could be that the bacteria can efficiently spread to other hosts because of diarrhoea.

On the other hand, among a family of ICE/CED-3 related cysteine proteases involved in programmed cell death, ICE might have particularly evolved as the trigger component of defensive apoptosis which upon stimulation specifically elicits a disease-limiting inflammation via release of mature IL-1β.

In recent years tremendous progress has been made in identifying molecular players of programmed cell death (Fraser & Evan, 1996). Mechanisms and regulation of apoptosis, however, still remain largely unsolved. Since IpaB induces apoptosis by directly binding to ICE or a related protease, this bacterial invasion might turn out to be a valuable tool in dissecting the delicate balance between the life and death of a cell.

ACKNOWLEDGEMENTS

This work was supported by a grant from the Swiss National Science Foundation (H.H.), by Public Health Service grant AI37720 from the National Institutes of Health (A.Z.) and by a collaborative research grant from the North Atlantic Treaty Organization (NATO) International Scientific Exchange Programs. The electron-micrograph was kindly provided by Station Centrale de Microscopie Electronique, Institut Pasteur.

REFERENCES

ALLAOUI, A., MOURNIER, J., PRÉVOST, M. C., SANSONETTI, P. J. & PARSOT, C. (1992). *icsB*: a *Shigella flexneri* virulence gene necessary for the lysis of protrusions during intercellular spread. *Molecular Microbiology* **6**, 1605–1616.

ALNEMRI, E. S., LIVINGSTON, D. J., NICHOLSON, D. W., SALVESEN, G., THORNBERRY, N. A., WONG, W. W. & YUAN, J. (1996). Human ICE/CED-3 protease nomenclature. *Cell* **87**, 171.

AMEISEN, J. C. (1996). The origin of programmed cell death. *Science* **272**, 1278–1279.

ARENDS, M. J. & WYLLIE, A. H. (1991). Apoptosis: mechanisms and roles in pathology. *International Review of Experimental Pathology* **32**, 223–254.

BARR, P. J. & TOMEI, L. D. (1994). Apoptosis and its role in human disease. *Biotechnology* **12**, 487–493.

BERNARDINI, M. L., MOUNIER, J., D'HAUTEVILLE, H., COQUIS-RONDON, M. & SANSONETTI, P. J. (1989). Identification of *icsA*, a plasmid locus of *Shigella flexneri* that governs bacterial intra and intercellular spread through interaction with F-actin. *Proceedings of the National Academy of Sciences, USA* **86**, 3867–3871.

BEUSCHER, H. U., RÖDEL, F., FORSBERG, A. & RÖLLINGHOFF, M. (1995). Bacterial evasion of host immune defense: *Yersinia enterocolitica* encodes a suppressor for tumor necrosis factor alpha expression. *Infection and Immunity* **63**, 1270–1277.

BHAKDI, S., MUHLY, M., KOROM, S. & SCHMIDT, G. (1990). Effects of *Escherichia coli* hemolysin on human monocytes. Cytocidal action and stimulation of interleukin 1 release. *Journal of Clinical Investigation* **85**, 1746–1753.

CERRETTI, D. P., KOZLOSKY, C. J., MOSLEY, B., NELSON, N., VAN NESS, K., GREENSTREET, T. A., MARCH, C. J., KRONHEIM, S. R., DRUCK, T., CANNIZARRO, L. A.,

HUEBNER, K. & BLACK, R. A. (1992). Molecular cloning of the interleukin-1β converting enzyme. *Science* **256**, 97–99.

CHEN, L. M., KANIGA, K. & GALAN, J. E. (1996*a*). *Salmonella* spp. are cytotoxic for cultured macrophages. *Molecular Microbiology* **21**, 1101–1115.

CHEN, Y., SMITH, M. R., THIRUMALAI, K. & ZYCHLINSKY, A. (1996*b*). A bacterial invasion induces macrophage apoptosis by directly binding ICE. *EMBO Journal* **15**, 3853–3860.

CHEN, Y. & ZYCHLINSKY, A. (1994). Apoptosis induced by bacterial pathogens. *Microbial Pathogenesis* **17**, 203–212.

CZUPRYNSKI, C. J. & WELCH, R. A. (1995). Biological effects of RTX toxins: the possible role of lipopolysaccharide. *Trends in Microbiology* **3**, 480–483.

D'HAUTEVILLE, H., LAGELOUSE, R. D., NATO, F. & SANSONETTI, P. J. (1996). Lack of cleavage of IcsA in *Shigella flexneri* causes aberrant movement and allows demonstration of a cross-reactive eukaryotic protein. *Infection and Immunity* **64**, 511–517.

D'HAUTEVILLE, H. & SANSONETTI, P. J. (1992). Phosphorylation of IcsA by cAMP-dependent protein kinase and its effects on intercellular spread of *Shigella flexneri*. *Molecular Microbiology* **6**, 833–841.

DINARELLO, C. A. (1992). Role of Interleukin-1 in infectious diseases. *Immunological Reviews* **127**, 119–146.

DINARELLO, C. A. & MARGOLIS, N. H. (1995). Stopping the cuts. *Current Biology* **5**, 587–590.

ELLIS, R. E., YUAN, J. & HORVITZ, H. R. (1991). Mechanisms and functions of cell death. *Annual Reviews in Cell Biology* **7**, 663–98.

FRASER, A. & EVAN, G. (1996). A licence to kill. *Cell* **85**, 781–784.

GOLDBERG, M. B. & THERIOT, J. A. (1995). *Shigella flexneri* surface protein IcsA is sufficient to direct actin-based motility. *Proceedings of the National Academy, USA* **92**, 6572–6576.

GUICHON, A. & ZYCHLINSKY, A. (1997). Clinical isolates of *Shigella* spp. induce apoptosis in macrophages. *Journal of Infectious Disease* (in press).

HAKANSSON, S., BERGMAN, T., VANOOTEGHEM, J. C., CORNELIS, G. & WOLF-WATZ, H. (1993). YopB and YopD constitute a novel class of *Yersinia* Yop proteins. *Infection and Immunity* **61**, 71–80.

HAKANSSON, S., SCHESSER, K., PERSSON, C., GALYOV, E. E., ROSQVIST, R., HOMBLÉ, F. & WOLF-WATZ, H. (1996). The YopB protein of *Yersinia pseudotuberculosis* is essential for the translocation of Yop effector proteins across the target cell plasma membrane and displays a contact-dependent membrane disrupting activity. *EMBO Journal* **15**, 5812–5823.

HANNA, P. C., ACOSTA, D. & COLLIER, R. J. (1993). On the role of macrophages in anthrax. *Proceedings of the National Academy of Sciences, USA* **90**, 10198–10201.

HIGH, N., MOUNIER, J., PREVOST, M. C. & SANSONETTI, P. J. (1992). IpaB of *Shigella flexneri* causes entry into epithelial cells and escape from the phagocytic vacuole. *EMBO Journal* **1992**, 1191–1999.

HOGQUIST, K. A., NETT, M. A., UNANUE, E. R. & CHAPLIN, D. D. (1991). Interleukin-1 is processed and released during apoptosis. *Proceedings of the National Academy of Sciences, USA* **88**, 8485–8489.

KANIGA, K., TUCKER, S., TROLLINGER, D. & GALAN, J. E. (1995). Homologs of the shigella IpaB and IpaC invasins are required for *Salmonella typhimurium* entry into cultured epithelial cells. *Journal of Bacteriology* **177**, 3965–3971.

KHELEF, N., ZYCHLINSKY, A. & GUISO, N. (1993). *Bordetella pertussis* induces apoptosis in macrophages: role of adenylate cyclase-hemolysin. *Infection and Immunity* **61**, 4064–4071.

KOCHI, S. K. & COLLIER, R. J. (1993). DNA fragmentation and cytolysis in U937 cells treated with diphtheria toxin or other inhibitors of protein synthesis. *Experimental Cell Research* **208**, 296–302.

KUIDA, K., LIPPKE, J. A., KU, G., HARDING, M. W., LIVINGSTON, D. J., SU, M. S.-S. & FLAVELL, R. A. (1995). Altered cytokine export and apoptosis in mice deficient in interleukin-1β converting enzyme. *Science* **267**, 2000–2003.

KUIDA, K., ZHENG, T. S., NA, S., KUAN, C., YANG, D., KARASUYAMA, H., RAKIC, P. & FLAVELL, R. A. (1996). Decreased apoptosis in the brain and premature lethality in CPP32-deficient mice. *Nature* **384**, 368–372.

LI, P., ALLEN, H., BANERJEE, S., FRANKLIN, S., HERZOG, L., JOHNSTON, C., MCDOWELL, J., PASKIND, M., RODMAN, L., SALFELD, J., TOWNE, E., TRACEY, D., WARDWELL, S., WEI, F.-Y., WONG, W., KAMEN, R. & SESHADRI, T. (1995). Mice deficient in IL-1β-converting enzyme are defective in production of mature IL-1β and resistance to endotoxic shock. *Cell* **80**, 401–411.

LINDGREN, S. W., STOJILKOVIC, I. & HEFFRON, F. (1996). Macrophage killing is an essential virulence mechanism of *Salmonella typhimurium*. *Proceedings of the National Academy of Sciences, USA* **93**, 4197–4201.

LOPES, M. F., DA VEIGA, V. F., SANTOS, A. R., FONSECA, M. E. F. & DOSREIS, G. A. (1995). Activation-induced CD4+ T cell death by apoptosis in experimental Chagas' disease. *Journal of Immunology* **154**, 744–752.

MANGAN, D. F., TAICHMAN, N. S., LALLY, E. T. & WAHL, S. M. (1991). Lethal effects of *Actinobacillus actinomycetemcomitans* leukotoxin on human T lymphocytes. *Infection and Immunity* **59**, 3267–3272.

MANTIS, N., PRÉVOST, M.-C. & SANSONETTI, P. J. (1996). Analysis of epithelial cell stress response during infection by *Shigella flexneri*. *Infection and Immunity* **64**, 2474–2482.

MÉNARD, R., DEHIO, C. & SANSONETTI, P. J. (1996a). Bacterial entry into epithelial cells: the paradigm of *Shigella*. *Trends in Microbiology* **4**, 220–226.

MÉNARD, R., PRÉVOST, M. C., GOUNON, P., SANSONETTI, P. J. & DEHIO, C. (1996b). The secreted Ipa complex of *Shigella flexneri* promotes entry into mammalian cells. *Proceedings of the National Academy of Sciences, USA* **93**, 1254–1258.

MÉNARD, R., SANSONETTI, P. J. & PARSOT, C. (1993). Nonpolar mutagenesis of the *ipa* genes defines IpaB, IpaC, and IpaD as effectors of *Shigella flexneri* entry into epithelial cells. *Journal of Bacteriology* **175**, 5899–5906.

MÉNARD, R., SANSONETTI, P. J. & PARSOT, C. (1994a). The secretion of the *Shigella flexneri* Ipa invasins is activated by epithelial cells and controlled by IpaB and IpaD. *EMBO Journal* **13**, 5293–5302.

MÉNARD, R., SANSONETTI, P. J., PARSOT, C. & VASSELON, T. (1994b). Extracellular association and cytoplasmic partitioning of the IpaB and IpaC invasins of *S. flexneri*. *Cell* **79**, 515–525.

MIURA, M., ZHU, H., ROTELLO, R., HARTWIEG, E. A. & YUAN, J. (1993). Induction of apoptosis in fibroblasts by IL-1β converting enzyme, a mammalian homolog of the *C. elegans* cell death gene *ced-3*. *Cell* **75**, 653–660.

MONACK, D. M., RAUPACH, B., HROMOCKYJ, A. E. & FALKOW, S. (1996). *Salmonella typhimurium* invasion induces apoptosis in infected macrophages. *Proceedings of the National Academy of Sciences, USA* **93**, 9833–9838.

MOORE, K. J. & MATLASHEWSKI, G. (1994). Intracellular infection by *Leishmania donovani* inhibits macrophage apoptosis. *Journal of Immunology* **152**, 2930–2937.

MORIMOTO, H. & BONAVIDA, B. (1992). Diphtheria toxin and pseudomonas A toxin mediated apoptosis. *Journal of Immunology* **149**, 2089–2094.

MOUNIER, J., VASSELON, T., HELLIO, R., LESOURD, M. & SANSONETTI, P. J. (1992). *Shigella flexneri* enters human colonic Caco-2 epithelial cells through the basolateral pole. *Infection and Immunity* **60**, 237–248.

PERDOMO, O. J. J., CAVAILLON, J. M., HUERRE, M., OHAYON, H., GOUNON, P. & SANSONETTI, P. J. (1994b). Acute inflammation causes epithelial invasion and mucosal destruction in experimental shigellosis. *Journal of Experimental Medicine* **180**, 1307–1319.

PERDOMO, J. J., GOUNON, P. & SANSONETTI, P. J. (1994a). Polymorphonuclear leukocyte transmigration promotes invasion of colonic epithelial monolayer by *Shigella flexneri*. *Journal of Clinical Investigation* **93**, 633–643.

RAFF, M. C. (1992). Social controls on cell survival and cell death. *Nature* **356**, 397–400.

RAGLAND, B. D., ASHLEY, L. S., VAUX, D. L. & PETRI, W. A. (1994). *Entamoeba histolytica*: target cells killed by trophozoites undergo DNA fragmentation which is not blocked by Bcl-2. *Experimental Parasitology* **79**, 460–467.

ROGERS, H. W., CALLERY, M. P., DECK, B. & UNANUE, E. R. (1996). *Listeria monocytogenes* induces apoptosis in infected hepatocytes. *Journal of Immunology* **156**, 679–684.

SANSONETTI, P. J., ARONDEL, J., CAVAILLON, J.-M. & HUERRE, M. (1995). Role of IL-1 in the pathogenesis of experimental shigellosis. *Journal of Clinical Investigation* **96**, 884–892.

SANSONETTI, P. J., ARONDEL, J., FONTAINE, A., D'HAUTEVILLE, H. & BERNARDINI, M. L. (1991). OmpB (osmo-regulation) and *icsA* (cell to cell spread) mutants of *Shigella flexneri*: vaccine candidates and probes to study the pathogenesis of shigellosis. *Vaccine* **9**, 416–422.

SANSONETTI, P. J., MOUNIER, J., PRÉVOST, M. C. & MÈGE, R.-M. (1994). Cadherin expression is required for the spread of *Shigella flexneri* between epithelial cells. *Cell* **76**, 829–839.

SIEBERS, A. & FINLAY, B. B. (1995). Models to study enteropathogenic bacteria: lessons learned from *Shigella*. *Trends in Microbiology* **3**, 207–209.

STELLER, H. (1995). Mechanism and genes of cellular suicide. *Science* **267**, 1445–1449.

THOMPSON, C. B. (1995). Apoptosis in the pathogenesis and treatment of disease. *Science* **267**, 1456–1462.

THORNBERRY, N. A., BULL, H. G., CALAYCAY, J. R., CHAPMAN, K. T., HOWARD, A. D., KOSTURA, M. J., MILLER, D. K., MOLINEAUX, S. M., WEIDNER, J. R., AUNINS, J., ELLISTON, K. O., AYALA, J. M., CASANO, F. J., CHIN, J., DING, J.-F., EGGER, L. A., GAFFNEY, E. P., LIMJUCO, G., RAJU, S. M., ROLANDO, A. M., SALLEY, J. P., YAMIN, T. T., LEE, T. D., SHIVELY, J. E., MACCROSS, M., MUMFORD, R. A., SCHMIDT, J. A. & TOCCI, M. J. (1992). A novel heterodimeric cysteine protease is required for interleukin-1β processing in monocytes. *Nature* **356**, 768–774.

TOURE-BALDE, A., SARTHOU, J. L., ARIBOT, G., MICHEL, P., TRAPE, J. F., ROGIER, C. & ROUSSILHON, C. (1996). *Plasmodium falciparum* induces apoptosis in human mononuclear cells. *Infection and Immunity* **64**, 744–750.

VAUX, D. L., HAECKER, G. & STRASSER, A. (1994). An evolutionary perspective on apoptosis. *Cell* **76**, 777–779.

WASSEF, J. S., KEREN, D. F. & MAILLOUX, J. L. (1989). Role of M cells in initial antigen uptake and in ulcer formation in rabbit intestinal loop model of shigellosis. *Infection and Immunity* **57**, 858–863.

WATARAI, M., TOBE, T., YOSHIKAWA, M. & SASAKAWA, C. (1995). Contact of *Shigella* with host cells triggers release of Ipa invasins and is an essential function of invasiveness. *EMBO Journal* **14**, 2461–2470.

WHYTE, M. (1996). ICE/CED-3 proteases in apoptosis. *Trends in Cell Biology* **6**, 245–248.

YUAN, J., SHAHAM, S., LEDOUX, S., ELLIS, H. M. & HORVITZ, H. R. (1993). The *C. elegans* cell death gene *ced-3* encodes a protein similar to mammalian interleukin-1β-converting enzyme. *Cell* **75**, 641–652.

ZYCHLINSKY, A., FITTING, C., CAVAILLON, J. M. & SANSONETTI, P. J. (1994a). Interleukin-1 is released by murine macrophages during apoptosis induced by *Shigella flexneri*. *Journal of Clinical Investigation* **94**, 1328–1332.

ZYCHLINSKY, A., KENNY, B., MÉNARD, R., PRÉVOST, M. C., HOLLAND, I. B. & SANSONETTI, P. J. (1994b). IpaB mediates macrophage apoptosis induced by *Shigella flexneri*. *Molecular Microbiology* **11**, 619–627.

ZYCHLINSKY, A., PERDOMO, J. J. & SANSONETTI, P. J. (1994c). Molecular and cellular mechanisms of tissue invasion by *Shigella flexneri*. *Annals of the New York Academy of Science* **730**, 197–208.

ZYCHLINSKY, A., PRÉVOST, M. C. & SANSONETTI, P. J. (1992). *Shigella flexneri* induces apoptosis in infected macrophages. *Nature* **358**, 167–168.

ZYCHLINSKY, A. & SANSONETTI, P. J. (1997). Apoptosis as a proinflammatory event or, what we can learn from bacterial induced cell death. *Trends in Microbiology*, **5**, 201–204.

ZYCHLINSKY, A., THIRUMALAI, K., ARONDEL, J., CANTEY, J. R., ALIPRANTIS, A. & SANSONETTI, P. J. (1996). *In vivo* apoptosis in *Shigella flexneri* infections. *Infection and Immunity* **64**, 5357–5365.

Virus encoded cytokines and cytokine receptors

M. BARRY[1] *and* G. McFADDEN[2]*

[1] *Department of Biochemistry, University of Alberta, Edmonton, Alberta, Canada T6G 2H7*
[2] *The John P. Robarts Research Institute, and Department of Microbiology and Immunology, University of Western Ontario, London, Ontario, Canada N6G 2V4*

SUMMARY

In order to replicate efficiently within the host, viruses have evolved multiple strategies to evade the host's immune system. In many cases viruses have actually hijacked various components of the host's immune system to ensure their own survival. One such strategy is the expression of virus encoded cytokines and cytokine receptors. Members of the poxvirus and herpesvirus families have been particularly successful with this strategy. The study of virus survival strategies provides important information regarding both virus biology as well as information about the immune system itself.

Key words: Virokines, viroceptors, herpes virus, immunosubversion, poxvirus.

INTRODUCTION

Large DNA viruses, such as the herpesviruses and poxviruses, encode upwards of 100 open reading frames that are predicted to express protein products. Many of these genes encode proteins that have been shown to counteract specifically the host's immune response thus facilitating survival of the virus within immunocompetent hosts (reviewed in Marrack & Kappler, 1994; Smith, 1994; Spriggs, 1996). A number of these proteins were identified by virtue of their homology to known cellular proteins (Murphy, 1993). Other virus open reading frames, however, do not have known corresponding cellular counterparts, suggesting that either these viral open reading frames are unique or the eukaryotic homologue remains unidentified. Although viruses have been viewed in the past simply as vehicles of plague and pestilence, it is now clear that viruses are extensively educated in regard to the innermost workings of the vertebrate immune system and therefore can be viewed as educational tools for the study of immunology. In fact, not only have viruses been adept pupils, but they have systematically exploited components of the host's immune system to overcome the hostile host environment in which they replicate. Since viruses have co-evolved with animal hosts for untold millennia, it is not surprising that they have much to teach us, and we are now realizing that viruses can be used as potential probes for understanding and manipulating the immune system.

* Corresponding author: The John P. Robarts Research Institute, 1400 Western Road, London, Ontario, Canada N6G 2V4. Telephone 519-663-3184. Fax: 519-663-3847. E-mail: mcfadden@rri.on.ca

VIRUS ENCODED CYTOKINES AND GROWTH FACTORS

One strategy of immune evasion by viruses is the capture and expression of host cytokines and growth factors. Currently, several viruses are known to express cytokines and growth factors, all of which are restricted to members of the herpesvirus and poxvirus families (Table 1) (also reviewed in Smith 1994; Spriggs, 1994, 1996; Evans, 1996). It has been proposed that the expression of these viral cytokines and growth factors is important for virus survival, although in most cases the rationale for this particular strategy remains poorly understood. The recent identification of additional virus encoded cytokines in molluscum contagiosum (Senkevich *et al.* 1996) and in Kaposi's sarcoma-associated herpesvirus (KSHV) (Moore *et al.* 1996) suggests that more virus encoded cytokines probably remain to be discovered.

Virus encoded IL-10

The first example of a virus encoded cytokine identified was the BCRF1 open reading frame of Epstein–Barr virus (EBV) which encodes a secreted polypeptide with significant homology to interleukin-10 (IL-10) (Baer *et al.* 1984; Moore *et al.* 1990; reviewed in Swaminathan & Kieff, 1995). Epstein–Barr virus is a member of the herpesvirus family and is the causative agent of infectious mononucleosis. In addition, EBV has been associated with various lymphomas such as Burkitt's lymphoma and nasopharyngeal carcinoma. The BCRF1 gene of EBV, designated viral IL-10 (vIL-10), is 70% homologous to murine IL-10 and 84% homologous to human IL-10, indicating that vIL-10 is more closely related to human IL-10 (Moore *et al.* 1990; Vieira *et al.* 1991). IL-10 is a multifunctional cytokine normally

Table 1. Virus encoded cytokines

Virus	Gene	Host homologue
Herpesviruses		
Epstein–Barr	BCRF1 (vIL-10)	IL-10
Herpesvirus saimiri	ORF13 (vIL-17)	CTLA-8/IL-17
Kaposi's sarcoma-associated virus	K4 and K6	MIP-1α
Kaposi's sarcoma-associated virus	K2 (vIL-6)	IL-6
Poxviruses		
Vaccinia virus[a]	VGF	EGF/TGFα
Myxoma virus[a]	MGF	EGF/TGFα
ORF virus	A2R	VEGF
Molluscum contagiosum	MC148R	MIP-1β

[a] Other poxviruses express VGF/MGF homologues.

produced by Th2 CD4[+] helper T cells, and from activated B cells and macrophages, and was originally classified as a cytokine synthesis inhibitory factor (CSIF) because of its ability to inhibit synthesis of a wide variety of cytokines, from macrophages, monocytes, natural killer cells, and T cells (reviewed in deWaal Malefyt *et al.* 1992; Howard & O'Garra, 1992; Moore *et al.* 1993). The high degree of amino acid identity between vIL-10 and the human protein suggests that EBV may have originally captured this gene from its human host in order to enhance virus propagation in IL-10 responsive leukocytes.

vIL-10 has retained many of the functions of IL-10, such as the inhibition of cytokine synthesis and the stimulation of B cell growth and differentiation (Hsu *et al.* 1990; Go *et al.* 1990; Rousset *et al.* 1992). The inhibitory effects of vIL-10 on the synthesis of both human and murine IFN-γ has been confirmed (Hsu *et al.* 1990), and recombinant EBVs devoid of the BCRF1 gene demonstrate directly that vIL-10 is necessary for the inhibition of IFN synthesis during virus infection (Swaminathan *et al.* 1993). Although vIL-10 has clearly retained several functions of its cellular counterpart, it has not retained the ability to stimulate mast cells (Vieira *et al.* 1991), murine thymocytes and T cells, or induce the expression of MHC class II antigens on murine B cells (Go *et al.* 1990; MacNeil *et al.* 1990). These observations clearly suggest that retention of cytokine synthesis inhibition and B cell stimulation is more important for virus survival. In theory, the ability of vIL-10 to inhibit the production of anti-viral cytokines such as IFN-γ and promote B cell growth should greatly benefit a B cell lymphotropic virus such as EBV. In support of this notion, two other herpesviruses, herpesvirus papio and equine herpesvirus type 2, have also been found to contain vIL-10 homologues (Rode *et al.* 1993; Swaminathan & Kieff, 1995).

Virus encoded CTLA-8/IL-17

Completion of the genomic sequence of Herpesvirus saimiri yielded the identification of 76 putative open

reading frames (Albrecht *et al.* 1992). A number of these were demonstrated to contain sequence similarities to cellular proteins, including those involved in host defence or immune modulation, such as complement regulatory proteins, and chemokine receptors (Albrecht *et al.* 1992). When first sequenced, no homologue for the ORF13 gene of Herpesvirus saimiri had been identified, suggesting that ORF13 encoded a virus-specific gene, or that the cellular counterpart was not yet characterized. It became apparent, however, that ORF13 was indeed homologous to a cellular open reading frame when CTLA-8 was cloned from a murine lymphocyte cDNA library (Rouvier *et al.* 1993). All cysteine residues were conserved between the two proteins, as well as a single glycosylation site and five putative phosphorylation sites. Overall, the percent amino acid identity was found to be 57% between CTLA-8 and the protein encoded by ORF13 of Herpesvirus saimiri (Rouvier *et al.* 1993). CTLA-8 is only expressed from a subset of activated T cells and the presence of AU rich repeats in the 3′ untranslated region of the mRNA suggest that expression of the protein is tightly controlled via mRNA degradation, a mechanism utilized for the controlled expression of many cytokines and growth factors (Rouvier *et al.* 1993). Although no function was initially associated with CTLA-8 or ORF13, the acquisition of this gene by Herpesvirus saimiri, which encodes other proteins important for immune evasion, suggested that the protein may be important to counteract host immune mechanisms.

Recently, Yao *et al.* have cloned and characterized a novel receptor that binds both CTLA-8 and the Herpesvirus saimiri homologue (Yao *et al.* 1995 *a*). The authors report that this receptor, now designated the interleukin 17 receptor (IL-17R), is widely distributed among various tissues and cell lines, and database searches have revealed no homology with previously characterized sequences, including members of the cytokine receptor family. The extensive distribution of the IL-17R is in direct contrast to the expression of CTLA-8, now referred to as IL-17, which reportedly is expressed only in activated T cells (Rouvier *et al.* 1993; Yao *et al.* 1995*a*). Interaction of IL-17 with the receptor results in the activation of NFκ-B, the secretion of IL-6 from fibroblasts and the enhanced proliferation of murine T cells (Yao *et al.* 1995*a*). The Herpesvirus saimiri protein,ORF13, has retained all of these functions, suggesting that these functional properties may confer a selective advantage for the virus.

The identification and cloning of the human IL-17 gene from a CD4+ T cell library revealed a greater degree of homology with ORF13, sharing 72% amino acid sequence identify (Yao *et al.* 1995*b*). Like its murine counterpart, human IL-17 was found to be expressed only in activated T cells. Human IL-17 mediates the secretion of both IL-6 and IL-8 from cells and increases the cell surface expression of the intracellular adhesion molecule ICAM-1 (Yao *et al.* 1995*b*). Studies to elucidate further the functional properties of IL-17 and its receptor should also provide information regarding the functional significance of the viral homologue ORF13, since the exact function of this protein during virus infection is currently unknown.

Virus encoded chemokines

Molluscum contagiosum virus is a relatively benign poxvirus that infects humans and causes dermal wart-like lesions. DNA sequence analysis of the genome has revealed the presence of 163 putative open reading frames of which 59 were found to be unique (not present in the genome of other poxviruses) (Senkevich *et al.* 1996). Different members of the poxvirus family are known to contain a diverse array of genes that function to evade the host immune response, but surprisingly many of these anti-immune genes have no obvious counterpart within the genome of molluscum contagiosum virus. Several of the open reading frames of molluscum contagiosum are, however, homologous to cellular genes with known function, including the open reading frame MC148R which shares amino acid identity with the C-C chemokine macrophage inflammatory protein 1β (MIP-1β) (Senkevich *et al.* 1996). Chemokines are a family of immune modulatory chemoattractant cytokines that mediate the chemotaxis of various leukocytes (reviewed in Baggiolini, Dewald & Moser; 1994; Schall & Bacon, 1994). The predicted protein encoded by MC148R has retained the correct cysteine positioning important for overall structure of the protein, but lacks a significant segment of the N-terminal region. Since studies have demonstrated that engineered chemokines that lack this N-terminal region can bind to chemokine receptors but are unable to trigger an intracellular signal (Clark-Lewis *et al.* 1995), the authors speculate that MC148R functions as a chemokine inhibitor by competing directly with

cellular chemokines for binding to chemokine receptors and preventing the subsequent activation of leukocytes (Senkevich *et al.* 1996). This type of a strategy is not foreign to members of the poxvirus family, since various other secreted viral proteins seems to function by preventing activation of the immune system, but in a very different manner (discussed in later sections of this review). However, molluscum contagiosum was the first virus, and currently the only poxvirus, reported to encode a gene with homology to a cellular chemokine which may contribute to virus survival.

A recent report has revealed the presence of two chemokine homologues in the genome of Kaposi's sarcoma-associated herpesvirus (KSHV), also designated human herpesvirus 8 (HHV-8) (Moore *et al.* 1996). KSHV is a member of the gamma herpesvirus family which also includes EBV and Herpesvirus saimiri, both of which encode multiple proteins involved in immune evasion. Open reading frames K6 and K4 of KSHV are predicted to encode 10·5 kDa proteins with approximately 40% amino acid identity to human MIP-1α. Although nothing is currently known about the function of K4, the K6 open reading frame encodes a functional chemokine which can interact with chemokine receptors and subsequently inhibit infection of HIV, which utilizes the CCR5 receptor for cellular entry (Moore *et al.* 1996). In contrast to the chemokine encoded by the MC148R open reading frame in molluscum contagiosum, the K4 and K6 open reading frames in KSHV seem to be intact and are not missing the N-terminal region. This suggests that the chemokines encoded by KSHV may in fact trigger a signal upon interaction with the cellular receptor, but this remains to be demonstrated experimentally. The rationale for the expression of a virus encoded functional chemokine is currently unknown, but since the virus has presumably acquired this gene from the host, it seems likely that it may somehow function to facilitate virus survival *in vivo*. It is possible that chemokine/chemokine receptor interaction may modify the host cell environment in order to benefit the virus. For example, the binding of a virus encoded chemokine to the normal cellular receptor may inappropriately induce chemotaxis or cellular activation that favours virus survival.

Virus encoded IL-6

In addition to the presence of virus encoded chemokines, genomic sequencing of KSHV revealed the presence of an open reading frame with homology to interleukin-6 (IL-6), including the presence of four cysteine residues which are necessary for structural conservation (Moore *et al.* 1996). IL-6 is a multifunctional cytokine that is an important growth factor for various cell types and stimulates B-cell differentiation (reviewed in Hirano *et al.* 1990).

The KSHV K2 open reading frame encodes a secreted 23 kDa protein, designated viral IL-6 (vIL-6), that has retained at least one of the functions of IL-6. Supernatants from Cos-7 cells that express vIL-6 have been shown to support growth of an IL-6-dependent myeloma cell line (Moore *et al.* 1996). Thus, despite the overall low sequence homology, the K2 open reading frame encodes a functional cytokine capable of stimulating leukocyte cell growth. Tissue sections from patients infected with KSHV demonstrated the presence of vIL-6 associated with haematopoietic lymph node cells, particularly in areas containing abundant B cells (Moore *et al.* 1996), suggesting that vIL-6 may play a role in KSHV pathogenesis by stimulating B cells *in vivo*. In fact, various reports suggest a link between KSHV infection of patients and B cell lymphomas (Cesarman *et al.* 1995; Gessain *et al.* 1995; Soulier *et al.* 1995; Corbellino *et al.* 1996). Since the IL-6 receptors is present on a wide variety of cell types (Taga *et al.* 1987), it is likely that vIL-6 interacts with this receptor and stimulates cell growth to enhance KSHV replication and survival.

Virus encoded growth factors

The capture and expression of cellular growth factors by numerous poxviruses has been well documented (reviewed in McFadden, Graham & Opgenorth, 1995*b*). To date, virus encoded proteins that mimic the activities of epidermal growth factor (EGF), transforming growth factor-α (TGFα) and vascular endothelial growth factor (VEGF) have been identified. The first example of a virus encoded growth factor was found serendipitously when the vaccinia virus 19K open reading frame, now designated vaccinia growth factor (VGF), was sequenced (Bloomquist, Hunt & Barker, 1984; Brown *et al.* 1985; Reisner, 1985). VGF displays significant homology to EGF and TGFα including the conservation of 6 essential cysteine residues. Additional secreted EGF-like proteins are encoded by other poxviruses, including variola virus (Massung *et al.* 1993*a*, 1994), Shope fibroma virus (Chang *et al.* 1987), and myxoma virus (Upton *et al.* 1987*b*). Virus encoded growth factors exploit the cellular system by binding to cellular growth factor receptors and inappropriately stimulating cell growth. Deletion of both VGF and the myxoma virus growth factor (MGF) results in attenuation *in vivo* (Buller *et al.* 1988; Opgenorth *et al.* 1992), indicating that the presence of VGF and MGF confer a selective advantage for the virus in the host. In order to restore virus virulence, the MGF gene of myxoma virus can be replaced with either VGF or rat TGFα, further supporting the theory that MGF and VGF are *bona fide* functional growth factors (Opgenorth *et al.* 1993). In addition to the EGF-like growth factors encoded by members of the poxvirus family, orf

virus, another member of the poxvirus family, expresses a protein with sequence homology to mammalian VEGF. Supernatants from orf infected cells, display mitogenic activity suggesting that the A2R open reading frame encodes a functional VEGF homologue (Lyttle *et al.* 1994). Since VEGF is an important regulator of angiogenesis, the expression of VEGF from orf virus infected cells is consistent with the presence of lesions at the primary site of infection that are characterized by capillary proliferation and oedema (Ferrara *et al.* 1992; Lyttle *et al.* 1994). To date, mimics of EGF and VEGF growth factors that favour virus replication have only been observed in members of the poxvirus family; however, the presence of these growth factors in other viruses is a distinct possibility.

VIRUS ENCODED CYTOKINE RECEPTORS

The acquisition of genes encoding cytokine receptors by poxviruses and their subsequent expression from infected cells has been extensively documented (reviewed in Pickup, 1994; Alcami & Smith, 1995*b*; McFadden *et al.* 1995*a*). The term 'viroceptor' was coined to describe soluble virus-encoded cytokine receptors that are expressed from virus infected cells and function by binding to host cell cytokines preventing interaction of host cytokines with cellular receptors (Upton *et al.* 1991). Currently, five such soluble viroceptors have been described in poxviruses (Table 2). In addition, both herpesviruses and poxviruses express proteins with significant homology to transmembrane cytokine receptors (reviewed in Ahuja, Gao & Murphy, 1994; Murphy, 1994; Schall *et al.* 1995) and this review will discuss both types of viroceptors.

Virus encoded TNF receptor homologues

The first example of a viroceptor was discovered in Shope fibroma virus (SFV), a poxvirus that causes benign subcutaneous fibromas in rabbits (Upton *et al.* 1987*a*; Smith *et al.* 1990; reviewed in Smith & Goodwin, 1995; McFadden, Schreiber & Sedger, 1996). Sequence analysis initially revealed significant homology between SFV-T2 and the nerve growth factor receptor. It later became apparent, when the type 1 and type 2 tumour necrosis factor (TNF) receptors were cloned and sequenced, that T2 was indeed a virus encoded homologue of the TNF type 1 receptor that was able to bind TNFα and lymphotoxin (Smith *et al.* 1990, 1991). Like its cellular homologue, the T2 protein contains a cysteine-rich region and a signal sequence, but no obvious transmembrane domain, suggesting that T2 is in fact a secreted TNF receptor capable of binding and sequestering TNF. Further studies revealed the

Table 2. Virus encoded cytokine receptors

Virus	Gene	Host homologue
Soluble cytokine receptors[a]		
Myxoma virus	M-T2	TNF receptor
Cowpox virus	CrmB, C	TNF receptor
Myxoma virus	M-T7	IFN-γ receptor
Vaccinia virus	B15R	IL-1β receptor
Vaccinia virus	B18R	IFN-αβ receptor
Myxoma virus	M-T1	Unknown
Membrane-associated receptors		
Herpesvirus samiri	ECRF3	Chemokine receptor
Human cytomegalovirus	US28	Chemokine receptor
Swinepox virus	K2R	Chemokine receptor
Capripox virus	Q2/3L	Chemokine receptor

[a] Related cytokine receptor homologues are encoded in other poxviruses.

presence of similar genes in myxoma virus and malignant rabbit fibroma virus (MRV) (Upton *et al.* 1991), two poxviruses that induce lethal diseases in European rabbits. In order to investigate the role of the myxoma virus T2 gene in virus virulence, both copies of the gene were disrupted. When the resulting recombinant virus was used to infect European rabbits, the majority of rabbits were able to recover from the infection in contrast to rabbits infected with the wild type virus (Upton *et al.* 1991), indicating that the M-T2 protein is necessary for virus pathogenesis *in vivo*. Related open reading frames have been observed in other poxviruses including variola virus (Massung *et al.* 1993*a*, 1994), and various strains of vaccinia virus, although the vaccinia genes appear to be disrupted (Goebel *et al.* 1990; Howard, Chan & Smith, 1991). In addition, the genome of cowpox virus contains two distinct open reading frames, crmB and crmC, with homology to TNF receptors (Hu, Smith & Pickup, Smith & Pickup, 1994; Smith *et al.* 1996). CrmB is the cowpox virus equivalent of M-T2, while crmC encodes a distinct TNF receptor homologue that is secreted late during infection and only interacts with TNF and not lymphotoxin (Smith *et al.* 1996).

Purified M-T2 protein, generated from a vaccinia virus engineered to over-express M-T2, inhibited rabbit TNFα-mediated lysis of L929-8 cells, but was unable to inhibit lysis mediated by either human or murine TNFα (Schreiber & McFadden, 1994). Thus, M-T2 exhibits narrow species specificity and specifically binds and inhibits rabbit TNFα. Furthermore, Scatchard analysis has demonstrated that M-T2 binds rabbit TNFα with an affinity similar to the mammalian TNF receptors (Schreiber, Rajarathnam & McFadden, 1996). M-T2 is secreted from infected cells as both a 80 kDa disulphide linked dimer and a 40 kDa monomer. Both forms can interact with TNFα, but the dimer appears to provide better protection against TNFα-mediated

cell lysis (Schreiber *et al.* 1996). A series of deleted and truncated M-T2-expressing recombinant viruses have demonstrated that the N-terminal cysteine-rich domain, the region homologous with other members of the TNF superfamily, is essential for binding rabbit TNFα (Schreiber & McFadden, 1996). In addition, it appears that the C-terminal 132 amino acids, which are unique to M-T2, are necessary for the efficient secretion of the protein, since deletion results in the intracellular retention of M-T2 (Schreiber & McFadden, 1996).

While M-T2 has definitively been shown to be important for the inhibition of cellular lysis due to TNF, it appears that M-T2 may have a dual function in infected cells. Although myxoma virus can replicate normally in infected CD4[+] T lymphocytes, infection of lymphocytes with the M-T2 mutant virus results in the induction of cellular DNA fragmentation, a characteristic of cellular apoptosis (Macen *et al.* 1996). Furthermore, experiments have shown that the ability of M-T2 to inhibit apoptosis does not correlate with TNF binding since truncated versions of the protein which cannot bind or inhibit TNF are still able to inhibit apoptosis in virus infected T cells (Schreiber, Sedger & McFadden, 1997). While many viruses encode proteins that are essential for protection against cellular apoptosis (reviewed in White, 1996), M-T2 appears to be the first protein with this unique dual function. Other poxvirus encoded M-T2 homologues have not been tested experimentally for this dual activity, but it is possible that other classes of viral immunomodulatory proteins are multifunctional as well.

Virus encoded IFN-γ receptor homologues

The protein encoded by the M-T7 open reading frame from myxoma virus was the first example of a secreted virus protein designed to circumvent the affects of IFN-γ prior to cytokine/receptor engagement (reviewed in Mossman *et al.* 1995*a*, 1997). Since IFN-γ was originally characterized as a cytokine with potent anti-viral activity, it is not surprising that viruses have evolved strategies to overcome the affects of this cytokine. N-terminal amino acid sequence analysis and subsequent DNA sequence analysis of the major 37 kDa secreted protein from myxoma virus infected cells revealed the existence of a virus encoded protein with approximately 30% amino acid identity to human and murine IFN-γ receptors (Upton, Mossman & McFadden, 1992). Homology is only evident within the ligand-binding domain and the positioning of eight cysteine residues throughout this domain.

The M-T7 protein functions as a *bona fide* soluble IFN-γ receptor that can both bind and inhibit the activity of rabbit IFN-γ (Upton *et al.* 1992). M-T7 binds rabbit IFN-γ with high affinity ($1 \cdot 2 \times 10^{-9}$ M), but not human or murine IFN-γ, reflecting the close evolutionary relationship of a rabbit virus with its natural host (Mossman, Upton & McFadden, 1995 c). An identical result was also observed for the myxoma virus TNF receptor homologue which also displays strict species specificity for ligand recognition (Schreiber & McFadden, 1994). In order to evaluate directly the role of M-T7 *in vivo*, a recombinant myxoma virus containing a deletion in the M-T7 open reading frame was used to infect European rabbits (Mossman *et al.* 1996). Rabbits infected with the recombinant virus remained essentially disease free and recovered fully from infection, in stark contrast to rabbits infected with the parental M-T7 containing virus. Thus, the presence of a virus encoded IFN-γ binding protein contributes directly to virus virulence *in vivo* (Mossman *et al.* 1996).

IFN-γ receptor homologues have been reported in a wide variety of poxviruses (Massung *et al.* 1994; Alcami & Smith, 1995 a; Mossman *et al.* 1995 b). In contrast to the myxoma virus encoded IFN-γ receptor, IFN-γ receptor homologues expressed by other poxviruses, particularly members of the orthopoxvirus genus, demonstrate a broader species specificity (Alcami & Smith, 1995 a; Mossman *et al.* 1995 b). The evolutionary origin of myxoma virus has been firmly established in the South American rabbit (Fenner & Ratcliffe, 1965); however, the evolutionary origin of most orthopoxviruses is not as clearly established, leading to the suggestion that the broad specificity demonstrated by these virus encoded IFN-γ binding proteins may reflect a varied evolutionary history involving the replication of virus in several species (Alcami & Smith, 1995 a, 1996 a; Mossman *et al.* 1995 b). Intriguingly, IFN-γ binding proteins encoded by members of the orthopoxviruses have retained only 6 of 8 conserved cysteine residues in the ligand-binding domain, which may explain the altered ligand binding properties displayed by these proteins.

Virus encoded IL-1 receptor homologues

Two groups have independently demonstrated that the B15R open reading frame in vaccinia virus (strain WR) encodes a secreted glycoprotein that specifically interacts with IL-1β (Alcami & Smith, 1992; Spriggs *et al.* 1992; reviewed in Alcami & Smith, 1995 c). Similar to the myxoma virus encoded TNF and IFN-γ receptor homologues, the B15R protein is homologous to the external portion of the cellular IL-1 type II receptor, and does not contain a transmembrane region. Although the cellular IL-1 receptor is able to bind both IL-1α, IL-1β and IL-RA (the naturally occurring receptor antagonist), the virus equivalent has clearly lost these functions and only retains the ability to interact with IL-1β (Alcami & Smith, 1992), suggesting that the inhibition of IL-1β is of particular importance during poxvirus

infection. Other poxviruses also have been shown to encode IL-1β-binding proteins, including cowpox virus and myxoma virus (Alcami & Smith, 1992; Spriggs *et al.* 1992; and K. Graham, K. Mossman & G. McFadden, unpublished). Alcami & Smith demonstrated that the vaccinia virus encoded IL-1β binding protein bound IL-1β with high affinity similar to that of cellular IL-1 receptors, and that cells infected with vaccinia virus (WR) secreted large quantities of the protein. In fact, the presence of B15R protein can effectively inhibit binding of IL-1β to cellular IL-1 receptors (Alcami & Smith, 1992), a result consistent with the proposed activity of a *bona fide* viroceptor.

Recombinant vaccinia viruses devoid of the B15R gene were constructed by 2 groups and were found to have conflicting pathogenic effects *in vivo*, probably due to variable routes of inoculation (Alcami & Smith, 1992; Spriggs *et al.* 1992). Mice infected by the intracranial route with the B15R deletion virus exhibited diminished mortality compared to infection with the parental virus indicating that B15R is important for virus virulence (Spriggs *et al.* 1992). In contrast, mice infected by the intranasal route with the B15R deletion virus displayed no alteration in mortality, but did display an accelerated disease profile with 70 % of infected mice dying earlier than those infected with the parental virus (Alcami & Smith, 1992). In a subsequent paper, Alcami and Smith report that mice infected with the B15R deletion virus respond to infection with increased fever (Alcami & Smith, 1996 b), which may account for the accentuated disease symptoms observed after intranasal inoculation of mice with the B15R deletion virus. In addition, virus-induced fever can be inhibited by antibodies specific for murine IL-1β (Alcami & Smith, 1996 b), clearly demonstrating IL-1β is responsible for the induction of fever and that the presence of B15R is important for virus modulation of fever during infection.

Virus encoded IFN-αβ receptor homologues

Although the importance of type I interferon for recovery from vaccinia virus infection has been well documented (Buller & Palumbo, 1991), the existence of a soluble poxvirus encoded protein that inhibits type I interferon was only recently identified (Colmanonici *et al.* 1995; Symons, Alcami & Smith, 1995). The B18R gene of vaccinia virus (strain WR) encodes a 60–65 kDa protein that is present both on the surface of infected cells and in extracellular supernatants (Ueda, Morikawa & Matsuura, 1990; Alcami & Smith, 1992; Morikawa & Ueda, 1993). The soluble B18R protein binds human IFNα2 with high affinity and inhibits both the binding of human IFN to its cellular receptor (Colmanonici *et al.* 1995; Symons *et al.* 1995) and subsequent signal transduction (Colmanonici *et al.* 1995). Since B18R does not contain any obvious transmembrane region, yet can be detected on the surface of uninfected cells, Colmanonici *et al.* suggest that B18R binds to the cell surface after secretion. B18R present on the cell surface of infected cells can also interact with type I IFN suggesting the existence of an additional protective mechanism (Colmanonici *et al.* 1995). In addition, B18R reportedly interacts with an unknown protein on the surface of cells with saturable kinetics (Colmanonici *et al.* 1995), suggesting that the interaction is specific. The elucidation of the B18R cellular binding partner may provide significant information in regards to IFN biology. This characteristic sets B18R apart from other described virus encoded cytokine receptor homologues, since none of the others seem to be present as both soluble and cell associated forms. In contrast to soluble cytokine receptors encoded by myxoma virus, the B18R protein exhibits broad species specificity and is able to bind bovine, rat, mouse, and rabbit interferon type 1 with varying affinities, in addition to human IFN type 1. Several other poxviruses secrete similar proteins that inhibit type 1 IFN, including various strains of vaccinia, cowpox, buffalopox, elephantpox, camelpox, rabbitpox (Symons *et al.* 1995), and ectromelia virus (Colmanonici *et al.* 1995). Additionally, variola virus contains an open reading frame which is homologous to B18R (Massung *et al.* 1993*a*, 1994). Despite the fact that B18R has a much lower affinity for murine IFNβ than human IFNβ (150-fold lower) the recombinant vaccinia virus devoid of the B18R gene is attenuated in mice, indicating that B18R is a functionally important immune modulator within infected animals (Symons *et al.* 1995).

Virus encoded chemokine receptors

DNA sequence analysis of the genomes of two herpesviruses, Herpesvirus saimiri and human cytomegalovirus, have revealed the presence of several putative seven-transmembrane serpentine receptors that are related to mammalian chemokine receptors (reviewed in Ahuja *et al.* 1994; Murphy, 1994; Schall *et al.* 1995). The ECRF3 gene of Herpesvirus saimiri and the US28 gene of human cytomegalovirus encode proteins that share approximately 30 % amino acid identity with identified chemokine receptors. Both genes have been shown to encode functional chemokine receptors that bind various members of the chemokine family and subsequently trigger calcium mobilization within the cell (Ahuja & Murphy, 1993; Neote *et al.* 1993; Gao & Murphy, 1994). The chemokine family can be divided into three subgroups based on the position of conserved cysteine residues. The α chemokines (C-X-C) include IL-8, NAP-2, and GRO/MGSA, while the β chemokines (C-C) include MCP-1, MIP-1α, MIP-1β, and RANTES, and the (C) chemokines contain only lymphotactin. ECRF3 has been shown to interact with only members of the α chemokine subgroup (Ahuja & Murphy, 1993), while US28 only interacts with members of the β chemokine subgroup (Gao & Murphy, 1994). The rationale behind the capture of a chemokine receptor that specifically interacts with only α chemokines by Herpesvirus saimiri and β chemokines by human cytomegalovirus is not known, and further studies to unravel this relationship will undoubtedly provide critical information in regards to both virus and chemokine biology. Since both ECRF3 and US28 are able to trigger a calcium flux in response to chemokine binding it seems likely the piracy and expression of these functional receptor genes may somehow enhance virus survival, however the precise function of virus encoded chemokine receptors is currently unknown. In support of this, a number of other viruses contain putative transmembrane serpentine molecules that may also interact with members of the chemokine family. These include two additional open reading frames in human cytomegalovirus (Chee *et al.* 1990), three open reading frames in equine herpesvirus type 2 (Telford *et al.* 1995), two open reading frames in human herpesvirus type 6 (Gompels *et al.* 1995), one open reading frame in KSHV (human herpesvirus type 8) (Cesarman *et al.* 1996) and two open reading frames in members of the poxvirus family, capripox and swinepox (Massung, Jayarama & Moyer, 1993*b*; Cao, Gershon & Black, 1995). The K2R open reading frame of swinepox virus and the Q2/3L open reading frame of capripox virus share significant homology with the US28 gene. Since many poxvirus encoded cytokine receptors act to simply bind and sequester cytokines, it will be particularly important to ascertain whether or not the poxvirus encoded transmembrane chemokine receptors simply act as a decoy or transduce a signal similar to the US28 and ECRF3 proteins. In addition, the role of these proteins in virus pathogenesis needs to be evaluated via gene disruption and *in vivo* analysis.

Virus encoded soluble chemokine-binding proteins

In addition to membrane-bound chemokine receptors, poxviruses also encode soluble chemokine-binding proteins. The first example of a virus encoded soluble chemokine-binding protein was observed when purified M-T7, the myxoma virus IFN-γ receptor homologue, was found to bind promiscuously to various chemokines from the C-C, C-X-C, and C subgroups in an *in vitro* chemical crosslinking assay (Lalani *et al.* 1997). The initial indication that M-T7 may have additional functions was observed when rabbits infected with the M-T7 deletion virus were found to contain elevated levels of reactive leukocytes in secondary lymphoid organs and a marked disruption of immune cell migration into infected lesions (Mossman *et al.* 1996). It was postulated that the M-T7 protein may possess additional biological activities since these functions have not been attributed to IFN-γ *in vitro*. Purified M-T7 interacts with chemokines from a broad range of species including rabbit, human and murine. Furthermore, experiments with various IL-8 analogues containing deletions demonstrated that the C-terminal heparin-binding domain of IL-8 was essential for binding to M-T7 (Lalani *et al.* 1997), indicating that M-T7 interacts non-specifically with chemokines via a common heparin-binding domain. Further evidence supports this observation since heparin can compete for the binding of M-T7 to chemokines, but not for IFN-γ (Lalani *et al.* 1997). In this respect M-T7 may function similarly to the Duffy antigen on erythrocytes which also binds promiscuously to members of at least two chemokine subgroups (Horuk *et al.* 1994). Thus it is proposed that M-T7 may function *in vivo* to disrupt chemokine gradients necessary for inflammatory cell migration, in addition to its ability to bind and sequester IFN-γ.

In a systematic approach to identify other secreted myxoma virus proteins that interact with cytokines, it was discovered that myxoma virus encodes a second distinct chemokine-binding protein encoded by the M-T1 open reading frame (Graham *et al.* 1997). Similar chemokine-binding activity was also observed in supernatants from a number of other poxviruses, including Shope fibroma virus, rabbitpox virus, cowpox virus, raccoonpox virus, and vaccinia virus (strain Lister). It seems likely that vaccinia virus (strain Copenhagen) also expresses a functional homologue, but not vaccinia virus strains Wyeth or Tian Tan (Patel *et al.* 1990). Since no detectable chemokine-binding activity could be observed in vaccinia virus (strain WR) and since vaccinia virus strain Lister contains a gene C23L/B19R that is truncated in vaccinia virus strain WR but is homologous to the M-T1 gene in myxoma virus, the M-T1 open reading frame provided a likely candidate for chemokine-binding activity. Supernatants from cells infected with a recombinant vaccinia virus strain WR engineered to express M-T1, exhibited chemokine-binding activity indicating that the secreted M-T1 protein was capable of chemokine binding (Graham *et al.* 1997). In addition, supernatants from cells infected with a recombinant rabbitpox virus containing a deletion in the homologous gene demonstrated no chemokine-binding activity (Graham *et al.* 1997). Purified M-T1 protein interacts with members of both the C-X-C and C-C chemokine families and binds RANTES with high affinity (Kd = 73 nM). Database searches have revealed no similarity between any non-viral genes and the M-T1/35 kDa family, indicating that these are either a unique set of virus proteins or the eukaryotic homologue has not yet been identified.

FUTURE CONSIDERATIONS

One intriguing recent development is the observation that some virus encoded proteins with immune modulation capabilities appear to have multiple functions. This is supported by recent evidence that several poxvirus proteins with previously characterized specificities are able to interact with other components of the immune system. The first documented case of this phenomenon is the presence of a single secreted protein from Tanapox infected cells that is able to bind multiple cytokines, including IL-2, IL-5, and IFN-γ (Essani *et al.* 1994). The second example of a dual functional protein was discovered when the purified myxoma virus IFN-γ homologue, M-T7, was able to interact with various members of the chemokine superfamily in addition to rabbit IFN-γ (Lalani *et al.* 1997). Finally, the observation that the presence of the myxoma virus TNF receptor homologue, M-T2 not only inhibits extracellular TNF but is also necessary for the intracellular inhibition of apoptosis in virus infected CD4[+] T cells supports the notion that many virus proteins may exhibit more than one function (Macen *et al.* 1996; Schreiber *et al.* 1997).

The realization that viruses are important for elucidating fundamental mechanisms of the immune system comes at a time when scientists are debating the future destruction of the remaining stocks of variola virus, the causative agent of smallpox (Massung *et al.* 1993*a*). Smallpox virus was systematically eradicated from the human population by a worldwide vaccination programme and the last known case occurred in 1977. In light of the fact that such highly pathogenic viruses, and particularly poxviruses, have so much still to teach us about the human immune system, it is instructive to consider whether the irrevocable loss of variola virus as a potential tool to investigate immunology might be a high price to pay for the presumptive security that its destruction provides.

ACKNOWLEDGEMENTS

Work in the author's laboratory is supported by grants from the Medical Research Council of Canada and the National Cancer Institute of Canada. G. M. is a Medical Research Council of Canada Senior Medical Scientist and M. B. is the recipient of a fellowship award from the Alberta Heritage Foundation for Medical Research. We thank members of the McFadden laboratory for dynamic discussions, and significant scientific contributions to the field. We also thank K. Graham, A. Lalani, K. Mossman and P. Nash for critically reviewing the manuscript.

REFERENCES

AHUJA, S. K., GAO, J. L. & MURPHY, P. M. (1994). Chemokine receptors and molecular mimicry. *Immunology Today* **15**, 281–287.

AHUJA, S. K. & MURPHY, P. M. (1993). Molecular piracy of

mammalian Interleukin-8 receptor type B by herpesvirus saimiri. *Journal of Biological Chemistry* **268**, 20691–20694.

ALBRECHT, J. C., NICHOLAS, J., BILLER, D., CAMERON, K. R., BIESINGER, B., NEWMAN, C., WITTMANN, S., CRAXTON, M. A., COLEMAN, H., FLECHENSTEIN, B. & HONESS, R. W. (1992). Primary structure of the Herpesvirus saimiri genome. *Journal of Virology* **66**, 5047–5058.

ALCAMI, A. & SMITH, G. L. (1992). A soluble receptor for interleukin-1β encoded by vaccinia virus: a novel mechanism of virus modulation of the host response to infection. *Cell* **71**, 153–167.

ALCAMI, A. & SMITH, G. L. (1995*a*). Vaccinia, cowpox, and camelpox viruses encode soluble gamma interferon receptors with novel broad species specificity. *Journal of Virology* **69**, 4633–4639.

ALCAMI, A. & SMITH, G. L. (1995*b*). Cytokine receptors encoded by poxviruses: a lesson in cytokine biology. *Immunology Today* **16**, 474–478.

ALCAMI, A. & SMITH, G. L. (1995*c*). Interleukin-1 receptors encoded by poxviruses. In *Viroceptors, Virokines and Related Immune Modulators Encoded by DNA Viruses* (ed. McFadden, G.), pp. 17–27. Austin, Texas. R. G. Landes Company.

ALCAMI, A. & SMITH, G. L. (1996*a*). Receptors for gamma-interferon encoded by poxviruses: implications for the unknown origin of vaccinia virus. *Trends in Microbiology* **4**, 321–326.

ALCAMI, A. & SMITH, G. L. (1996*b*). A mechanism for the inhibition of fever by a virus. *Proceedings of the National Academy of Sciences, USA* **93**, 11029–11034.

BAER, R., BANKIER, A. T., BIGGIN, M. D., DEININGER, P. L., FARRELL, P. J., GIBSON, T. J., HATFULL, G., HUDSON, G. S., SATCHWELL, S. C., SEQUIN, C., TUFFNELL, P. S. & BARRELL, B. G. (1984). DNA sequence and expression of the B95-8 Epstein–Barr virus genome. *Nature* **310**, 207–211.

BAGGIOLINI, M., DEWALD, B. & MOSER, B. (1994). Interleukin-8 and related chemotactic cytokines: CXC and CC chemokines. *Advances in Immunology* **55**, 97–179.

BLOOMQUIST, M. C., HUNT, L. T. & BARKER, W. C. (1984). Vaccinia virus 19-kilodalton protein: relationship to several mammalian proteins, including two growth factors. *Proceedings of the National Academy of Sciences, USA* **81**, 7363–7367.

BROWN, J. P., TWARDSIK, D. R., MARQUARDT, H. & TODARO, G. J. (1985). Vaccinia virus encodes a polypeptide homologous to epidermal growth factor and transforming growth factor. *Nature* **313**, 491–492.

BULLER, R. M., CHARKRABARTI, S., COOPER, J. A., TWARDZIK, D. R. & MOSS, B. (1988). Deletion of the vaccinia virus growth factor gene reduces virus virulence. *Journal of Virology* **62**, 866–874.

BULLER, R. M. L. & PALUMBO, G. J. (1991). Poxvirus Pathogenesis. *Microbiological Reviews* **55**, 80–122.

CAO, J. X., GERSHON, P. D. & BLACK, D. N. (1995). Sequence analysis of *Hind* III Q2 fragment of capripox virus reveals a putative gene encoding a G-protein-coupled chemokine receptor homologue. *Virology* **209**, 207–212.

CESARMAN, E., CHANG, Y., MOORE, P. S., SAID, J. W. & KNOWLES, D. M. (1995). Kaposi's sarcoma-associated herpesvirus-like DNA sequences are present in AIDS-related body cavity based lymphomas. *New England Journal of Medicine* **332**, 1186–1191.

CESARMAN, E., NADOR, R. G., BAI, F., BOHENZKY, R. A., RUSSO, J. J., MOORE, P. S., CHANG, Y. & KNOWLES, D. M. (1996). Kaposi's sarcoma-associated herpesvirus contains G protein-coupled receptor and cyclin D homologs which are expressed in Kaposi's sarcoma and malignant lymphoma. *Journal of Virology* **70**, 8218–8223.

CHANG, W., UPTON, C., HU, S., PURCHIO, A. F. & McFADDEN, G. (1987). The genome of Shope fibroma virus, a tumorigenic poxvirus, contains a growth factor gene with sequence similarity to those encoding epidermal growth factor and transforming growth factor alpha. *Molecular Cell Biology* **7**, 535–540.

CHEE, M. S., SATCHWELL, S. C., PREDDIE, E., WESTON, K. M. & BARRELL, B. G. (1990). Human cytomegalovirus encodes three G protein-coupled receptor homologues. *Nature* **344**, 774–777.

CLARK-LEWIS, I., KIM, K. S., RAJARATHNAM, K., GONG, J. H., DEWALD, B., MOSER, B., BAGGIOLINI, M. & SYKES, B. D. (1995). Structure-activity relationships of chemokines. *Journal of Leucocyte Biology* **57**, 703.

COLAMONICI, O. R., DOMANSKI, P., SWEITZER, S. M., LARNER, A. & BULLER, R. M. (1995). Vaccinia virus B18R gene encodes a type I interferon-binding protein that blocks interferon alpha transmembrane signalling. *Journal of Biological Chemistry* **270**, 15974–15978.

CORBELLINO, M., POIREL, L., BESTETTI, G., AUBIN, J. T., CAPRA, M., BERTI, E., GALLI, M. & PARRAVICINI, B. (1996). Human herpesvirus-8 in AIDS-related and unrelated lymphomas. *AIDS* **10**, 545–546.

DEWAAL MALEFYT, R., YESSEL, H., RONCAROLO, M. G., SPITS, H. & DEVRIES, J. E. (1992). Interleukin-10. *Current Opinion in Immunology* **4**, 314–320.

ESSANI, K., CHALASANI, S., EVERSOLE, R., BEUVING, L. & BIRMINGHAM, L. (1994). Multiple anti-cytokine activities secreted from tanapox virus-infected cells. *Microbial Pathogenesis* **17**, 347–353.

EVANS, C. H. (1996). Cytokines and viral anti-immune genes. *Stem Cells* **14**, 177–184.

FENNER, F. & RATCLIFFE, F. N. (1965). *Myxomatosis.* Cambridge, Cambridge University Press.

FERRARA, N., HOUCK, K., JAKEMAN, L. & LEUNG, D. W. (1992). Molecular and biological properties of the vascular endothelial growth factor family of proteins. *Endocrine Review* **13**, 18–32.

GAO, J. L. & MURPHY, P. M. (1994). Human cytomegalovirus open reading frame US28 encodes a functional β chemokine receptor. *Journal of Biological Chemistry* **269**, 28539–28542.

GESSAIN, A., SUDAKA, A., BRIERE, J., FOUCHARD, N., NICOLA, M. A., RIO, B., ARBORIO, M., TROUSSARD, X., AUSOUIN, J., DIEBOLD, J. & DE THE, G. (1995). Kaposi sarcoma associated herpes-like virus (human herpesvirus type 8) DNA sequences in multicentric Castleman's disease: is there any relevant association in non-human immunodeficiency virus-infected patients? *Blood* **87**, 414–416.

GO, N. F., CASTLE, B. E., BARRETT, R., KASTELEIN, R., DANG, W., MOSMANN, T. R., MOORE, K. W. & HOWARD, M. (1990). Interleukin 10, a novel B cell stimulatory factor: unresponsiveness of X chromosome-linked immunodeficiency B cells. *Journal of Experimental Medicine* **172**, 1625–1631.

GOEBEL, S. J., JOHNSON, G. P., PERKUS, M. E., DAVIS, S. W., WINSLOW, J. P. & PAOLETTI, E. (1990). The complete DNA sequence of vaccinia virus. *Virology* **179**, 247–266.

GOMPELS, U. A., NICHOLAS, J., LAWRENCE, G., JONES, M., THOMSON, B. J., MARTIN, M. E., EFSTATHIOU, S., CRAXTON, M. & MACAULAY, H. A. (1995). The DNA sequence of human herpesvirus-6: structure, coding content, and genome evolution. *Virology* **209**, 29–51.

GRAHAM, K. A., LALANI, A. S., MACEN, J. L., NESS, T. L., BARRY, M., LIU, Y. L., LUCAS, A., CLARK-LEWIS, I., MOYER, R. W. & McFADDEN, G. (1997). The T1/35 kDa family of poxvirus secreted proteins bind chemokines and modulate leucocyte influx into virus infected tissues. *Virology* **229**, 12–24.

HIRANO, T., AKIRA, S., TAGA, T. & KISHIMOTO, T. (1990). Biological and clinical aspects of interleukin 6. *Immunology Today* **11**, 443–449.

HORUK, R., WANG, Z. X., PEIPER, S. C. & HESSELGESSER, J. (1994). Identification and characterization of a promiscuous chemokine-binding protein in a human erythroleukemic cell line. *Journal of Biological Chemistry* **269**, 17730–17733.

HOWARD, M. & O'GARRA, A. (1992). Biological properties of interleukin 10. *Immunology Today* **13**, 198–200.

HOWARD, S. T., CHAN, Y. S. & SMITH, G. L. (1991). Vaccinia virus homologues of the shope fibroma virus inverted terminal repeat proteins and a discontinuous ORF related to the tumor necrosis factor receptor family. *Virology* **180**, 633–647.

HSU, D. H., DEWAAL MALEFYT, R., FIORENTINO, D. F., DANG, M. N., VIEIRA, P., DEVRIES, J., SPITS, H., MOSMANN, T. R. & MOORE, K. W. (1990). Expression of interleukin-10 activity by Epstein–Barr virus protein BCRF1. *Science* **250**, 830–832.

HU, F., SMITH, C. A. & PICKUP, D. (1994). Cowpox virus contains two copies of an early gene encoding a soluble secreted form of the type II TNF receptor. *Virology* **204**, 343–356.

LALANI, A. S., GRAHAM, K., MOSSMAN, K., RAJARATHNAM, K., CLARK-LEWIS, I., KELVIN, D. & McFADDEN, G. (1997). The purified myxoma virus IFN-γ receptor homolog, M-T7, interacts with the heparin binding domains of chemokines. *Journal of Virology* **71**, 4356–4363.

LYTTLE, D. J., FRASER, K. M., FLEMING, S. B., MERCER, A. A. & ROBINSON, A. J. (1994). Homologs of vascular endothelial growth factor are encoded by the poxvirus ORF virus. *Journal of Virology* **68**, 84–92.

MACEN, J. L., GRAHAM, K. A., LEE, S. F., SCHREIBER, M., BOSHKOV, L. K. & McFADDEN, G. (1996). Expression of the myxoma virus tumor necrosis factor receptor homologue and M11L genes is required to prevent virus-induced apoptosis in infected rabbit T lymphocytes. *Virology* **218**, 232–237.

MACNEIL, I. A., TAKASHI, S., MOORE, K. W. & MOSMANN, T. R. (1990). IL-10, a novel growth cofactor for mature and immature T cells. *Journal of Immunology* **145**, 4167–4173.

MARRACK, P. & KAPPLER, J. (1994). Subversion of the immune system by pathogens. *Cell* **76**, 323–332.

MASSUNG, R. F., ESPOSITO, J. J., LUI, L., QI, J., UTTERBACK, T. R., KNIGHT, J. C., AUBIN, L., YUARAN, T. E., PARSONS, J. M., LOPAREV, V. N., SELIVANOV, N. A., CAVALLARO, K. F., KERLAVAGE, A. R., MAHY, B. W. J. & VENTER, J. C. (1993*a*). Potential virulence determinants in terminal regions of variola smallpox virus genome. *Nature* **366**, 748–751.

MASSUNG, R. F., JAYARAMA, V. & MOYER, R. W. (1993*b*). DNA sequence analysis of conserved regions of swinepox virus: identification of genetic elements supporting phenotypic observations including a novel G protein-coupled receptor homologue. *Virology* **197**, 511–528.

MASSUNG, R. F., LIU, L., QI, J., KNIGHT, J. C., YURAN, T. E., KERLAVAGE, A. R., PARSONS, J. M., VENTER, J. C. & ESPOSITO, J. J. (1994). Analysis of the complete genome of smallpox variola major virus strain Bangladesh-1975. *Virology* **201**, 215–240.

McFADDEN, G., GRAHAM, K., ELLISON, K., BARRY, M., MACEN, J., SCHREIBER, M., MOSSMAN, K., NASH, P., LALANI, A. & EVERETT, H. (1995*a*). Interruption of cytokine networks by poxviruses: lessons from myxoma virus. *Journal of Leucocyte Biology* **57**, 731–738.

McFADDEN, G., GRAHAM, K. & OPGENORTH, A. (1995*b*). Poxvirus growth factors. In *Viroceptors, Virokines and Related Immune Modulators Encoded by DNA Viruses* (ed. McFadden, G.), pp. 1–15. Austin, Texas: R. G. Landes Company.

McFADDEN, G., SCHREIBER, M. & SEDGER, L. (1997). Myxoma T2 protein as a model for poxvirus TNF receptor homologs. *Journal of Neuroimmunology* **72**, 119–126.

MOORE, K. W., O'GARRA, A., DEWAAL MALEFYT, R., VIEIRA, P. & MOSMANN, T. R. (1993). Interleukin-10. *Annual Review of Immunology* **11**, 165–190.

MOORE, K. W., VIEIRA, P., FIORENTINO, D. F., TROUNSTINE, M. L., KHAN, T. A. & MOSMANN, T. R. (1990). Homology of cytokine synthesis inhibitory factor (IL-10) to the Epstein–Barr virus gene BCRF1. *Science* **248**, 1230–1234.

MOORE, P. S., BOSHOFF, C., WEISS, R. A. & CHANG, Y. (1996). Molecular mimicry of human cytokine and cytokine response pathway genes in KSHV. *Science* **274**, 1739–1744.

MORIKAWA, S. & UEDA, Y. (1993). Characterization of vaccinia surface antigen expressed by recombinant baculovirus. *Virology* **193**, 753–761.

MOSSMAN, K., BARRY, M. & McFADDEN, G. (1995*a*). Interferon-γ receptors encoded by poxviruses. In *Viroceptors, Virokines and Related Immune Modulators Encoded by DNA Viruses* (ed. McFadden, G.), pp. 41–54. Austin, Texas: R. G. Landes Company.

MOSSMAN, K., BARRY, M. & McFADDEN, G. (1997). Regulation of interferon-γ gene expression and extracellular ligand function by immunomodulatory viral proteins. In *Gamma Interferon in Antiviral Defence* (ed. Karupiah, G.), pp. 177–188. Austin, Texas: R. G. Landes Company.

MOSSMAN, K., NATION, P., MACEN, J., GARBUTT, M., LUCAS, A. & McFADDEN, G. (1996). Myxoma virus M-T7, a secreted homolog of the interferon-γ receptor, is a critical virulence factor for the development of myxomatosis in European rabbits. *Virology* **215**, 17–30.

MOSSMAN, K., UPTON, C., BULLER, R. M. L. & McFADDEN, G. (1995 b). Species specificity of ectromelia virus and vaccinia virus interferon-γ binding proteins. *Virology* **208**, 762–769.

MOSSMAN, K., UPTON, C. & McFADDEN, G. (1995 c). The myxoma virus-soluble interferon-γ receptor homolog, M-T7, inhibits interferon-γ in a species-specific manner. *Journal of Biological Chemistry* **270**, 3031–3038.

MURPHY, P. M. (1993). Molecular mimicry and the generation of host defense protein diversity. *Cell* **72**, 823–826.

MURPHY, P. M. (1994). Molecular piracy of chemokine receptors by herpesviruses. *Infectious Agents and Disease* **3**, 137–154.

NEOTE, K., DIGREGORIO, D., MAK, J. Y., HORUK, R. & SCHALL, T. J. (1993). Molecular cloning, functional expression, and signalling characteristics of a C-C chemokine receptor. *Cell* **72**, 415–425.

OPGENORTH, A., GRAHAM, K., NATION, N., STRAYER, D. & McFADDEN, G. (1992). Deletion analysis of two tandemly arranged virulence genes in myxoma virus, MIIL and myxoma growth factor. *Journal of Virology* **66**, 4720–4731.

OPGENORTH, A., NATION, N., GRAHAM, K. & McFADDEN, G. (1993). Transforming growth factor alpha, Shope fibroma growth factor, and vaccinia growth factor can replace myxoma growth factor in the induction of myxomatosis in rabbits. *Virology* **192**, 701–709.

PATEL, A. H., GAFFNEY, D. F., SUBAK-SHARPE, J. H. & STOW, N. D. (1990). DNA sequence of the gene encoding a major secreted protein of vaccinia virus, strain Lister. *Journal of General Virology* **71**, 2013–2021.

PICKUP, D. J. (1994). Poxviral modifiers of cytokine responses to infection. *Infectious Agents and Disease* **3**, 116–127.

REISNER, A. H. (1985). Similarity between the vaccinia virus 19K early protein and epidermal growth factor. *Nature* **313**, 801–803.

RODE, H. J., JANSSEN, W., ROSEN-WOLFF, A., BUGERT, J. J., THEIN, P., BECKER, Y. & DARAI, G. (1993). The genome of equine herpesvirus type 2 harbors an interleukin 10 (IL-10)-like gene. *Virus Genes* **7**, 111–116.

ROUSSET, F., GARCIA, E., DEFRANCE, T., PERONNE, C., VEZZIO, N., HSU, D. H., KASTELEIN, R., MOORE, K. W. & BANCHEREAU, J. (1992). Interleukin 10 is a potent growth and differentiation factor for activated human B lymphocytes. *Proceedings of the National Academy of Sciences, USA* **89**, 1890–1893.

ROUVIER, E., LUCIANI, M. F., MATTEI, M. G., DENIZOT, F. & GOLSTEIN, P. (1993). CTLA-8, cloned from an activated T cell, bearing AU-rich messenger RNA instability sequences, and homologous to a herpesvirus saimiri gene. *Journal of Immunology* **150**, 5445–5456.

SCHALL, T. J. & BACON, K. B. (1994). Chemokines, leucocyte trafficking, and inflammation. *Current Opinion in Immunology* **6**, 865–873.

SCHALL, T. J., STEIN, B., GORGONE, G. & BACON, K. B. (1995). Cytomegalovirus encodes a functional receptor for C-C chemokines. In *Viroceptors, Virokines and Related Immune Modulators Encoded by DNA Viruses* (ed. McFadden, G.), pp. 201–214. Austin, Texas: R. G. Landes Company.

SCHREIBER, M. & McFADDEN, G. (1994). The myxoma virus TNF-receptor homologue (T2) inhibits tumor necrosis factor-α in a species-specific fashion. *Virology* **204**, 692–705.

SCHREIBER, M. & McFADDEN, G. (1996). Mutational analysis of the ligand binding domain of M-T2 protein, the tumor necrosis factor receptor homologue of myxoma virus. *Journal of Immunology*, **157**, 4486–4495.

SCHREIBER, M., RAJARATHNAM, K. & McFADDEN, G. (1996). Myxoma virus T2 protein, a tumor necrosis factor (TNF) receptor homolog, is secreted as a monomer and dimer that each bind rabbit TNFα, but the dimer is a more potent TNF inhibitor. *Journal of Biological Chemistry* **271**, 13333–13341.

SCHREIBER, M., SEDGER, L. & McFADDEN, G. (1997). Distinct domains of M-T2, the myxoma virus TNF receptor homolog, mediate extracellular TNF binding and Intracellular apoptosis inhibition. *Journal of Virology*, **71**, 2171–2181.

SENKEVICH, T. G., BUGERT, J. J., SISLER, J. R., KOONIN, E. V., DARAI, G. & MOSS, B. (1996). Genome sequence of a human tumorigenic poxvirus: prediction of specific host response-evasion genes. *Science* **273**, 813–816.

SMITH, C. A., DAVIS, T., ANDERSON, D., SOLAM, L., BECKMANN, M. P., JERZY, R., DOWER, S. K., COSMAN, D. & GOODWIN, R. G. (1990). A receptor for tumor necrosis factor defines an unusual family of cellular and viral proteins. *Science* **248**, 1019–1023.

SMITH, C. A., DAVIS, T., WIGNALL, J. M., DIN, W. S., FARRAH, T., UPTON, C., McFADDEN, G. & GOODWIN, R. G. (1991). T2 open reading frame from the Shope fibroma virus encodes a soluble form of the TNF receptor. *Biochemical and Biophysical Research Communications* **176**, 335–342.

SMITH, C. A., FANG-QI, H., DAVIS SMITH, T., RICHARDS, C. L., SMOLAK, P., GOODWIN, R. G. & PICKUP, D. J. (1996). Cowpox virus genome encodes a second soluble homologue of cellular TNF receptors, distinct from CrmB, that binds TNF but not LTα. *Virology* **223**, 132–147.

SMITH, C. A. & GOODWIN, R. G. (1995). Tumor necrosis factor receptors in the poxvirus family: biological and genetic implications. In *Viroceptors, Virokines and Related Immune Modulators Encoded by DNA Viruses* (ed. McFadden, G.), pp. 29–40. Austin, Texas: R. G. Landes Company.

SMITH, G. L. (1994). Virus strategies for evasion of the host response to infection. *Trends in Microbiology* **2**, 81–88.

SOULIER, J., GROLLET, L., OKSENHENDLER, E., CACOUB, P., CAZALS-HATEM, D., BABINET, P., D'AGAY, M. F., CLAUVEL, J. P., RAPHAEL, M., DEGOS, L. & SIGAUX, F. (1995). Kaposi's sarcoma-associated herpesvirus-like DNA sequences in multicentric Castleman's disease. *Blood* **86**, 1275–1280.

SPRIGGS, M. K. (1994). Cytokine and cytokine receptor genes 'captured' by viruses. *Current Opinion in Immunology* **6**, 526–529.

SPRIGGS, M. K. (1996). One step ahead of the game: viral immunomodulatory molecules. *Annual Review of Immunology* **14**, 101–131.

SPRIGGS, M. K., HRUBY, D. E., MALISZEWSKI, C. R., PICKUP, D. J., SIMS, J. E., BULLER, R. M. L. & VANSLYKE, J. (1992).

Vaccinia and cowpox viruses encode a novel secreted interleukin-1-binding protein. *Cell* **71**, 145–152.

SWAMINATHAN, S., HESSELTON, R., SULIVAN, J. & KIEFF, E. (1993). Epstein–Barr virus recombinants with specifically mutated BCRF1 genes. *Journal of Virology* **67**, 7406–7413.

SWAMINATHAN, S. & KIEFF, E. (1995). The role of BCRF1/vIL-10 in the life cycle of Epstein–Barr virus. In *Viroceptors, Virokines and Related Immune Modulators Encoded by DNA Viruses* (ed. McFadden, G.), pp. 111–125. Austin, Texas: R. G. Landes Company.

SYMONS, J. A., ALCAMI, A. & SMITH, G. L. (1995). Vaccinia virus encodes a soluble type 1 interferon receptor of novel structure and broad species specificity. *Cell* **81**, 551–560.

TAGA, T., KAWANISHI, Y., HARDY, R. R., HIRANO, T. & KISHIMOTO, T. (1987). Receptors for B cell stimulatory factor 2. Quantitation specificity, distribution, and regulation of their expression. *Journal of Experimental Medicine* **166**, 967–981.

TELFORD, E. A., WATSON, M. S., AIRD, H. C. & DAVISON, P. J. (1995). The DNA sequence of equine herpesvirus. 2. *Journal of Molecular Biology* **249**, 520–528.

UEDA, Y., MORIKAWA, S. & MATSUURA, Y. (1990). Identification and nucleotide sequence of the gene encoding a surface antigen induced by vaccinia virus. *Virology* **177**, 588–594.

UPTON, C., DeLANGE, A. M. & McFADDEN, G. (1987 a). Tumorigenic poxviruses: genomic organization and DNA sequence of the telomeric region of the Shope fibroma virus genome. *Virology* **160**, 20–30.

UPTON, C., MACEN, J. L. & McFADDEN, G. (1987 b). Mapping and sequencing of a gene from myxoma virus that is related to those encoding epidermal growth factor and transforming growth factor alpha. *Journal of Virology* **61**, 1271–1275.

UPTON, C., MACEN, J. L., SCHREIBER, M. & McFADDEN, G. (1991). Myxoma virus expresses a secreted protein with homology to the tumor necrosis factor receptor gene family that contributes to viral virulence. *Virology* **184**, 370–382.

UPTON, C., MOSSMAN, K. & McFADDEN, G. (1992). Encoding of a homolog of the IFN-γ receptor by myxoma virus. *Science* **258**, 1369–1372.

VIEIRA, P., DEWAAL MALEFYT, R., DANG, M. N., JOHNSON, K. E., KASTELEIN, R., FIORENTINO, D. F., DEVRIES, J. E., RONCAROLO, M. G., MOSMANN, T. R. & MOORE, K. W. (1991). Isolation and expression of human cytokine synthesis inhibitor factor cDNA clones: homology to Epstein–Barr virus open reading frame BCRF1. *Proceedings of the National Academy of Sciences, USA* **88**, 1172–1176.

WHITE, E. (1996). Life, death, and the pursuit of apoptosis. *Genes and Development* **10**, 1–15.

YAO, Z., FANSLOW, W. C., SELDIN, M. F., ROUSSEAU, A. M., PAINTER, S. L., COMEAU, M. R., COHEN, J. I. & SPRIGGS, M. J. (1995 a). Herpesvirus saimiri encodes a new cytokine, IL-17, which binds to a novel cytokine receptor. *Immunity* **3**, 811–821.

YAO, Z., PAINTER, S. L., FANSLOW, W. C., ULRICH, D., MACDUFF, B. M., SPRIGGS, M. K. & ARMITAGE, R. J. (1995 b). Human IL-17: a novel cytokine derived from T cells. *Journal of Immunology* **155**, 5483–5486.

Production of an interferon-gamma homologue by an intestinal nematode: functionally significant or interesting artefact?

R. K. GRENCIS *and* G. M. ENTWISTLE

School of Biological Sciences, Stopford Building, University of Manchester, Oxford Road, Manchester, M13 9PT, UK

SUMMARY

Chronic infection is a prominent feature of many intestinal nematode infections in man and animals. It is also clear that in such situations host immunity is activated but is unable to induce a protective response. A great deal of work has shown that genetic control of host immunity contributes to the variation in worm burdens often observed in the field. There is increasing appreciation, however, of the capability of infectious agents themselves to modulate the host immune response and potentiate their own survival. Using an immunologically well defined model of intestinal nematode infection in mice (*Trichuris muris*) we have shown that parasite derived molecules share cross reactive epitopes with the host cytokine interferon-γ using cytokine specific monoclonal antibodies in ELISA, Western blotting and immunoprecipitation assays. Furthermore, the parasite molecules can be shown to bind to the interferon-γ receptor and induce change in lymphoid cells similar to those induced by murine interferon-γ. The functional activity of the molecule *in vivo* remains to be determined. Previous studies have established that interferon-γ is critical for progression to chronic *T. muris* infection in mice and, therefore, it raises the distinct possibility that the production of an interferon-γ homologue by the worm may be one mechanism whereby the parasite is able to interfere with the regulation of the host immune response and potentiate its own survival.

Key words: Nematode, immunity, cytokines, immunomodulation.

INTRODUCTION

An established feature of many intestinal nematode infections is chronicity. This is particularly true for many species of parasite which infect the human gut and is at least in part responsible for the high prevalence of such infections observed in the field. This situation presents a paradox in that the host immune response is highly adapted to recognise, respond to and eliminate foreign antigen or damage caused by pathogens (Matzinger, 1994). Thus it is clear that in a chronic infection either a true symbiotic relationship has evolved between parasite and host or, more likely, that the infectious agent is actively evading being destroyed by the host immune system. Studies in both human and laboratory rodent systems have shown that the host immune response does recognise intestinal nematode antigens and responds vigorously to them. This is most easily demonstrated by the detection of parasite-specific antibody in the circulation of infected individuals (see review by Maizels *et al.* 1993). Thus the immune system responds during chronic infection but, by definition, is not efficient at promoting resistance.

A number of approaches has been used to investigate the basis of chronic infection with the most common being based based on studies which have analysed mechanisms of *protective* immunity to intestinal nematodes especially in laboratory models. Systems in rodents (*Nippostrongylus brasileinsis, Trichinella spiralis, Strongyloides venezuelensis* and *Heligmosomoides polygyrus*) have provided a great deal of information on the underlying regulation of resistance to intestinal nematodes, although definition of effector mechanisms is still lacking for many (reviewed in Grencis, 1997*b*). Based upon these studies it is quite clear that resistance is mediated by CD4$^+$ T cells through the secretion of defined sets of cytokines (reviewed in Finkelman *et al.* 1997; Grencis, 1997*b*).

It is broadly accepted that two major subsets of CD4$^+$ T cells, Th$_1$ and Th$_2$, regulate qualitatively and quantitatively the immune response generated against pathogens/antigens. Furthermore, it is clear that Th$_1$ and Th$_2$ subsets can regulate each other reciprocally through production of their characteristic cytokines. IL-4 promotes the production of Th$_2$ cells and down-regulates the production of Th$_1$ cells. Interferon (IFN)-γ produced by Th$_1$ cells potentiates Th$_1$ cell responses and down-regulates the Th$_2$ subset and hence IL-4 production (Mosmann & Coffman, 1989; Mosmann & Subach, 1996). With regards to intestinal nematode infection T cells of the Th$_2$ type are associated with resistance, i.e. cells secreting IL-4, IL-5, IL-6, IL-9, IL-10 and IL-13 (Finkelman *et al.* 1996; Grencis 1997*a*). Indeed, it is clear that in all systems studied potentiation of a Th$_1$ response is beneficial to the parasite in terms of survival, but detrimental to the host in terms of resistance (Grencis 1997*a*; Finkelman *et al.* 1997).

Parasitology (1997), **115**, S101–S105.   © 1997 Cambridge University Press

THE *TRICHURIS MURIS* MODEL

One particular rodent system which has been particularly useful and has extended our understanding of resistance and susceptibility to intestinal nematodes is that of whipworm (*Trichuris muris*) infection in the mouse. This is a naturally-occurring parasite of the mouse and shares many features with its human counterpart (*Trichuris trichiura*). One of the values of this model system is that different inbred strains of mouse present different response phenotypes under similar infection conditions. The majority of inbred strains of mouse actively expel their parasite burdens although a few strains are unable to do so and exhibit high levels of chronic infection (Else & Wakelin, 1988). This spectrum of infection reflects the overdispered distribution of infection commonly seen in gut nematode infections of outbred naturally-infected host populations including humans (Bundy, 1995).

It is quite clear that in those strains which expel *T. muris* resistance is mediated by Th_2 cells with particular importance for the cytokines, IL-4, IL-9 and IL-13 (see review by Grencis, 1997*a*). Interestingly, however, in those strains which do not expel their parasites (susceptible) a Th_1-dominated response develops which is critically dependent upon the production of IFN-γ. This is most readily demonstrated by *in vivo* experiments in which IFN-γ is neutralized in naturally susceptible infected mice using IFN-γ-specific monoclonal antibodies (Else *et al.* 1994). Depression of a Th_1 response allows such animals to mount a Th_2 response and to expel their worms through a Th_2-dependent response (Else *et al.* 1994). The importance of IFN-γ is further supported by experiments in which a Th_1 response was induced by *in vivo* administration of IL-12. This is a non-T cell-derived cytokine which promotes a Th_1 response through upregulation of IFN-γ. Treatment of normally resistant mice with IL-12 induced a strong Th_1 response and the mice did not expel their worms. Moreover, the change in response phenotype was entirely dependent upon IFN-γ as demonstrated by co-incident administration of anti-IFN-γ antibody to IL-12-treated animals (Bancroft *et al.* 1997). Thus it is clear that for *T. muris* infections IFN-γ plays a critical role in the establishment of a chronic infection.

The reasons why some strains of mouse mount a strong Th_2 response, whereas others allow a Th_1 response to become dominant are unknown. It is reasonable that the host plays a major role. It is well established that multiple factors contribute to the initiation and development of either a Th_1 and Th_2 response including cytokine environment, antigen presenting cell type, antigen dose, MHC/TCR ligand density, MHC etc. (Fitich *et al.* 1993). Host genetic heterogeneity in any of these can influence the qualitative development of the Th cell response

and this is undoubtedly true for *T. muris* infections (Grencis, 1997*a*).

However, there is also considerable evidence to support an active role for the parasite itself in influencing the development of a Th_1 response and susceptibility. In normally susceptible strains of mouse abbreviation of patent *T. muris* infections (> day 33 post infection, p.i.) by anthelmintic treatment does not change the mode of response to a challenge infection, i.e. a secondary infection readily goes through to patency. If, however, primary infections are abbreviated at times up until approximately day 21 p.i. then upon challenge, the mice are relatively resistant (Else, Wakelin & Roach, 1989). These data suggest that exposure of the host to the parasite for differing lengths of time or to different life cycle stages is important in the development of either resistance or susceptibility.

More convincing evidence comes from studies of *T. muris* infections in 'differential responder' strains of mouse. There are some strains of mouse which show a split response phenotype (i..e. between individuals within the inbred strain). In these strains some individuals expel their parasites and become resistant and others do not expel their parasites and become susceptible. The common feature of the strains of mouse that exhibit this split response is that they all begin to expel their parasites around days 17–21 p.i. (Lee & Wakelin, 1982). Nevertheless, the individual mice within such strains which expel their worms mount a Th_2 response and those that do not expel mount a Th_1 response (Else, Entwistle & Grencis, 1993). Because all mice within an inbred strain are genetically identical, there can be no difference between them in terms of host genes, molecules or cells involved in immune responses. Furthermore all mice received the same level of infection. Thus for the *T. muris* system at least, there is compelling evidence to suggest that the parasite itself is inducing susceptibility, presumably by redirecting the host immune response towards one which is inappropriate for mediating worm expulsion.

Recent data from our laboratory have raised one possible mechanism whereby *T. muris* may be interfering with the generation of a host protective Th_2-mediated immune response. The data initially stemmed from the observation that excretory/secretory (ES) products of adult *T. muris* generated a positive signal in a capture ELISA assay to detect murine IFN-γ (Entwistle *et al.* unpublished observations; see Fig. 1). Such assays utilize two cytokine-specific monoclonal antibodies to detect cytokines via a solid phase. Briefly, one cytokine-specific monoclonal antibody is bound to a plastic plate. Following blocking of spare protein binding sites the test solution is added (containing cytokine, or in this case parasite molecules). Folllowing washing, a second cytokine-specific monoclonal

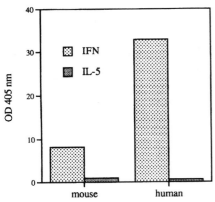

Fig. 1. Determination of cross-reactivity between
T. muris antigens in excretory/secretory antigens (ES)
and IFN-γ. A capture ELISA using either anti-mouse
or human IFN-γ specific or anti-mouse or human
interleukin (IL)-5 specific monoclonal antibodies were
used. Results are expressed as 'equivalent' units of
cytokine per mg of ES protein when compared to a
standard curve.

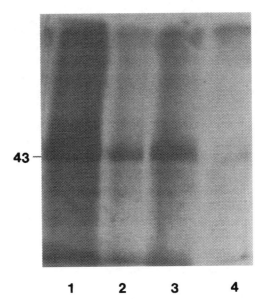

Fig. 3. Immunoprecipitation of [³⁵S]methionine
labelled *T. muris* pseudocoelomic (PC) fluid using anti-
human IFN-monoclonal antibodies (lanes 2 and 3).
Lane 1 shows non-precipitated PC fluid, lane 4 shows a
precipitation using an isotyped matched control
monoclonal antibody of irrelevant specificity.

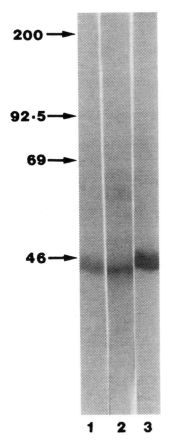

Fig. 2. Western blot of *T. muris* ES products probed
with anti-human IFN-γ monoclonal antibodies (lane 3)
or post infection sera from naturally infected animals
(lane 2). Lane 1 shows Ponceau-S staining of blotted ES
products. Isotype matched control monoclonal
antibodies showed no staining of ES products (data not
shown).

antibody is added. This will bind to any cytokine
'captured' by the first monoclonal antibody. This
interaction is detected using standard ELISA metho-

dology. The assays are highly specific detecting
distinct epitopes on the cytokine in question and
used routinely for detection of numerous cytokines
in the laboratory and clinic. The observation that
T. muris ES products were detected in an IFN-γ-
specific ELISA was a quite unexpected but a
consistent one and has been extended to several
IFN-γ-specific monoclonal antibodies for the mouse,
rat and human cytokine (see Fig. 1). IFN-γ from
different species share some homology (see Gray,
1994) although a number of the monoclonal antibodies used are believed to be epitope- and species-
specific. Moreover, Western Blot analysis of ES
products using IFN-γ-specific monoclonal antibodies (Fig. 2) and immunoprecipitation assays of
[Se³⁵]-methionine labelled ES products (see Fig. 3)
identified a major parasite-derived protein/glyco-
protein of approximately 43 kDa molecular weight
recognized by these antibodies. Interestingly this
molecule is known to be a major secreted antigen of
T. muris and is also known to be a major component
of the pseudocoelomic fluid (PC) of the worm (Else,
1989). Its relationship to the pore-forming molecule
of a similar molecular weight isolated from *Trichuris*
species (Drake *et al.* 1994) is unknown, although
recent studies may suggest that a multi-gene family
exists for these molecules (G. Barker, personal
communication).

Further analysis of ES products has shown their
capacity to bind to the murine IFN-γ receptor. This
was assayed using a solid phase ELISA in which a
soluble construct of the murine IFN-γ receptor
(Fernando *et al.* 1991) was bound to a plastic plate
followed by ES products or PC fluid and detection

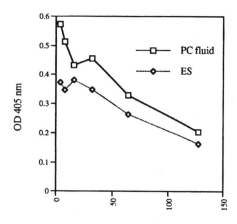

Fig. 4. Parasite ES and PC fluid binding to a plate bound soluble construct of the murine IFN-γ receptor. Binding was detected using a mono-specific anti-*T. muris* 43 kDa antigen rabbit antiserum.

Table 1. Spleen and mesenteric lymph node cells from BALB/c mice were cultured for 24 h under the conditions described. The cells were then stained with a biotinylated anti-Ly6A/E followed by phycoerythrin conjugated streptavidin and either a FITC conjugated anti-CD3 monoclonal antibody to detect T cells or FITC conjugated anti-B220 monoclonal antibody to detect B cells. Cells were analysed by two colour flow cytometry

	% cells bearing Ly6A/E	
	T cells	B cells
Untreated	15	18
LPS (20 μg/ml)	29	37
IFN-γ (2 U/ml)	26	40
T. muris ES (25 μg/ml)	30	42

using a monospecific polyclonal anti-43 kDa rabbit antiserum (see Fig. 4). This analysis clearly suggests that the 43 kDa molecule is capable of binding to the cytokine receptor and therefore may have a functional effect. The latter has been examined by investigation of the regulation of lymphocyte cell surface molecules by parasite products.

It is well known that *in vitro* IFN-γ can induce the up-regulation of expression of the Ly 6 molecule on murine lymphocytes (Snapper *et al.* 1991). It is clear from our own studies that parasite products are also capable of doing so (Table 1). This again strengthens the possibility that *T. muris* produces a molecule which shares some of the functional characteristics of host IFN-γ.

It also follows that if *T. muris* has a major antigen which shares epitopes with IFN-γ then it is reasonable to suggest that *T. muris*-infected mice

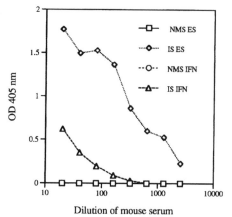

Fig. 5. Cross-reactivity in *T. muris* post-infection serum between IFN-γ and parasite ES products. A mouse IFN-γ specific monoclonal antibody was used as a capture antibody. Recombinant mouse IFN-γ (IFN) or *T. muris* ES (ES) was bound and probed with normal mouse serum (NMS) or immune mouse serum (IMS) taken on day 35 post infection. N.B. The NMS IFN combination gave an identical curve to that of NMS ES but is not depicted for clarity of presentation.

may make antibodies against such an antigen which show cross-reactivity with IFN-γ. This is indeed the case and can be demonstrated using a capture ELISA in which recombinant IFN-γ is captured by a cytokine specific monoclonal and probed by *T. muris* post infection sera (Fig. 5). Thus, following infection mice make antibodies to parasite antigens which share crossreactivity with IFN-γ. If such an antibody had a relatively low affinity for IFN-γ it could conceivably act as a carrier for the cytokine extending its biological activity. Certainly this kind of approach has been used to great effect experimentally in extending the half like of cytokines *in vivo* following coupling to cytokine specific monoclonal antibodies (Finkelman *et al.* 1993).

The preceding data lend strong support to the hypothesis that *T. muris* shares an epitope (functional?) with IFN-γ. It is tempting to speculate that it may be one mechanism whereby the parasite interferes with the immune response of the host. If such a molecule is able to potentiate a Th₁ response against the parasite it will promote its own survival. Whilst we have preliminary evidence that *T. muris* may possess an IFN-γ homologue we have yet to prove it plays a functional role in potentiating Th₁ responses *in vivo*. Further studies are needed to consolidate these observations. Cloning of the molecule is indeed a priority. It must be remembered, however, that even though the two molecules share epitopes, there may be little sequence homology. Whilst it is relatively easy to appreciate how viruses can capture parts of the host genome, e.g. cytokine genes (see review by Spriggs, 1996), large metazoan parasites are unable to do this. Nevertheless any molecule evolved by the parasite which gives it an advantage is likely to be selected. Indeed such a strategy for potentiating helminth infection has been previously hypothesized (Behnke, Barnard & Wakelin, 1992). Furthermore, it is quite apparent from studies of other infectious agents that structural similarity between pathogen and cytokine can occur in the absence of sequence homology (Matthews *et al.* 1994).

CONCLUSION

The *T. muris* system has, therefore, provided a model system whereby a helminth-derived cytokine homologue may play a role in promoting parasite survival. In the mouse, the availability of such a molecule may function to potentiate an inappropriate (Th$_1$-dominated) host immune response. Other mechanisms will also undoubtedly contribute including host genetics and parasite effects such as infection level (Bancroft, Else & Grencis, 1994). Whether other species of *Trichuris* or other intestinal nematodes have IFN-γ homologues remains to be seen. Whilst Th$_2$-dominated responses do appear to be responsible for mediating resistance to intestinal nematodes it is clear that in some animal model or human systems where chronic infection occurs, Th$_2$ responses are evident (see review by Maizels *et al.* 1993). In these systems, chronic infection is therefore not the result of or unlikely to result in a complete shift from a Th$_2$ to a Th$_1$ response. Rather, a more selective down-regulation of some particular aspects of the Th$_2$ response may be operating. Evidence to support this hypothesis comes from the *H. polygyrus* system in which chronic infection is associated with depression of only some Th$_2$ cytokines and certain Th$_2$-mediated effector mechanisms (Behnke *et al.* 1993). Certainly, different intestinal nematodes appear to be controlled by different Th$_2$-controlled effector mechanisms (see Grencis, 1997*b*). Further study should elucidate whether other helminth parasites mimic host cytokines to interefere with the regulation of anti-parasite immunity. Such an evolutionary strategy is not without precedent in other infectious agents and is compatible with a subtle and stable interaction with the host, often a hallmark of successful parasitism.

REFERENCES

BANCROFT, A. J., ELSE, K. J. & GRENCIS, R. K. (1994). Low level infection of *Trichuris muris* significantly affects the polarisation of the CD4 response. *European Journal of Immunology* **24**, 3113–3118.

BANCROFT, A. J., ELSE, K. J., SYPEK, J. P. & GRENCIS, T. K. (1997). IL-12 promotes a chronic intestinal nematode infection. *European Journal of Immunology* **27**, 866–870.

BEHNKE, J. M., BARNARD, C. J. & WAKELIN, D. (1992). Understanding chronic nematode infections:

evolutionary considerations, current hypotheses and the way forward. *International Journal for Parasitology* **22**, 861–907.

BUNDY, D. A. P. (1995). Epidemiology and transmission of intestinal helminths. In *Enteric Infection 2. Intestinal Helminths*. (Ed. Farthing M. J. C., Keusch G. T. & Wakelin D.), pp. 5–24. London: Chapman & Hall.

DRAKE, L., KORCHEV, Y., BASHFORD, L., DJAMGOZ, M., WAKELIN, D., ASHALL, F. & BUNDY, D. (1994). The major secreted product of the whipworm *Trichuris* is a pore-forming protein. *Proceedings of the Royal Society of London, B* **257**, 255–261.

ELSE, K. J. (1989). Immunogenetics of *Trichuris muris* infection. PhD thesis, University of Nottingham.

ELSE, K. J., ENTWISTLE, G. M. & GRENCIS, R. K. (1993). Correlations between worm burden and markers of Th1 and Th2 cell subset induction in an inbred strain of mouse infected with *Trichuris muris*. *Parasite Immunology* **10**, 595–600.

ELSE, K. J. & WAKELIN, D. (1988). The effects of H-2 and non-H-2 genes on the expulsion of the nematode *Trichuris muris* from inbred and congenic strains of mouse. *Parasitology* **96**, 543–550.

ELSE, K. J., FINKELMAN, F. D., MALISZEWSKI, C. R. & GRENCIS, R. K. (1994). Cytokine mediated regulation of chronic intestinal helminth infection. *Journal of Experimental Medicine* **179**, 347–351.

ELSE, K. J., WAKLEIN, D. & ROACH, T. I. A. R. (1989). Host predisposition to trichuriasis: the mouse – *T. muris* model. *Parasitology* **98**, 275–282.

FERNANDO, L. P., LeCLAIRE, R. D., OBICI, S., ZAVODNY, P. J., RUSSELL, S. W. & PACE, J. L. (1991). Stable expression of a secreted form of the mouse IFN-γ receptor by rat cells. *Journal of Immunology* **147**, 541–547.

FINKELMAN, F. D., MADDEN, K. B., MORRIS, S. C., HOLMES, J. M., BOIANI, N., KATONA, I. M. & MALISZEWSKI, C. R. (1993). Anti-cytokine antibodies as carrier proteins: prolongation of *in vivo* effects of exogenous cytokines by injection of cytokine-anti-cytokine antibody complexes. *Journal of Immunology* **151**, 1235–1244.

FINKELMAN, F. D., SHEA-DONAHUE, T., GOLDHILL, J., SULLIVAN, C. A., MORRIS, S. C., MADDEN, K. B., GAUSE, W. C. & URBAN, J. F. (1997). Cytokine regulation of host defence against parasitic gastrointestinal nematodes: lessons from studies with rodent models. *Annual Review of Immunology* **15**, 505–534.

FITCH, F. W., McKLSIC, D. W., LANCKI, D. W. & GAJEWSKI, T. F. (1993). Differential regulation of murine T lymphocyte subsets. *Annual Review of Immunology* **11**, 29–48.

GRAY, P. W. (1994). Interferon gamma. In *Guidebook to Cytokines and their Receptors* (Ed. Nicola, N. A.), pp. 118–119. Oxford, Oxford University Press.

GRENCIS, R. K. (1997*a*). Enteric helminth infection: immunopathology and resistance during intestinal nematode infection. *Chemical Immunology* **66**, 41–61.

GRENCIS, R. K. (1997*b*). Th2 mediated host protective immunity to intestinal nematode infections. *Philosophical transactions of the Royal Society B: Biological Sciences.* **352**, 1377–1384.

LEE, T. D. G. & WAKELIN, D. (1982). The use of host strain variation to assess the significance of mucosal mast cells in the spontaneous cure response of mice in the nematode *Trichuris muris*. *International Archives of Allergy and Applied Immunology* **67**, 302–307.

MATTHEWS, S., BARLOW, P., BOYD, J., BARTON, G., RUSSELL, R., MILLS, H., CUNNINGHAM, M., MEYWERS, N., BURNS, N., CLARK, N., KINGSMAN, S., KINGSMAN, A. & CAMPBELL, I. (1994). Structural similarity between p17 matrix protein of HIV-1 and Interferon-γ. *Nature* **370**, 666–668.

MATZINGER, P. (1994). Toelerance, danger and the extended family. *Annual Review of Immunology* **12**, 991–1045.

MAIZELS, R. M., BUNDY, D. A. P., SELKIRK, M. E., SMITH, D. F. & ANDERSON, R. M. (1993). Immunological modulation and evasion by helminth parasites in human populations. *Nature* **365**, 797–805.

MOSMANN, T. R. & COFFMAN, R. L. (1989). Th1 and Th2 cells: different patterns of lymphokine secretion lead to different functional properties. *Annual Review of Immunology* **7**, 145–173.

MOSMANN, T. R. & SUBACH, S. (1996). The expanding universe of T cell subsets: Th1, Th2 and more. *Immunology Today* **68**, 18–23.

SNAPPER, C. M., YAMAGUCHI, H., URBAN, J. F. & FINKELMAN, F. D. (1991). Induction of Ly-6AE expression by murine lymphocytes after *in vivo* immunisation is strictly dependent upon the action of IFN-αβ and/or IFN-γ. *International Immunology* **3**, 845–852.

SPRIGGS, M. K. (1996). One step ahead of the game: viral immunomodulatory molecules. *Annual Review of Immunology* **14**, 101–130.

Cytokine induction and exploitation in schistosome infections

J. H. McKERROW

Departments of Pathology and Pharmaceutical Chemistry, University of California, San Francisco,
c/o Pathology, 4150 Clement St., 113B, San Francisco, CA 94121

SUMMARY

Schistosome parasites, despite being multicellular organisms several millimetres in length, can survive in the bloodstream of mammalian hosts for decades. The remarkable and complex adaptation exemplified in the host–parasite relationship in schistosomiasis may include not only immune evasion by the parasite, but also immune exploitation. While the developmental and adult stages of the parasite are by and large invisible to the immune response, the parasite egg induces a granulomatous reaction which not only protects the host from a diffusible parasite toxin, but also is required for normal transmission of parasite eggs from the host to the external environment. Other possible mechanisms of immune exploitation by schistosomes are discussed including skewing of cytokine responses, effects of cytokines on worm fecundity, exploitation of endothelial cell adherence, and induction of IgE.

Key words: Schistosome, host, cytokine, immune, evasion, granuloma, egg.

INTRODUCTION

Trematode helminth parasites of the genus *Schistosoma* have complex life-cycles, and yet are some of the most successful parasites of man. The relationship between the human host and the schistosome parasite is an ancient one. The three major species of schistosomes, *S. haematobium*, *S. japonicum* and *S. mansoni*, appear to have had their origins in roughly the same location as three of the great cradles of human civilization – the Nile River Delta, the Yangtze River, and the African Lake Plateau (Warren, 1984). It is not unreasonable to assume that schistosome parasites co-evolved with their human hosts.

In contrast to many other infectious organisms, the relationship between human host and schistosome parasite can extend for years, or even decades. While significant morbidity and occasional mortality is associated with a small percentage of infections, most individuals harbour schistosome parasites and transmit their eggs while exhibiting only modest clinical signs and symptoms. This occurs despite the fact that schistosome adults are complex multicellular organisms that reach 9–12 mm in length. The observation that these relatively large organisms can exist in the bloodstream of the human host for up to 3 decades has intrigued immunologists and parasitologists for some time. Several mechanisms have been proposed for how the schistosome manages to elude the immune response, and more recently, intriguing results from several studies suggest that the parasite may go a step further and actually exploit the immune response for its own replication and transmission (Fig. 1).

THE CIRCUMOVAL GRANULOMA: PROTECTOR OF THE HOST, CHAPERONE OF THE PARASITE

One of the first observations that led to the hypothesis that schistosome parasites may subvert the human immune response for their own benefit came from studies of granuloma formation in response to schistosome eggs in T cell-deprived mice. Doenhoff *et al.* (1978) noted that the rate of egg excretion from hosts infected with either *S. bovis* or *S. mansoni* was reduced by immunosuppression. Injection of serum from immunologically intact mice, either chronically infected or immunized with egg antigens, partly restored the rate of egg excretion in *S. mansoni*-infected immunosuppressed mice (Doenhoff *et al.* 1981; Dunne *et al.* 1983). If mesenteric lymph node cells or splenic lymphocytes were used to reconstitute the immunosuppressed mice, the number of eggs excreted per worm pair was restored to nearly normal levels by cell transfer (Doenhoff, Hassounah & Lucas, 1985).

When infected with *S. mansoni*, severe combined immunodeficient (SCID) mice, which lack functional T and B cells, were found not only to be deficient in granuloma formation, but also not to excrete eggs (Amiri *et al.* 1992). Egg excretion was restored in parallel with granuloma formation by injection of splenic lymphocytes, a T lymphocyte clone, or even cell-free cytokine-containing supernatant from the T cell clone. Taken together, these experiments support the notion that egg transmission from infected hosts is actually dependent on the host immune response. On the one hand this may reflect an immune influence on the number of eggs produced (see discussion below), but the formation of granulomas around eggs is also critical for egg

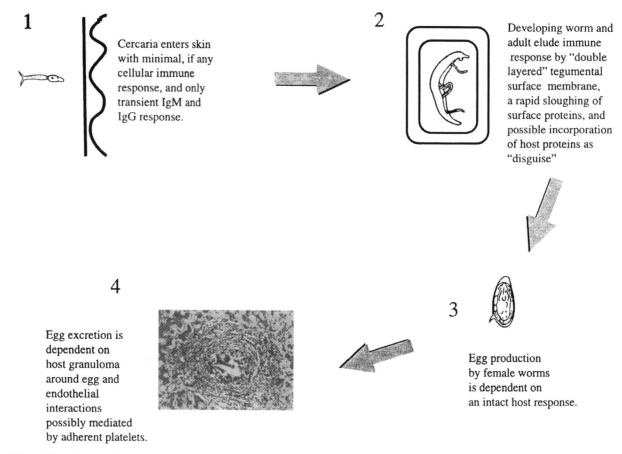

Fig. 1. Evasion and exploitation of the host immune response in schistosomiasis.

passage through intestinal or bladder wall. How granulomas aid in egg excretion is still a matter of speculation. One possibility is that the granuloma 'chaperones' the egg through tissue. Activated macrophages, or other cells within the granuloma, may provide hydrolytic enzymes which could aid in creating a path through the extracellular matrix. The increased volume that a granuloma gives an egg may also be a factor in aiding physical expulsion during a wave of peristalsis. In support of this notion is the observation that there is a strong correlation between the mean granuloma diameter and the number of eggs excreted (Doenhoff *et al.* 1985).

While the necessity of granuloma formation for egg excretion, coupled with the clinical consequences of granuloma formation for the host, makes this immune response seem disadvantageous, granuloma formation does protect the host from a more severe disease. In T-cell deprived mice or SCID mice, the absence of granuloma formation results in an acute severe liver damage apparently caused by diffusible hepatotoxin released from the eggs (Byram *et al.* 1979; Cheever, Byram & von Lichtenberg, 1985; Doenhoff *et al.* 1985; Amiri *et al.* 1992). The damaged liver cells have an accumulation of lipid vacuoles. In SCID mice, this leads to a mortality of up to 30 % (Amiri *et al.* 1992). A granulomatous reaction to eggs in the liver therefore serves to wall off diffusion of the hepatotoxin to protect the host

from a more acute, and a potentially more fatal, disease. The 'trade off' is the consequence of chronic schistosomiasis.

TEMPORAL EXPRESSION OF CYTOKINES DURING SCHISTOSOME INFECTION

While invading schistosome cercariae are usually 'invisible' to the host immune response, maturing adult worms stimulate both the humoral and cellular arms of the host immune response. Adult worms induce a Th1 type cytokine profile with elevated levels of interferon gamma, IL-12, and TNFα (Grzych *et al.* 1991; Leptak & McKerrow, 1997). With the onset of parasite egg laying between 5 and 6 weeks post infection, this predominantly Th1-like response is modulated by Th2 cytokines, such as IL-10 and IL-4. While the specific cytokines induced, and their cellular sources, vary with the experimental model used (lung *versus* liver *versus* peritoneal egg injection), the overall shift in cytokine profile before and after egg laying is dramatic. One of the key questions that remains unanswered is how much of this cytokine response is in fact beneficial to the host in terms of controlling infection? Riffkin *et al.* (1996) suggest that the skewing of Th1 and Th2 cytokine responses may in fact be a strategy evolved by the parasite to limit an effective host immune response.

DO SCHISTOSOMES EXPLOIT CYTOKINE SIGNALS TO INDUCE EGG LAYING IN THE PORTAL VENOUS SYSTEM?

Aside from the role of granulomas in facilitating egg excretion, a second remarkable adaptation of the schistosome parasite is reflected by the need for an intact immune response for normal egg production. Reduced fecundity of *S. mansoni* females was first noted in T cell-deprived mice as early as 1983 (Dunne *et al.* 1983). However, the most dramatic support for this concept came from studies of schistosome infection in SCID mice. While it was not surprising to find that granuloma formation was severely muted in schistosome-infected SCID mice, a profound effect on egg production by female worms that was observed was not expected (Amiri *et al.* 1992). The ability of SCID mice to receive immune 'transplants' allowed reconstitution of the mice first with dissociated splenocytes, and then more specifically with a T cell clone, or supernatant from that cultured T cell clone. In each instance, egg production was restored to normal levels. Because the T cell clone used had been characterized phenotypically for cytokine production, the role of individual cytokines could then be assessed. Antibodies to each of the cytokines known to be present in the supernatant were used to remove that cytokine before reconstitution of schistosome-infected SCID mice. Interferon gamma, IL-4, IL-5, or IL-6 depletion had no significant effect on either granuloma formation or egg production. However, when TNFα was removed from the supernatant, there was no reconstitution of either granuloma formation or normal egg production. Were female schistosomes responding to TNFα as a signal for egg production? To address this question, female worms were placed in culture, and egg laying monitored in the presence and absence of TNFα. A single dose of TNFα could increase egg laying by females in a dose-dependent manner. If the TNFα was then removed from the culture, the worms washed and monitored again, egg production fell to baseline levels. If TNFα was again added, egg production was again induced. It appears, therefore, that in the evolution of the host–parasite relationship in schistosomiasis, the parasite had exploited not only granuloma formation, but a specific cytokine signal as a means of ensuring transmission. TNFα appears to be key both to granuloma formation, as well as to egg production.

If female worms are responding to TNFα as a stimulus to egg laying, the implication is that they have a means of sensing the cytokine molecule. TNFα-like molecules have been identified in invertebrates (Beck *et al.* 1989). It is possible, therefore, that the response of the schistosome to host TNFα may have evolved from a more primitive invertebrate receptor–ligand interplay where a TNFα-like molecule acted as a growth factor. To date, no specific

TNFα receptor or binding protein has been identified on the surface of schistosome females. However, detergent extracts of female worms do contain a 70 kDa protein which cross-reacts with antibodies to both the mouse and human p60 TNFα receptor (D. M. Ritter & J. H. McKerrow, unpublished observations).

WHEN AND HOW IS TNFα INDUCED DURING SCHISTOSOME INFECTION?

Another unexpected observation came from studies of the induction of TNFα messenger RNA in the livers of schistosome-infected mice. Injection of eggs alone into the portal vein of mice failed to induce either TNFα message expression or granuloma formation (Leptak & McKerrow, 1997). However, if mice were infected by male worms only, expression of TNFα message was noted at 5 weeks, at levels comparable to that seen in natural infections. If eggs were injected into animals that had previously been infected by males only, or injected with recombinant TNFα, normal granuloma formation was restored. This experiment confirmed that the injected eggs failed to induce granulomas not because they were 'defective', but because the adult worms induce cytokine signals that are required for complete granuloma formation in the liver. Cheever, Lewis & Wynn (1997) also found accelerated and augmented reactions to eggs injected into lungs of mice harbouring a unisexual infection. A remarkable adaptation of parasite to host may thus follow the scenario of schistosome worms inducing TNFα upon their migration to the liver, and TNFα in turn inducing egg laying and granuloma formation, ultimately facilitating transmission of the infection to a new host.

The role of the adult worms in inducing TNFα, and the subsequent role of TNFα in enhancing transmission, may explain in part the observations of Hirata *et al.* (1996), who found small or poorly formed granulomas around eggs in less susceptible hosts like rats and quails, and no granulomas at all in guinea-pigs. In these unnatural hosts, worms appear to be stunted, and failure to mature completely may result in diminution of a worm-derived protein required for induction of granuloma formation.

PARASITE STIMULATION OF ENDOTHELIAL CELLS AND ENDOTHELIAL CELL–EGG INTERACTIONS

Endothelial cells of the hepatic sinusoids respond to schistosome egg deposition by the up-regulation of ICAM-1, the adhesion molecule responsible for recruitment of certain leucocytes from the blood-stream (Ritter & McKerrow, 1996). Secor *et al.*

(1994) showed that in human infections, there are significant levels of circulating soluble ICAM-1, presumably released from the large surface area of the hepatic sinusoidal vasculature. In fact, circulating soluble ICAM-1 appears to be more abundant in hepatosplenic schistosomiasis than in uncomplicated intestinal schistosomiasis of humans. These authors proposed that the soluble ICAM-1 may be a means of immunomodulation. By binding to its cognate LFA-1 receptor on lymphocytes, soluble ICAM-1 cellular could down-regulate an immune response by decreasing cell adhesion.

Another possible benefit for the parasite of inducing endothelial cell adhesion molecules is suggested by the observations of Ngaiza, Doenhoff & Jaffe (1993) and File (1995), that there is a close association between eggs and endothelial cells prior to egg migration out of the vessel and into the bowel wall. Adherence of endothelial cells to eggs *in vitro* results in movement of the endothelial cells over the egg (D. M. Ritter, unpublished observations). *In vivo*, this might be one mechanism for breaching the endothelial cell barrier as the egg leaves the vessel lumen. Prior to granuloma formation, the egg could reach a subendothelial location through migration of endothelial cells over the surface of the egg. The interaction of eggs and endothelial cells *in vivo* may occur via platelet coating of the eggs, an issue discussed in another chapter of this monograph (see Doenhoff, this volume).

IGE: A KEY TO PARASITE ELIMINATION OR A TOOL FOR PARASITE SURVIVAL?

Another controversial, but intriguing, relationship between the schistosome parasite and the host immune response involves the high levels of IgE and eosinophilia observed during schistosome infection. An elevated IgE antibody response is one of the hallmarks of infection with many parasitic helminths. IgE is thought to be key in the host defence against helminths (King *et al.* 1997). B cell depletion prevents development of immunity and, in some cases, effective immunity could be adoptively transferred by IgE. *In vitro*, IgE can directly mediate parasite killing through eosinophil, macrophage or platelet-mediated cytotoxicity (Capron *et al.* 1984). On the other hand, in many parasitic infections, including schistosomiasis, much of the IgE induced is nonspecific. Amiri *et al.* reported that *suppression* of IgE levels by a specific soluble IgE monoclonal antibody actually resulted in a *lower* number of eggs in infected mice (Amiri *et al.* 1994). On the other hand, targeted *disruption* of the IgE gene resulted in a *higher* worm burden in primary schistosome infections of mice (King *et al.* 1997). Further studies may help to confirm whether in fact high levels of IgE are a boon or an anathema to the parasite.

RECENT STUDIES USING TARGETED GENE DISRUPTION, OR 'CHEMICAL KNOCKOUT', THAT ARE RELEVANT TO CYTOKINE INDUCTION IN SCHISTOSOME INFECTION

Mice lacking interleukin 5 (IL-5) showed normal granuloma formation in response to schistosome ova (Sher *et al.* 1990). A more interesting pattern of response was seen in mice in which the interleukin 4 (IL-4) gene was disrupted (Pearce *et al.* 1996). When granuloma formation was accessed in the 'natural' liver venue, no effect on granuloma size or induction was seen. However, when the experimental lung granuloma model was examined, there was a marked reduction in granuloma size. This study helps to re-emphasize the importance of tissue-specific immune responses to infectious agents. It serves as a caveat to the interpretation of studies on immune responses like granuloma formation when such responses are evaluated in different organs or tissue locations.

Studies evaluating granuloma formation in mice in which the gene for tumour necrosis factor α (TNFα) receptor p55 is disrupted have been carried out. Hepatic granuloma formation is markedly diminished in response to *Propionibacterium*, *Mycobacterium* and *Corynebacterium* infections. Interestingly, in parallel with the observations of hepatotoxicity in the absence of granuloma formation around schistosome eggs, a more acute bacterial disease, often resulting in death, occurs in the granuloma-defective mice (Tsuji *et al.* 1997). These studies support the role of TNFα in hepatic granuloma formation in response to a number of different infectious agents. Results evaluating the response of these same TNFα receptor knockout mice to schistosome eggs have been less revealing. Two unpublished studies (C. Leptak; A. Sher; personal communications) found no effect on hepatic granuloma formation in TNFα p55 knockout mice. A third study (P. Amiri, personal communication) reported delayed and diminished granuloma formation. To date, no analyses of granuloma formation in mice which have the TNFα gene disrupted have been reported. It will be interesting to see whether TNFα is absolutely required for hepatic granuloma formation in schistosomiasis, or whether another cytokine pathway can 'cover' in its absence. For example, while ICAM-1 appears to be the key adhesion molecule up-regulated in the liver of schistosome-infected mice, ICAM-1 knockout mice had normal granuloma formation. However, in the absence of ICAM-1 in the liver, VCAM-1 was up-regulated, and could account for the maintenance of normal leucocyte recruitment (Ritter & McKerrow, 1996).

Schistosome granuloma induction differs qualitatively in some ways from that occurring in response to *Mycobacterium* in that a strong egg-induced Th2 response follows the initial worm-induced Th1 response. Perhaps in schistosome infections, both 'Th1' and 'Th2' granulomas may be induced under different conditions. Disruption of either pathway alone may only partly modify the final outcome. Further analysis of TNFα or TNFα receptor knockout mice may also provide more clues as to whether the influence on worm fecundity by TNFα is a direct or an indirect effect.

More consistent data have come from the use of 'chemical knockout' of TNFα. For example, a chimaeric soluble TNFα receptor-Fc protein (STNFr:Fc) produced by Immunex Corporation (Seattle, Washington, USA) produced a dramatic inhibition of granuloma formation and fibrosis in schistosome-infected mice. A parallel reduction in the number of eggs was also noted (C. Leptak, thesis, University of California, San Francisco, 1997 and C. Leptak, J. McKerrow and G. Labib, unpublished). In experiments with the soluble STNFr:Fc, the problems associated with redundant pathways being induced, as can occur in gene knockout mice, can be circumvented. Indeed, consistent, confirmatory results using this type of reagent have come from studies of granuloma formation in response to *Mycobacterium* and *Corynebacterium* (Senaldi *et al.* 1996).

REFERENCES

AMIRI, P., HAAK-FRENDSCHO, M., ROBBINS,K., McKERROW, J. H., STEWART, T. & JARDIEU. P. (1994). Anti-immunoglobulin E treatment decreases worm burden and egg production in *Schistosoma mansoni*-infected normal and interferon gamma knockout mice. *Journal of Experimental Medicine* **180**, 43–51.

AMIRI, P., LOCKSLEY, R. M., PARSLOW, T. G., SADICK, M., RECTOR, E., RITTER, D. & McKERROW, J. H. (1992). Tumor necrosis factor α restores granulomas and induces parasite egg laying in schistosome-infected *scid* mice. *Nature* **356**, 604–607.

BECK, G., VASTA, G. R., MARCHALONIS, J. J. & HABICHT, G. S. (1989). Characterization of interleukin-1 activity in tunicates. *Comparative Biochemistry and Physiology* **92B**, 93–98.

BYRAM, J. E., DOENHOFF, M. J., MUSALLAM, R., BRINK, L. H. & VON LICHTENBERG, F. (1979). *Schistosoma mansoni* infections in T-cell deprived mice, and the ameliorating effect of administering homologous chronic infection serum. II. Pathology. *American Journal of Tropical Medicine and Hygiene* **28**, 274–285.

CAPRON, M., SPIEGELBERG, H. L., PRIN, L., BENNICH, H., BUTTERWORTH, A. E., PIERCE, R. J., OUAISSI, M. A. & CAPRON, A. (1984). Role of IgE receptors in effector function of human eosinophils. *Journal of Immunology* **132**, 462–468.

CHEEVER, A. W., BYRAM, J. E. & VON LICHTENBERG, F. (1985). Immunopathology of *Schistosoma japonicum* infection in athymic mice. *Parasite Immunology* **7**, 387–398.

CHEEVER, A. W., LEWIS, F. A. & WYNN, T. A. (1997).

Schistosoma mansoni: unisexual infections sensitize mice for granuloma formation around intravenously injected eggs. *Parasitology Research* **83**, 57–59.

DOENHOFF, M. J., DUNNE, D. W., BAIN, J., LILLYWHITE, J. E. & McLAREN, M. L. (1985). Serodiagnosis of schistosomias is mansoni with CEF6, a cationic antigen fraction of *Schistosoma mansoni* eggs. *Development of Biological Standards* **62**, 63–73.

DOENHOFF, M. J., HASSOUNAH, O. A. & LUCAS, S. B. (1985). Does the immunopathology induced by schistosome eggs potentiate parasite survival? *Immunology Today* **6**, 203–206.

DOENHOFF, M., MUSALLAM, R., BAIN, J. & McGREGOR, A. (1978). Studies on the host–parasite relationship in *Schistosoma mansoni*-infected mice: the immunological dependence of parasite egg excretion. *Immunology* **35**, 771–778.

DOENHOFF, M. J., PEARSON, S., DUNNE, D. W., BICKLE, Q., LUCAS, S., BAIN, J., MUSALLAM, R. & HASSOUNAH, O. (1981). Immunological control of hepatotoxicity and parasite egg excretion in *Schistosoma mansoni* infections: stage specificity of the reactivity of immune serum in T-cell deprived mice. *Transactions of the Royal Society of Tropical Medicine and Hygiene* **75**, 41–53.

DUNNE, D. W., HASSOUNAH, O., MUSALLAM, R., LUCAS, S., PEPYS, M. B., BALTZ, M. & DOENHOFF, M. (1983). Mechanisms of *Schistosoma mansoni* egg excretion: parasitological observations in immunosuppressed mice reconstituted with immune serum. *Parasite Immunology* **5**, 47–60.

FILE, S. (1995). Interaction of schistosome eggs with vascular endothelium. *Journal of Parasitology* **81**, 234–238.

GRZYCH, J. M., PEARCE, E., CHEEVER, A., CAULADA, Z. A., CASPAR, P., HEINY, S., LEWIS, F. & SHER, A. (1991). Egg deposition is the major stimulus for the production of Th2 cytokines in murine schistosomiasis mansoni. *Journal of Immunology* **146**, 1322–1327.

HIRATA, M., KAGE, M., HABE, S., AGATSUMA, T. & FUKUMA, T. (1996). *In vivo* and *in vitro* cellular response to *Schistosoma japonicum* eggs in hosts with differing susceptibilities. *Parasite Immunology* **18**, 431–438.

KING, C., XINALI, J., MALHORTRA, I., LIU, S., MAHMOUD, A. A. F. & OETTGEN, H. C. (1997). Mice with a targeted deletion of the IgE gene have increased worm burdens and reduced granulomatous inflammation following primary infection with *Schistosoma mansoni*. *Journal of Immunology* **158**, 294–300.

LEPTAK, C. L. & McKERROW, J. H. (1997). Schistosome egg granulomas and hepatic expression of TNFα are dependent on immune priming during parasite maturation. *Journal of Immunology* **158**, 301–307.

NGAIZA, J. R., DOENHOFF, M. J. & JAFFE, E. A. (1993). *Schistosoma mansoni* egg attachment cultured human umbilical vein endothelial cells: an *in vitro* model of an early step of parasite egg excretion. *Journal of Infectious Diseases* **168**, 1576–1580.

PEARCE, E. J., CHEEVER, A., LEONARD, S., COVALESKY, M., FERNANDEZ-BOTRAN, R., KOHLER, G. & KOPF, M. (1996). *Schistosomiasis mansoni* in IL-4 deficient mice. *International Immunology* **8**, 435.

RIFFKIN, M., SEOW, H.-F., JACKSON, D., BROWN, L. &

WOOD, P. (1996). Defence against the immune barrage: helminth survival strategies. *Immunology and Cell Biology* **74**, 564–574.

RITTER, D. M. & McKERROW, J. H. (1996). Intracellular adhesion molecule 1 is the major adhesion molecule expressed during schistosome granuloma formation. *Infection and Immunity* **64**, 4706–4713.

SECOR, W. E., DOS REIS, M. G., RAMOS, E. A., MATOS, E. P., REIS, E. A., DO CARMO, T. M. & HARN, D. A., JR (1994). Soluble intercellular adhesion molecules in human schistosomiasis: correlations with disease severity and decreased responsiveness to egg antigens. *Infection and Immunity* **62**, 2695–2701.

SENALDI, G., YIN, S., SHAKLEE, C. ., PIGUET, P. F., MAK, T. W. & ULICH, T. R. (1996). *Corynebacterium parvum*- and *Mycobacterium bovis* Bacillus Calmette–Guérin-induced granuloma formation is inhibited in TNF receptor 1 (TNF-R1) knockout mice and by treatment with soluble TNF-R1. *Journal of Immunology* **157**, 5022.

SHER, A., COFFMAN, R. L., HIENY, S., SCOTT, P. & CHEEVER, A. W. (1990). Interleukin 5 is required for the blood and tissue eosinophilia but not granuloma formation induced by infection with *Schistosoma mansoni*. *Proceedings of the National Academy of Sciences, USA* **87**, 61–65.

TSUJI, H., HARADA, A., MUKAIDA, N., NAKANUMA, Y., BLUETHMANN, H., KANEKO, S., YAMAKAWA, K., NAKAMURA, S. I., KOBAYASHI, K. I. & MATSUSHIMA, K. (1997). Tumor necrosis factor receptor P55 is essential for intrahepatic granuloma formation and hepatocellular apoptosis in a murine model of bacterium-induced fulminant hepatitis. *Infection and Immunity* **65**, 1892–1898.

WARREN, K. S. (1984). Water-poison disease. *World Health* December, 5–6.

A role for granulomatous inflammation in the transmission of infectious disease: schistosomiasis and tuberculosis

M. J. DOENHOFF

School of Biological Sciences, University of Wales, Bangor, Gwynedd LL57 2UW, UK

SUMMARY

The relationship between cell-mediated granulomatous inflammation and transmission of disease in schistosomiasis and tuberculosis has been explored. In 2 experiments involving *Schistosoma mansoni*-infected normal and T cell-deprived mice, and infected deprived mice that had been variously reconstituted with immune or normal lymphocytes or immune serum, there was a significant positive numerical correlation between mean liver granuloma diameters and faecal egg counts in individual animals. Lymphocytes from donors with recently patent infections were more active than cells from chronically infected or uninfected donors in reconstituting egg excretion rates in deprived recipients, and mesenteric lymph node (MLN) cells were more active than spleen cells. Modulation of granulomatous activity with increasing chronicity of infection in the donors, resulting in a decrease in granuloma size around freshly produced tissue-bound eggs, was paralleled by a waning of the capacity of transferred lymph node cells to reconstitute egg excretion in the recipients. Serum taken from chronically infected donor mice over the same period and transferred to infected deprived recipients became more active in enhancing egg excretion in the recipients as the cell-mediated activity declined. A recent study in Kenya has found that *S. mansoni*-infected patients with concurrent human immunodefficiency virus (HIV) infection excrete fewer eggs than patients exposed to the same levels of schistosome infection, but who are not HIV-infected, thus indicating that schistosome egg excretion in humans is also immune-dependent. Attention is drawn to an apparently parallel situation in human tuberculosis, another pathogen which induces a cell-mediated granulomatous immune response. Several studies have shown that patients with tuberculosis who are also HIV-seropositive tend to have fewer tubercle bacilli detectable in their saliva than those with tuberculosis, but who are HIV-negative. This discrepancy, associated with differences in lung pathology in HIV-positive patients, suggests that in tuberculosis immune cell-mediated granulomatous inflammation causes the destruction of host tissue in a manner which facilitates onward transmission of the bacterial pathogen.

Key words: Schistosomiasis, tuberculosis, *Schistosoma mansoni*, *Mycobacterium tuberculosis*, egg granuloma, granulomatous inflammation, immunopathology, disease transmission.

INTRODUCTION

Continuity of the life-cycles of disease-causing organisms requires that the respective causative agents have evolved mechanisms which allow them to be effectively transmitted to new hosts. Schistosomes, in common with many other helminths, depend on their eggs leaving the definitive host in excreted stools or urine, while tubercle bacilli are released from diseased lung tissue and expectorated in aerosol droplets. A feature common to both schistosomiasis and tuberculosis is the extensive granulomatous inflammation induced in diseased tissue, and it has previously been indicated that in schistosome-infected animals this immunopathology is important in facilitating the egg excretion process (Doenhoff, Hassounah & Lucas, 1985, 1986; Damian, 1987). An objective of this review is to draw attention to the possibility that the immunological reactivity induced by tubercle bacilli is likewise important in the release of the infectious agent from the body.

With respect to schistosomes, in contrast to many other species of helminth, the adult parasites reside inside blood vessels of the host (the hepatic portal system in the case of *Schistosoma mansoni* and *S. japonicum*, the vesicle plexus and other vessels of the urino-genital tract for *S. haematobium*), and many of the eggs produced by female worms are not in fact excreted. Instead they embolize in capillary beds of host tissues, particularly those downstream of the sites of oviposition. Once entrapped in a capillary the egg induces an inflammatory reaction or granuloma which occludes blood flow, resulting in lesions to which many of the disease symptoms of schistosomiasis are attributable (Warren, 1973).

While it is clear that the symptoms of schistosomiasis have an immunopathological aetiology (Warren, 1975), the function of the granuloma in an adaptive or evolutionary context has been less well defined. It has been suggested that the granuloma protects host tissues against toxic schistosome egg products (von Lichtenberg, 1964), and indeed, 'nude' mice (Byram & von Lichtenberg, 1977) or T-cell depleted mice (Buchanan, Fine & Colley, 1973; Byram *et al.* 1979; Doenhoff *et al.* 1981) which cannot form granulomas around schistosome eggs, have been found to suffer from hepatotoxicity putatively induced by *S. mansoni* eggs. However, reconstitution of *S. mansoni*-infected T-deprived

Parasitology (1997) **115**, S113–S125.   © 1997 Cambridge University Press

mice with immune lymphocytes, which augmented the host's capacity for granuloma formation, did not protect against egg-induced hepatocyte damage as well as passive immunization with immune serum. Furthermore, serum from infected mice acquired hepatoprotective activity early after patency when granulomatous activity was still maximal (Doenhoff *et al.* 1986; Hassounah & Doenhoff, 1993).

Experiments with *S. bovis* and *S. haematobium* also are not entirely consistent with the notion that the sole role of the granuloma is host protection, since livers of immunosuppressed mice heavily infected with these two schistosome species showed no evidence of a toxic reaction despite the absence of granulomatous inflammation (Murare *et al.* 1987; Agnew, Lucas & Doenhoff, 1988).

As well as suffering from an egg-induced hepatotoxicity reaction, *S. mansoni*-infected immunosuppressed mice are incapable of excreting parasite eggs in normal numbers, a defect which could in part be rectified by injections of serum from normal infected donor animals (Doenhoff *et al.* 1978, 1981; Dunne *et al.* 1983). In subsequent experiments involving adoptive transfer of immune cells or serum there was a positive relationship between the numerical values for the mean size of granulomas in the liver and the number of eggs being excreted in individual recipient animals (Doenhoff *et al.* 1985, 1986). While it was understood that liver-bound eggs were terminally incapacitated with respect to further transmission of the parasite, it was suggested that the immune effector mechanisms responsible for granuloma formation may also be involved in the processes of egg excretion.

From his observations on non-human primates Damian (1987) arrived at a similar conclusion concerning promotion of schistosome egg excretion by the egg granuloma. The experimental observations, first made some 20 years ago, are now supported by the finding that *S. mansoni* egg excretion rates are impaired in subjects with concomitant human immunodefficiency virus (HIV) infection in Kenya (Karanja *et al.* 1997). Insight about the molecular mediators involved in the schistosome egg excretion process has also recently been gained with the observation that tumour necrosis factor alpha (TNF-α) promotes both *S. mansoni* egg granuloma formation and egg excretion (Amiri *et al.* 1992; McKerrow, this volume).

The experimental results reported here are from studies in which the respective roles of cellular and humoral immune responses in the processes of *S. mansoni* egg excretion in mice have been further investigated. It is shown that as schistosome infections become increasingly chronic, egg-induced granuloma formation and the capacity of immune cells to mediate in the processes of egg excretion are modulated downwards in parallel, while the capacity of immune serum to promote egg excretion increases.

The activity in serum reaches a lower, but constantly maintained level of potency than that expressed by immune cells during acute infection.

In the discussion of these results below attention is drawn to the possibility that the cell-mediated immune responses generated by *Mycobacterium tuberculosis* may have a similar role to the schistosome egg granuloma; namely, to promote transmission of the infectious agent.

MATERIALS AND METHODS

Parasite and host

A Puerto Rican strain of *Schistosoma mansoni* was maintained in laboratory life-cycle in random-bred TO strain mice and albino *Biomphalaria glabrata* snails. Inbred CBA-H/*T6T6* strain mice were used for experimentation.

T cell-deprived mice were prepared as described previously by a combination of adult thymectomy and injections of rabbit anti-mouse thymocyte serum (Doenhoff *et al.* 1978, 1979). The mice were rested for approximately 30 days before use, and then infected percutaneously with *S. mansoni* cercariae (Smithers & Terry, 1965). Adult-thymectomized, anti-thymocyte serum-treated mice have approximately 10 % of the normal level of circulating T cells (Doenhoff & Leuchars, 1977). Donors of immune lymphocytes and serum were infected with 25 cercariae, and recipients of the cells or serum and the appropriate control groups were given 200 cercariae.

Cell and serum transfers

Mesenteric lymph node and spleen cells and serum were obtained from donors that had been infected between 1 and 20 weeks previously. Cell suspensions for injection into recipient mice were prepared as described (Doenhoff *et al.* 1986; Hassounah & Doenhoff, 1993). Immune serum was obtained from infected immune-intact donors between 2 and 16 weeks after infection with 25 cercariae.

Recipient T cell-deprived mice and the appropriate control groups were infected with 200 cercariae. Approximately 40 days after infection lymphocyte recipients were given intravenous injections of cells in medium: if more than one injection was given, each injection was separated by 24 h. The number of cells transferred is specified in the text relating to respective experiments. Recipients of immune serum were injected intraperitoneally once daily from day 40 after infection until the day before perfusion. Experiments were terminated approximately 47 days after infection of the recipients.

Perfusions, tissue egg counts and faecal egg counts

On the day of termination of the experiment a 40–50 mg faecal pellet was obtained from each mouse

Table 1. *Granuloma diameters and faecal egg counts in mice given serum, spleen or mesenteric lymph node cells from normal and infected donors*

A group of normal mice and the T cell-deprived mice which were to act as cell or serum recipients or unreconstituted controls were infected with 200 *S. mansoni* cercariae on day 0. Recipients were injected with 38×10^6 cells on each of days 38 and 39 after infection. Spleen cells (SpC) or mesenteric lymph node cells (LNC) were taken from uninfected mice (N) or from mice infected with 25 cercariae 8 weeks (AI), or 16 weeks (CI) previously. Recipients of serum (IS) were given daily intraperitoneal injections of serum (0·5 ml/mouse, the serum having been obtained from mice with patent 16 week-old infections) from day 40 to 46, and the experiment was terminated on day 47. Figures in parentheses to the right of values of faecal egg counts and granuloma diameters indicate position in numerical rank order with (1) the highest and (9) the lowest.

Group	No. of mice	No. of worms	Total gut eggs ($\times 10^{-3}$)	No. of eggs/ 100 mg faeces	Granuloma diameter
Normal	7	70·4 ± 17·1	59·7 ± 20·1	216·8 ± 103·7 (1)	354 ± 79 (1)
Deprived	8	91·3 ± 19·3	43·8 ± 17·4	5·0 ± 6·5 (9)	142 ± 44 (9)
Dep. + IS	8	88·6 ± 14·5	32·3 ± 13·8	16·0 ± 11·3 (6)	196 ± 50 (7)
Dep. + AI-LNC	8	84·9 ± 14·5	38·3 ± 9·1	93·9 ± 62·0 (2)	282 ± 65 (2)
Dep. + AI-SpC	8	73·5 ± 9·5	31·6 ± 8·0	36·7 ± 26·0 (3)	232 ± 73 (3)
Dep. + CI-LNC	8	75·7 ± 10·4	31·0 ± 16·1	26·3 ± 13·3 (4)	228 ± 65 (4)
Dep. + CI-SpC	8	83·3 ± 14·9	40·0 ± 13·9	12·6 ± 10·0 (8)	196 ± 57 (8)
Dep. + N-LNC	8	76·3 ± 11·9	34·9 ± 6·6	20·5 ± 25·4 (5)	220 ± 56 (5)
Dep. + N-SpC	8	90·5 ± 10·2	39·7 ± 12·7	13·8 ± 11·0 (7)	212 ± 56 (6)

and processed to display ninhydrin-stained eggs as previously described (Doenhoff *et al.* 1978). Adult worms were retrieved from infected mice by perfusion via the hepatic portal vein (Smithers & Terry, 1965) and they were counted on the same day. The intestine from the duodenum to the rectum was removed and kept at $-20\,°C$ for later digestion in 5 % KOH for tissue egg counting (Cheever, 1968).

Measurement of granuloma diameters

After perfusion of the mouse the two ventral median lobes of the liver were removed, fixed in formal saline, embedded in wax, and three $5\,\mu m$-thick sections cut at intervals of $200\,\mu m$. Following haematoxylin and eosin staining and mounting the sections were scanned and the diameters of granulomas around single separate eggs measured with an ocular micrometer as described by von Lichtenberg (1962). The mean liver granuloma diameter for each mouse was estimated after a minimum of 20 lesions had been scored, and the group mean value subsequently calculated from the individual mouse means.

Statistics

The Student *t*-test was used to determine the probability of the difference between group mean values being significant, with values of $P < 0.05$ being considered significant.

RESULTS

Results from preliminary unpublished experiments indicated that the degree to which lymphoid cells from *S. mansoni*-infected mice enhanced the rate of

egg excretion in infected immunosuppressed recipients varied with the source organ of the transferred cells (spleen or mesenteric lymph nodes) and with the age of the schistosome infections in the donors. Table 1 gives results from an experiment examining the effect of these two parameters.

The results in Table 1 are in agreement with earlier observations in that: (1) heavily infected T cell-deprived mice have liver egg granuloma diameters which are smaller than those in the normal animals, and the mean number of eggs excreted by the deprived mice is reduced by 98 % compared with only 27 % reduction in the intestinal egg count in the deprived animals; and (2) the egg excretion rate and granuloma size were increased marginally in the group of deprived mice that received immune serum injections. The results in the other groups in Table 1 indicate that with respect to enhancement of egg excretion rate and liver egg granuloma size: (i) cells from infected donors were more effective than cells from non-infected normal donors; (ii) transferred mesenteric lymph node cells were more effective than spleen cells; and (iii) cells from donors infected for 8 weeks (AI) were more effective than cells from donors infected for 16 weeks (CI).

The group mean results for faecal egg count and liver egg granuloma diameter in Table 1 showed, with one exception, the same numerical rank order, and regression analysis of the results from individual animals gave a significant positive linear correlation between the two numerical values ($P < 0.001$, result not illustrated).

The effect of chronicity of infection on the ability of donor mouse lymphoid cells to facilitate egg excretion was investigated in more detail. Mesenteric lymph node cells were taken from groups of donor

Table 2. Experiment 1. Granuloma diameters and faecal egg counts in mice reconstituted with lymph node cells from groups of donors infected for different times. Experimental design as in Table 1. Mesenteric lymph node cells were collected from groups of mice infected with 25 cercariae 1, 4, 6, 8, 10, 15 or 20 weeks previously. Each recipient received one injection of 44×10^6 lymphocytes on day 39 after infection with 200 cercariae. Perfusion etc. was on day 47. Experiment 2. Egg excretion rates in deprived mice reconstituted with immune serum. Groups of deprived mice infected with 200 cercariae were given serum from donor mice that had been infected with 25 cercariae between 2 and 16 weeks previously. Each recipient was injected with 0·5 ml serum/day between days 40 and 45 after infection, and the mice were perfused on day 46.

Exp.	Duration of infection in cell/serum donors	No. of mice	No. of worms	Total gut eggs ($\times 10^{-3}$)	No. of eggs/100 mg faeces	Granuloma diameter in recipients (μm)	Granuloma diameter in donors (μm)
1	—	7	112·7 ± 11·0	55·0 ± 15·9	7·5 ± 6·9	160 ± 69	—
	1	6	109·0 ± 22·7	43·0 ± 12·9	37·9 ± 38·2	208 ± 77	—
	4	7	111·8 ± 13·4	50·9 ± 9·2	81·8 ± 56·6	196 ± 90	—
	6	7	117·0 ± 16·2	58·2 ± 9·0	115·9 ± 101·3	192 ± 95	—
	8	6	116·2 ± 15·9	54·1 ± 7·0	291·0 ± 198·6	218 ± 94	305 ± 79
	10	7	87·4 ± 26·1	45·6 ± 8·4	97·6 ± 61·4	211 ± 130	357 ± 36
	15	6	103·2 ± 17·7	46·7 ± 13·2	79·3 ± 105·6	192 ± 81	318 ± 32
	20	7	95·0 ± 29·7	41·7 ± 5·3	37·9 ± 28·7	180 ± 86	258 ± 31
2	—	6	56·5 ± 11·8	31·5 ± 9·8	8·2 ± 8·5	ND*	
	2	6	52·5 ± 9·6	24·1 ± 10·9	4·4 ± 5·3	ND	
	4	6	55·5 ± 10·1	36·2 ± 9·3	13·2 ± 10·5	ND	
	6	6	52·8 ± 16·4	31·5 ± 13·4	10·7 ± 13·2	ND	
	8	6	65·3 ± 13·7	37·7 ± 6·5	47·4 ± 23·1	ND	
	10	6	55·8 ± 6·2	34·1 ± 5·6	92·9 ± 50·9	ND	
	13	6	61·8 ± 11·1	37·5 ± 10·3	74·6 ± 60·0	ND	
	16	6	44·0 ± 10·0	25·9 ± 5·5	64·9 ± 34·8	ND	

*ND = not determined.

mice that had been infected with *S. mansoni* for periods of between 1 and 20 weeks. The cells were transferred to a series of 7 groups of infected deprived mice, with inclusion of an eighth infected, but unreconstituted deprived group.

The results (Table 2, Exp. 1) indicate that the capacity of a given number of cells maximally to enhance egg excretion in the recipients coincided temporally with their ability to generate inflammation around eggs in the recipients' livers, and that the peak response was obtained from 8-week-infected donors. Maximum liver granuloma size in the donor animals was, however, found in mice with 10-week-old infections (last column, Table 2, Exp. 1). To facilitate visualization of these results the group mean values in Table 2 have been plotted in Fig. 1a (liver granuloma diameters in cell donors), 1b (liver granuloma diameters in recipients) and 1c (egg excretion rates in recipients).

As in the preceding experiment, there was a significant positive correlation between the pooled individual mouse values for faecal egg count and mean liver egg granuloma diameter, but at a lower probability level ($P < 0.025$, result not illustrated).

To determine at what time after infection serum acquired the capacity to promote egg excretion, blood from a group of normal donor mice was taken at intervals between 2 and 16 weeks after infection

with 25 cercariae. The results of transferring 7 pools of serum are given in Table 2 (Exp. 2) and the mean egg excretion rates in the recipient groups have been included in Fig. 1c. The cell-dependent effector mechanisms which facilitate egg excretion thus show maximum activity at an earlier time than humoral effector mechanisms, and the serological activity plateaus out at a lower, though subsequently constantly maintained level than the cellular activity.

DISCUSSION

Induction and regulation of schistosome egg-associated immunopathology

Granulomas induced by *S. mansoni* eggs in mice have many of the characteristics of a thymus- or T-dependent cell-mediated delayed-type hypersensitivity reaction (Warren, Domingo & Cowan, 1964). Earlier evidence suggested that naive animals could be sensitized for granuloma formation only with egg antigens (Warren & Domingo, 1970), but recent reports indicate that enhanced responses to injected eggs occur also in mice with unisexual worm infections (Cheever, Lewis & Wynn, 1997; Leptak & McKerrow, 1997). Granulomas induced by eggs of *S. haematobium* and the related species *S. bovis* share with *S. mansoni* granulomas characteristic of T

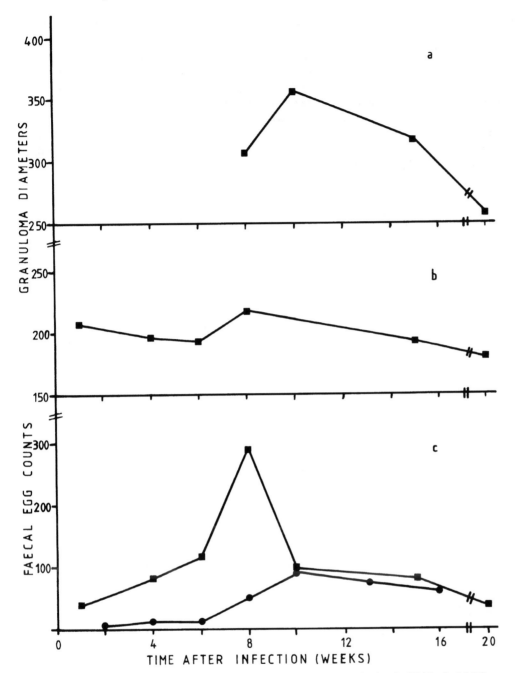

Fig. 1. Liver egg granuloma diameters and egg excretion rates of mice in Table 2. (*a*) Mean granuloma diameters of cell donor mice (last column, Table 2). (*b*) Granuloma diameters of cell recipient mice (6th column, Table 2). (*c*) Group mean egg excretion rates of mice reconstituted with cells (■: Table 2, Exp. 1) or serum, (●: Table 2, Exp. 2).

cell dependence (Murare *et al.* 1987; Agnew *et al.* 1988, 1989), but considerably less work has been done on the former species. Because the aetiology of *S. japonicum* egg granulomas is still not very well defined, this species will be ignored in this review. Space has also precluded exhaustive citation of the literature on *S. mansoni* immunopathology, but see recent reviews by Boros (1994), Lukacs *et al.* (1994), Wynn & Cheever (1995) and Pearce *et al.* (1996*a*.)

Schistosome egg granulomas are dynamic, both in terms of their growth and decay around individual eggs (von Lichtenberg, 1962), and because the

immune effector mechanisms mediating granuloma formation undergo modulation as the infection progresses (see below). In contrast to the protoypical granulomatous inflammatory lesion induced by tubercle bacilli, which consist almost entirely of macrophage-like cells (see below), many different cell types are found in schistosome egg granulomas, including eosinophils, macrophages, mast cells, neutrophils, T and B lymphocytes, plasma cells and fibroblasts (Stenger, Warren & Johnson, 1967; von Lichtenberg, Erickson & Sadun, 1973; Moore, Grove & Warren, 1977; Smith, 1977; Epstein *et al.*

1979). The proportions of each cell type present in an egg granuloma depend on the tissue in which the granuloma is located (Weinstock & Boros, 1983), and the time after infection (Chensue & Boros, 1979).

Leptak & McKerrow (1997) have shown that tumour necrosis factor-α (TNF-α), the production of which can be induced in liver cells by infecting with worms, is involved in priming the host for egg granuloma development. Granuloma formation is also dependent on CD4$^+$ T helper cells, and there is a consensus that the activities of Th2 cells, and the cytokines they produce (especially IL-4), are involved in granuloma formation during the early stages of infection (Lukacs & Boros, 1993; Wynn & Cheever, 1995; Jankovic & Sher, 1996). However, the results from IL-4 gene-knockout mice, which give diminished Th2 responses and enhanced Th1 responses, are inconsistent, with the intensity of hepatic pathology having been found to be either unaffected (Pearce et al. 1996b) or reduced (Metwali et al. 1996). Furthermore, SEA-specific Th0 and Th1 clones, as well as Th2 clones, have granuloma formation capability (Chikunguwo et al. 1991; Zhu, Lukacs & Boros, 1994). Th1 cell activity may thus not be completely inconsequential in granuloma formation during the early stages of patent infection, but its IgE-dependence (King et al. 1997) and the presence in these complex lesions of numerous eosinophils that are IL-5 dependent (Sher et al. 1990) argue strongly for the activity being dominated by Th2 cells. Hepatic fibrosis also appears to be particularly dependent on Th2 cell activity (Czaja et al. 1989; Cheever et al. 1994; Wynn et al. 1995a), and is perhaps a consequence of a failure to down-regulate hepatic TNF-α production (Adewusi et al. 1996) and the lack of anti-egg antibody of appropriate idiotype (Montesano et al. 1997).

After 10 weeks of infection in mice mean liver granuloma size begins to diminish (Andrade & Warren, 1964; Cheever, 1965), a process termed 'endogenous desensitization' (Domingo & Warren, 1968) or 'spontaneous modulation' (Boros, Pelley & Warren, 1975). Modulated lesions have a reduced ratio of CD4$^+$ helper T cells (Ragheb & Boros, 1989) and a higher percentage of B cells (Chensue & Boros, 1979), as well as an intrinsic capacity to synthesize immunoglobulins and specific antibody (Boros, Amsden & Hood, 1982). Infected mice with modulated granulomas show decreased activity in tests of cell-mediated immunity such as antigen-stimulated lymphocyte blastogenesis, delayed hypersensitivity responses to egg antigens and MIF production, and increased evidence of humoral immunity (Boros, Warren & Pelley, 1973; Boros et al. 1975; Colley, 1975).

Cytokines associated with Th1 cell differentiation and activity, particularly IL-12 and interferon-γ, suppress Th2 cell-mediated granuloma formation (Chensue et al. 1992, 1994; Wynn et al. 1994, 1995b), implicating Th1 cells in down modulation of granulomatous activity. CD8$^+$ T cells have also been implicated (Chensue & Boros, 1979; Chensue et al. 1993), and although S. mansoni egg granuloma formation cannot be suppressed by antibody with specificity for egg antigens (Colley, 1976), anti-idiotypic antibodies appear to exert a regulatory influence in infections of both humans (Montesano et al. 1990) and mice (Bosshardt, Nix & Colley, 1996).

The biological role of schistosome egg-induced immunopathology: host protection or parasite survival?

The pathological consequences of schistosome egg-induced inflammation have been well documented (Warren, 1973), and as the above brief review demonstrates, considerable progress has been made in defining the cellular and molecular factors which induce and regulate granuloma formation. Less consideration has been given to defining the adaptive advantages of granuloma formation for the host and/or the parasite?

The intuitive assumption is that the egg granuloma is an immune response which, like most others, is induced to defend the host (von Lichtenberg, 1964) and that the pathological sequelae are but a consequence of immunological over-reactivity. There is, however, no consistency between schistosome species with regard to the apparent relative toxicity of their eggs. Thus, while evidence of hepatotoxic potential has been found for S. mansoni eggs in immunosuppressed mice (Buchanan et al. 1973; Byram & von Lichtenberg, 1977; Byram et al. 1979; Lucas et al. 1980), a factor which may in turn contribute to the infected immunosuppressed animals dying earlier than comparably infected, immunologically intact controls (Doenhoff et al. 1979; Lucas et al. 1980), the same does not hold for S. bovis or S. haematobium eggs. Granuloma formation around eggs of the latter two species is as T cell-dependent as for S. mansoni eggs, but no evidence of toxicity was seen around S. bovis and S. haematobium eggs in livers of immunosuppressed mice, even when the eggs were present in relatively large numbers (Murare et al. 1987; Agnew et al. 1988). T cell-deprived mice infected with S. bovis in fact survived longer than comparably infected intact controls (Murare et al. 1987), a result which is commensurate with the assertion first made by Ken Warren, that schistosomiasis is an 'immunologic disease' (Warren, 1975). Furthermore, S. mansoni egg-induced hepatotoxicity is more effectively neutralized by transfer of immune serum from homologously-infected intact donors than by transfer of immune lymphocyte populations, even though the serum acts without concomitant restoration of granuloma formation (Byram et al. 1979; Doenhoff *et al.* 1981; Hassounah & Doenhoff, 1993). The hepatoprotective activity in infected donor serum has been attributed to antibodies which react with ω-1, a 31 kDa *S. mansoni* egg glycoprotein (Dunne *et al.* 1981; Dunne, Jones & Doenhoff, 1991) that is stage- and species-specific (Dunne *et al.* 1984; Murare *et al.* 1992). The activity is transferable with serum generated as early as 8 weeks after infection of the serum donors (Doenhoff *et al.* 1981) i.e. at the time when, in terms of lesion size, granulomatous activity around eggs is still at a maximum level in the mouse experimental system.

In contrast to the discrepancy between schistosome species with respect to the hepatotoxic potential of their eggs, rates of egg excretion of both *S. mansoni* (Doenhoff *et al.* 1978) and *S. bovis* (Murare & Doenhoff, 1987) are severely impaired in immunosuppressed mice. *S. mansoni* egg excretion rates can be partly restored in infected T cell-deprived mice by transfer of serum or lymphocyte populations from infected normal donor mice (Doenhoff *et al.* 1978, 1985, 1986). In these earlier experiments, and in the experiments reported above, there was a significant correlation between the respective numerical values for the number of schistosome eggs found in the faeces and the mean liver granuloma diameter in individual animals.

The results in Table 2 and Figs. 1*b* and 1*c* confirm the efficacy of immune cells in mediating *S. mansoni* egg excretion in homologously infected immunosuppressed recipients, with the group of mice that received cells from 8-week-infected donors showing both the highest rate of egg excretion and greatest mean granuloma size in the liver. However, maximum granuloma size occurred at a later time in the donor mice than in the recipients. This is probably due to the lymphocytes which mediate granuloma formation having become sensitized in lymphoid organs in the infected donor mice some time before they were removed and induced to participate in formation of the lesions around liver-bound eggs in the recipients.

Granuloma size in recipients given cells from donors with 1-week-old infections was enhanced when compared with unreconstituted controls, but this was without concomitant induction of an ability to enhance egg excretion rates. This may be due to the recently incoming larvae bearing some antigens that are cross-reactive with egg antigens (Dunne & Bickle, 1987), but which sensitize for inflammatory activity around eggs that is irrelevant to egg excretion.

The results in Table 1 indicate that transferred syngeneic mesenteric lymph node cells are more effective than an equivalent number of spleen cells in mediating schistosome egg excretion. This may not be due solely to the presence of a proportionately greater number of T cells in lymph nodes than in spleen, as in a separate unpublished experiment groups of infected deprived mice given twice as many spleen as lymph node cells still excreted fewer eggs. The discrepancy between spleen and lymph node cells is noteworthy since the spleen has been used almost exclusively as the source of lymphoid cells in investigations into the immune aetiology of schistosome egg granulomas. Cells involved in the egg excretion process and which mediate in granulomatous activity may, however, be preferentially sensitized and/or retained in the mesenteric lymph nodes of infected mice.

Cell-mediated immunological activity induced by schistosome eggs appears therefore to have a pivotal role in the process of schistosome egg excretion, at least during early infection patency. Extensive histological observations on *S. mansoni*-infected baboons likewise led Damian (1987) to propose that 'the granuloma is the agent of translocation (of a schistosome egg) from the site of oviposition...to the intestinal lumen'. Evidently, eggs in the liver and the granulomas formed around them play no direct role in parasite transmission. The liver is, however, the site where schistosome eggs are first found in substantial numbers after infection, and antigens from these early eggs are likely to initiate sensitization of the host for granuloma formation.

There is still much to elucidate about the egg excretion process at the cell and molecular levels. The identity and properties of the antigens which sensitize for granuloma formation and egg excretion have to be determined, and the manner in which the sensitized lymphocytes and the subsequently generated specific antibodies mediate egg excretion also has to be worked out. Are the sensitizing antigens the same for both the cell- and antibody-mediated events? It also remains to be established whether the post-8-week down-turn in cell-mediated activity noted in Fig. 1*c*, and the concomitantly increased serum-borne activity, have their counterparts in human infection. While the time-spans of this infection in mice and humans bear no comparison, faecal eggs counts do routinely reduce from their maxima as patients age in endemic areas.

Host factors other than those responsible for adaptive immunity are also involved in schistosome egg excretion. Thus, schistosome eggs are potent aggregators of blood platelets, and excretion rates are impaired in platelet-depleted mice (Ngaiza & Doenhoff, 1990). Platelets may aggregate around eggs as soon as they emerge from the female into the bloodstream, but it is not known whether they just serve to anchor the eggs to the endothelial surface against blood flow, or actively help the eggs penetrate through. Schistosome eggs interact actively with vascular endothelial cells (Freedman & Ottesen, 1988; Ngaiza, Doenhoff & Jaffe, 1993; File, 1995), but again we do not yet know much about the nature of this interaction, nor how the integrity of the endothelium is breached. And how does the granuloma which forms around the egg as it traverses the intestine wall act a 'translocation' device (Damian, 1987)?

Justification for a continuation of the experimental research on immune-dependent schistosome egg excretion is provided by the novel results of Karanja *et al.* (1997) who showed that *S. mansoni*-infected people in Kenya that were concurrently sero-positive for HIV had significantly (approximately 3-fold) lower *S. mansoni* faecal egg counts than HIV sero-negative people exposed to similar levels of schistosome infection. Control for variation in schistosome infection intensity levels between the 2 patient groups was provided by analysis of concentrations of a parasite worm-derived circulating antigen (Barsoum *et al.* 1991). The pathogenic effects of HIV infection are complex and heterogeneous, and perhaps more difficult to control for. However, the HIV positive group had a significantly lower mean CD4[+] blood cell count and a significant positive correlation between faecal egg counts and the percent of CD4[+] cells in peripheral blood was found in this group.

A role for granulomatous inflammation in the transmission of tuberculosis

Schistosome egg granulomas were initially called 'pseudotubercles' in acknowledgement of their similarity with the lesions which occur in host tissue infested with *Mycobacterium tuberculosis* (MTB). In contrast to the mixed cell nature of schistosome egg granulomas, however, MTB granulomas are comprised of macrophages or macrophage-derived epithelioid cells alone.

Lung tissue is the commonest site for mycobacterial growth and pathogenesis in tuberculosis patients. Although growth of MTB and ensuing pathology can occur in tissues other than the lungs, extrapulmonary lesions do not contribute to onward transmission of infection (similarly to those schistosome eggs which become trapped in tissues other than those from which they can be excreted). Transmission of the infection depends on bacteria being released from lung tissue into the airways and subsequent expectoration from the body in aerosol droplets, particularly during bouts of coughing. Transmission from index cases to secondary contacts occurs mainly within households or between individuals living at close quarters (e.g. in hostels or hospital wards) (Murray, 1990), and it is estimated that a minimum density of approximately 5000 bacterial cells per mL of saliva is necessary for transmission to occur (Grange, 1996). In one study it was estimated that children living in a household with a tuberculosis sputum smear-positive index case had approximately 8 times more chance of becoming infected than children in a household with a smear-negative source case (Geuns, van Meijer & Styblo, 1975).

Four to five distinguishable stages have been identified in the disease process of tuberculosis in a rabbit model system, namely: onset, symbiosis, initial caseous necrosis, liquefaction and cavity formation (Dannenberg, 1991; Dannenberg & Rook, 1994). In the first stage of infection the inhaled bacterium is ingested by a macrophage and the process may end there if, as is often the case, the micro-organism is destroyed. The 'symbiotic' second stage occurs if the bacterium survives and replicates, and a lesion consisting of pathogen-loaded macrophages develops. The third stage begins when logarithmic growth of the microbe is halted, assumedly as a result of control by induced cell-mediated immune (CMI) and delayed-type hypersensitivity (DTH) responses. Lesions may regress with destruction of bacteria or become necrotic, depending on the infection resistance/susceptibility status of the host. If the disease progresses to the final stages the lesions liquefy and host lung tissue cavitates, and these extremes may occur even if host CMI responses are well developed. In tuberculosis, as in schistosomiasis, the complicated interplay between the pathogen and cells of the host immune system, mediated to a large extent by host cell-derived cytokines (Dannenberg, 1994; Rook & Bloom, 1994) is still being intensively investigated.

Exit of mycobacteria from the body occurs particularly when the disease has progressed to give liquefaction and cavitation of lung tissue. Despite host immunity, the liquefied lung tissue occurring in the final stages of the disease process appears to provide an especially good environment for growth of the bacteria (Dannenberg, 1991) and following tissue necrosis and rupture (cavitation) the micro-organisms are discharged into the airways.

Two counterintuitive aspects of tuberculosis pathogenesis are noteworthy in the present context: (1) the frequency of liquefied caseous foci and cavitation tends to be less in immunosuppressed individuals (Dannenberg, 1991) and in those infected with HIV (Pitchenik & Rubinson, 1985; Pozniak *et al.* 1995; Haramati, JennyAvital & Alterman, 1997) than in tuberculous subjects with otherwise uncompromised immune responsiveness; and (2) rabbits selectively bred for enhanced resistance against infection with MTB suffered more severely from tissue liquefaction and cavitation than rabbits selected for increased susceptibility (Lurie & Dannenberg, 1965). Furthermore, delayed-type hypersensitivity to MTB antigens has been shown to be involved in the liquefaction reaction (Yamamura *et al.* 1974).

In view of the above two observations it is perhaps less surprising that several studies have shown that patients with tuberculosis and who were sero-positive for HIV infection were more likely to have sputum that was smear- or culture-negative for MTB than those with tuberculosis, but who were

HIV sero-negative (Klein *et al.* 1989; Elliott *et al.* 1990, 1993 *a*, *c*; Long *et al.* 1991 *a*; Pozniak *et al.* 1995). The concentration of bacteria in sputum of HIV +ve patients is also lower than in that of HIV −ves (Elliott *et al.* 1990, 1993 *a*). In some other studies concurrent HIV infection had either no or only a marginal effect on MTB sputum smear positivity (Githui *et al.* 1992; Long *et al.* 1991 *b*; Houston *et al.* 1994; Smith *et al.* 1994), but in these instances the HIV +ve and HIV −ve groups had been included in the study only if there was previous clinical evidence, including sputum smear- and culture-positivity, which indicated that they had tuberculosis.

Although the heterogeneity of HIV-induced pathology and CD4$^+$ cell counts were generally not controlled for in these studies, there seems to be little published work in which the prevalence of tuberculosis-positive sputum smears was found to be markedly higher in HIV +ve than in HIV −ve subjects, nor any evidence that the rate of growth of mycobacteria is inhibited in the lung tissue of the former patients to account for these results. Indeed, the increased mortality rates suffered by late-stage HIV patients with tuberculosis is assumed to be due to the opportunistic bacterial infection (de Cock *et al.* 1992).

A practically important consequence of the failure of mycobacteria to enter the airways as a result of immunosuppression is the delay in diagnosis of tuberculosis in HIV-infected patients which has been noted (Kramer *et al.* 1990), and attributed in part to a reduced sensitivity of sputum-smear examinations in these subjects.

At least 2 studies have shown that tuberculosis patients with AIDS or HIV infection were less infectious (with respect to transmission of the bacterial infection) than HIV −ve tuberculosis patients (Elliott *et al.* 1993 *b*; Cauthen *et al.* 1996). In a third study investigating secondary transmission (Klausner *et al.* 1993) there was no evidence for such a difference, but in this instance again only sputum smear-positive individuals had been selected as index patients.

That cell-mediated immune responses play a vital role in controlling the growth of mycobacteria (Dannenberg, 1994) is not being questioned here. The studies summarized above do, however, suggest that, as with schistosomiasis, the tubercle bacillus gains advantage from host-defensive cell-mediated granulomatous inflammation in terms of facilitated transmission to new hosts.

ACKNOWLEDGEMENTS

Assistance from Dr O. A. Hassounah and Ms J. Bain and financial support from the Wellcome Trust, London, are gratefully acknowledged with respect to completion of the experimental schistosome work described in the results section above. Critical evaluations of the manuscript by Dr S. Gillespie (Royal Free Hospital, London) and Dr S. Lucas (UMDS, London) helped considerably to improve it. This paper is dedicated to the memory of the late Dr Kenneth S. Warren in appreciation of his considerable positive influence on the course of schistosomiasis research, first as an experimentalist and subsequently as administrator and scientometrist.

REFERENCES

ADEWUSI, O. I., NIX, N. A., LU, X., COLLEY, D. G. & SECOR, E. (1996). *Schistosoma mansoni*: relationship of tumor necrosis factor-α to morbidity and collagen deposition in chronic experimental infection. *Experimental Parasitology* **84**, 115–123.

AGNEW, A. M., LUCAS, S. B. & DOENHOFF, M. J. (1988). The host–parasite relationship of *Schistosoma haematobium* in the mouse. *Parasitology* **97**, 403–424.

AGNEW, A. M., MURARE, H. M., LUCAS, S. B. & DOENHOFF, M. J. (1989). *Schistosoma bovis* as an immunological analogue of *Schistosoma haematobium*. *Parasite Immunology* **11**, 329–340.

AMIRI, P., LOCKSLEY, R. M., PARSLOW, T. G., SADICK, M., RITTER, D. & McKERROW, J. H. (1992). Tumour necrosis factor α restores granulomas and induces parasite egg laying in schistosome-infected *scid* mice. *Nature* **356**, 604–607.

ANDRADE, Z. A. & WARREN, K. S. (1964). Mild prolonged schistosomiasis in mice: alterations in host response with time and the development of portal fibrosis. *Transactions of the Royal Society of Tropical Medicine and Hygiene* **58**, 53–58.

BARSOUM, I. S., KAMAL, K. A., BASSILLY, S., DEELDER, A. M. & COLLEY, D. G. (1991). Diagnosis of human schistosomiasis by detection of circulating cathodic antigen with a monoclonal antibody. *Journal of Infectious Diseases* **164**, 1010–1013.

BOROS, D. L. (1994). The role of cytokines in the formation of the schistosome egg granuloma. *Immunobiology* **191**, 441–450.

BOROS, D. L., AMSDEN, A. F. & HOOD, A. T. (1982). Modulation of granulomatous hypersensitivity. IV. Immunoglobulin and antibody production by vigorous and immunomodulated liver granulomas of *Schistosoma mansoni*-infected mice. *Journal of Immunology* **128**, 1050–1053.

BOROS, D. L., PELLEY, R. P. & WARREN, K. S. (1975). Spontaneous modulation of granulomatous hypersensitivity in schistosomiasis mansoni. *Journal of Immunology* **114**, 1437–1441.

BOROS, D. L., WARREN, K. S. & PELLEY, R. P. (1973). The secretion of migration inhibitory factor by intact schistosome egg granulomas maintained *in vitro*. *Nature* **246**, 224–226.

BOSSHARDT, S. C., NIX, N. A. & COLLEY, D. G. (1996). Reactive idiotypes expressed on antibodies to soluble egg antigens during *Schistosoma mansoni* infection of mice. *European Journal of Immunology* **26**, 272–275.

BUCHANAN, R. D., FINE, D. P. & COLLEY, D. G. (1973). *Schistosoma mansoni* infection in mice depleted of thymus-dependent lymphocytes. II. Pathology and altered pathogenesis. *American Journal of Pathology* **71**, 207–218.

BYRAM, J. E., DOENHOFF, M. J., MUSALLAM, R., BRINK, L. H. & VON LICHTENBERG, F. (1979). *Schistosoma mansoni* infections in T-cell deprived mice, and the ameliorating effect of administering chronic infection serum. II. Pathology. *American Journal of Tropical Medicine and Hygiene* **28**, 274–285.

BYRAM, J. E. & VON LICHTENBERG, F. (1977). Altered schistosome granuloma formation in nude mice. *American Journal of Tropical Medicine and Hygiene* **26**, 944–956.

CAUTHEN, G. M., DOOLEY, S. W., ONORATO, I. M., IHLE, W. W., BURR, J. M., BIGLER, W. J., WITTE, J. & CASTRO, K. G. (1996). Transmission of *Mycobacterium tuberculosis* from tuberculosis patients with HIV-infection or AIDS. *American Journal of Epidemiology* **144**, 69–77.

CHEEVER, A. W. (1965). A comparative study of *Schistosoma mansoni* infections in mice, gerbils, multimammate rats and hamsters. I. The relationship of portal hypertension to size of hepatic granulomas. *American Journal of Tropical Medicine and Hygiene* **14**, 211–226.

CHEEVER, A. W. (1968). Conditions affecting the accuracy of potassium hydroxide digestion techniques for counting *Schistosoma mansoni* eggs in tissues. *Bulletin of the World Health Organization* **39**, 328–331.

CHEEVER, A. W., LEWIS, F. A. & WYNN, T. A. (1997). *Schistosoma mansoni*: unisexual infections sensitize mice for granuloma formation around eggs. *Parasitology Research* **83**, 57–59.

CHEEVER, A. W., WILLIAMS, M. E., WYNN, T. A., FINKELMAN, F. D., SEDER, R. A., COX, T. M., HIENY, S., CASPAR, P. & SHER, A. (1994). Anti-interleukin-4 treatment of *Schistosoma mansoni*-infected mice inhibits development of T-cells and non-B, non-T cells expressing Th2 cytokines while decreasing egg-induced hepatic fibrosis. *Journal of Immunology* **153**, 753–759.

CHENSUE, S. W. & BOROS, D. L. (1979). Population dynamics of T and B lymphocytes in the lymphoid organs, circulation, and granulomas of mice infected with *Schistosoma mansoni*. *American Journal of Tropical Medicine and Hygiene* **28**, 291–299.

CHENSUE, S. W., TEREBUH, P. D., WARMINGTON, K. S., HERSHEY, S. D., EVANOFF, H. L., KUNKEL, S. L. & HIGASHI, G. I. (1992). Role of interleukin-4 and gamma interferon in *Schistosoma mansoni* egg-induced hypersensitivity granuloma formation. Orchestration, relative contribution and relationship to granuloma function. *Journal of Immunology* **148**, 900–906.

CHENSUE, S. W., WARMINGTON, K. S., HERSHEY, S. D., TEREBUH, P. D., OTHMAN, M. & KUNKEL, S. L. (1993). Evolving T cell responses in murine schistosomiasis: Th2 cells mediate secondary granulomatous hypersensitivity and are regulated by CD8+ T cells *in vivo*. *Journal of Immunology* **151**, 1391–1400.

CHENSUE, S. W., WARMINGTON, K. S., RUTH, J., LINCOLN, P. M. & KUNKEL, S. L. (1994). Cross-regulatory role of interferon-gamma (IFN-γ), IL-4 and IL-10 in schistosome egg granuloma formation: *in vivo* regulation of Th activity and inflammation. *Clinical and Experimental Immunology* **98**, 395–400.

CHIKUNGUWO, S. M., KANAZAWA, T., DAYAL, Y. & STADECKER, M. J. (1991). The cell-mediated response to schistosomal antigens at the clonal level: *in vivo* function of cloned murine egg antigen-specific CD4+ T helper type 1 lymphocytes. *Journal of Immunology* **147**, 3921–3925.

COLLEY, D. G. (1975). Immune responses to a soluble schistosomal egg antigen preparation during chronic primary infection with *Schistosoma mansoni*. *Journal of Immunology* **115**, 150–156.

COLLEY, D. G. (1976). Adoptive suppression of granuloma formation. *Journal of Experimental Medicine* **143**, 696–700.

CZAJA, M. J., WEINER, F. R., TAKAHASHI, S., GIAMBRONE, M.-A., VAN DER MEIDE, P. H., SCHELLEKENS, H., BIEMPKA, L. & ZERN, M. A. (1989). γ-Interferon treatment inhibits collagen deposition in murine schistosomiasis. *Hepatology* **10**, 795–800.

DAMIAN, R. T. (1987). The exploitation of host immune responses by parasites. *Journal of Parasitology* **73**, 3–13.

DANNENBERG, A. M. (1991). Delayed-type hypersensitivity and cell-mediated immunity in the pathogenesis of tuberculosis. *Immunology Today* **12**, 228–233.

DANNENBERG, A. M. (1994). Roles of cytotoxic delayed-type hypersensitivity and macrophage-activating cell-mediated immunity in the pathogenesis of tuberculosis. *Immunobiology* **191**, 461–473.

DANNENBERG, A. M. & ROOK, G. A. W. (1994). Pathogenesis of pulmonary tuberculosis: an interplay of tissue-damaging and macrophage-activating immune response – dual mechanisms that control bacillary multiplication. In *Tuberculosis: Pathogenesis, Protection and Control* (ed. Bloom, B. R.) American Society for Microbiology, Washington DC, USA.

DE COCK, K. M., SORO, B., COULIBALY, I. M. & LUCAS, S. B. (1992). Tuberculosis and HIV infection in sub-Saharan Africa. *Journal of the American Medical Association* **268**, 1581–1587.

DOENHOFF, M. J., HASSOUNAH, O. A. & LUCAS, S. B. (1985). Does the immunopathology induced by schistosome eggs potentiate parasite survival? *Immunology Today* **6**, 203–206.

DOENHOFF, M. J., HASSOUNAH, O., MURARE, H., BAIN, J. & LUCAS, S. (1986). The schistosome egg granuloma: immunopathology in the cause of host protection or parasite survival? *Transactions of the Royal Society of Tropical Medicine and Hygiene* **80**, 503–514.

DOENHOFF, M. J. & LEUCHARS, E. (1977). Effects of irradiation, anti-thymocyte serum and corticosteroids on PHA and LPS responsive cells of the mouse. *International Archives of Allergy and Applied Immunology* **53**, 505–514.

DOENHOFF, M., MUSALLAM, R., BAIN, J. & McGREGOR, A. (1978). Studies on the host–parasite relationship in *Schistosoma mansoni*-infected mice: the immunological dependence of parasite egg excretion. *Immunology* **35**, 771–778.

DOENHOFF, M., MUSALLAM, R., BAIN, J. & McGREGOR, A. (1979). *Schistosoma mansoni* infections in T-cell deprived mice, and the ameliorating effect of administering chronic infection serum. I. Pathogenesis. *American Journal of Tropical Medicine and Hygiene* **28**, 260–273.

DOENHOFF, M. J., PEARSON, D., DUNNE, D. W., BICKLE, Q., LUCAS, S., BAIN, J., MUSALLAM, R. & HASSOUNAH, O. (1981). Immunological control of hepatotoxicity and parasite egg excretion in *Schistosoma mansoni* infections: stage specificity of the reactivity of immune serum in T-cell deprived mice. *Transactions of the Royal Society of Tropical Medicine and Hygiene* **75**, 41–53.

DOMINGO, E. O. & WARREN, K. S. (1968). Endogenous desensitization: changing host granulomatous response to schistosome eggs at different stages of infection. *American Journal of Pathology* **52**, 369–377.

DUNNE, D. W., BAIN, J., LILLYWHITE J. & DOENHOFF, M. J. (1984). The stage-, strain- and species-specificity of a *Schistosoma mansoni* egg antigen fraction with serodiagnostic potential. *Transactions of the Royal Society of Tropical Medicine and Hygiene* **78**, 460–470.

DUNNE, D. W. & BICKLE, Q. D. (1987). Identification and characterization of a polysaccharide-containing antigen from *Schistosoma mansoni* eggs which cross-reacts with the surface of schistosomula. *Parasitology* **94**, 255–268.

DUNNE, D. W., HASSOUNAH, O., MUSALLAM, R., LUCAS, S., PEPYS, M. B., BALTZ, M. & DOENHOFF, M. J. (1983). Mechanisms of *Schistosoma mansoni* egg excretion: parasitological observations in immunosuppressed mice reconstituted with immune serum. *Parasite Immunology* **5**, 47–60.

DUNNE, D. W., JONES, F. M. & DOENHOFF, M. J. (1991). The purification, characterization, serological activity and hepatotoxic properties of two cationic glycoproteins (α-1 and ω-1) from *Schistosoma mansoni* eggs. *Parasitology* **103**, 225–236.

DUNNE, D. W., LUCAS, S., BICKLE, Q., PEARSON, S., MADGWICK, L., BAIN, J. & DOENHOFF, M. J. (1981). Identification and partial purification of an antigen (ω-1) from *Schistosoma mansoni* eggs which is putatively hepatotoxic in T cell-deprived mice. *Transactions of the Royal Society of Tropical Medicine and Hygiene* **75**, 54–71.

ELLIOTT, A. M., HAYES, R. J., HALWINDI, B., LUO, N., TEMBO, G., POBEE, J. O. M., NUNN, P & McADAM, K. P. W. J. (1993*b*). The impact of HIV on infectiousness of pulmonary tuberculosis – a community study in Zambia. *AIDS* **7**, 981–987.

ELLIOTT, A. M., HALWINDI, B., HAYES, R. J., LUO, N., TEMBO, G., MACHIELS, L., BEM, C., STEENBERG, G., POBEE, J. O. M., NUNN, P. P. & McADAM, K. P. W. J. (1993*c*). The impact of human immunodefficiency virus on presentation and diagnosis of tuberculosis in a cohort study in Zambia. *Journal of Tropical Medicine* **96**, 1–11.

ELLIOTT, A. M., LUO, N., TEMBO, G., HALWINDI, B., STEENBERGEN, G., MACHIELS, L., POBEE, J., NUNN, P., HAYES, R. & McADAM, K. P. W. J. (1990). Impact of HIV on tuberculosis in Zambia: a cross-sectional study. *British Medical Journal* **301**, 412–415.

ELLIOTT, A. M., NAMAAMBO, K., ALLEN, B. W., LUO, N., HAYES, R. J., POBEE, J. O. M. & McADAM, K. P. W. J. (1993*a*). Negative sputum smear results in HIV-positive patients with pulmonary tuberculosis in Lusaka, Zambia. *Tubercle and Lung Disease* **74**, 191–194.

EPSTEIN, W. L., FUKUYAMA, K., DANNO, K. & KWAN-WONG, E. (1979). Granulomatous inflammation in normal and athymic mice infected with *Schistosoma mansoni*: an ultrastructural study. *Journal of Pathology* **127**, 207–216

FILE, S. (1995). Interaction of schistosome eggs with

vascular endothelium. *Journal of Parasitology* **81**, 234–238.

FREEDMAN, D. O. & OTTESEN, E. A. (1988). Eggs of *Schistosoma mansoni* stimulate endothelial cell proliferation. *Journal of Infectious Diseases* **158**, 556–562.

GEUNS, H. A., VAN MEIJER, J. & STYBLO, K. (1975). Results of contact examination in Rotterdam. *Bulletin of the International Union Against Tuberculosis* **50**, 107–121.

GITHUI, W., NUNN, P., JUMA, E., KARIMI, F., BRINDLE, R., KAMUNYI, R., GATHUA, S., GICHEHA, C., MORRIS, J. & OMWEGA, M. (1992). Cohort study of HIV-positive and HIV-negative tuberculosis, Nairobi, Kenya: comparison of bacteriological results. *Tubercle and Lung Disease* **73**, 203–209.

GRANGE, J. M. (1996). *Mycobacteria and Human Disease* 2nd edn. London, Arnold.

HARAMATI, L. B., JENNYAVITAL, E. R. & ALTERMAN, D. D. (1987). Effect of HIV status on chest radiographic and CT findings in patients with tuberculosis. *Clinical Radiology* **52**, 31–35.

HASSOUNAH, O. A. & DOENHOFF, M. J. (1993). Comparison of the hepatoprotective effects of immune cells and serum in *Schistosoma mansoni*-infected immunosuppressed mice. *Parasite Immunology* **15**, 657–661.

HOUSTON, S., RAY, S., MAHARI, M., NEILL, P., LEGG, W., LATIFF, A. S., EMMANUEL, J., BASSETT, M., POZNIAK, A., TSWANA, S. & FLOWERDEW, G. (1994). The association of tuberculosis and HIV infection in Harare, Zimbabwe. *Tubercle and Lung Disease* **75**, 220–226.

JANKOVIC, D. & SHER, A. (1996). Initiation and regulation of CD4+ T-cell function in host–parasite models. *Chemical Immunology* **63**, 51–65.

KARANJA, D. M. S., COLLEY, D. G., NAHLEN, B. L., OUMA, J. H. & SECOR, W. E. (1997). Studies on schistosomiasis in Western Kenya. I. Evidence for immune facilitated excretions of schistosome eggs from patients with *Schistosoma mansoni* and Human Immunodifficiency Virus co-infections. *American Journal of Tropical Medicine and Hygiene* **56**, 515–521.

KING, C. L., XIANLI, J., MALHOTRA, I., LIU, S., MAHMOUD, A. A. F. & OETTGEN, H. C. (1997). Mice with a targeted deletion of the IgE gene have increased worm burdens and reduced granulomatous inflammation following primary infection with *Schistosoma mansoni*. *Journal of Immunology* **158**, 294–300.

KLAUSNER, J. D., RYDER, R. W., BAENDE, E., LELO, U., WILLIAME, J.-C., NGAMBOLI, K., PERRIENS, J. H., KABOTO, M. & PRIGNOT, J. (1993). *Mycobacterium tuberculosis* in household contacts of Human Immunodeficiency Virus Type 1-seropositive patients with active pulmonary tuberculosis in Kinshasa, Zaire. *Journal of Infectious Diseases* **168**, 106–111.

KLEIN, N. C., DUNCANSON, F. P., LENOX, T. H., PITTA, A., COHEN, S. C. & WORMSER, G. P. (1989). Use of mycobacterial smears in the diagnosis of pulmonary tuberculosis in AIDS/ARC patients. *Chest* **95**, 1190–1192.

KRAMER, F., MODILEVSKY, T., WALIANY, A. R., LEEDOM, J. M. & BARNES, P. F. (1990). Delayed diagnosis of tuberculosis in patients with human immunodeficiency virus infection. *American Journal of Medicine* **89**, 451–456.

LEPTAK, C. L. & MCKERROW, J. H. (1997). Schistosome egg granulomas and hepatic expression of TNFα are dependent on immune priming during parasite maturation. *Journal of Immunology* **158**, 301–307.

LONG, R., SCALCINI, M., MANFREDA, J., CARRE, G., PHILIPPE, E., HERSCHFIELD, E., SEKLA, L. & STACKIW, W. (1991*b*). Impact of Human Immunodeficiency Virus Type 1 on tuberculosis in rural Haiti. *American Review of Respiratory Disease* **143**, 69–73.

LONG, R., SCALCINI, M., MANFREDA, J., JEAN-BAPTISTE, M. & HERSCHFIELD, E. (1991*a*). The impact of HIV on the usefulness of sputum smears for the diagnosis of tuberculosis. *American Journal of Public Health* **81**, 1326–1328.

LUKACS, N. W. & BOROS, D. L. (1993). Lymphokine regulation of granuloma formation in murine schistosomiasis mansoni. *Clinical Immunology and Immunopathology* **68**, 57–63.

LUKACS, N., KUNKEL, S. L., STRIETER, R. M. & CHENSUE, S. W. (1994). The role of chemokines in *Schistosoma mansoni* granuloma formation. *Parasitology Today* **10**, 322–324.

LUCAS, S. B., MUSALLAM, R., BAIN, J., HASSOUNAH, O., BICKLE, Q. & DOENHOFF, M. J. (1980). The pathological effects of immunosuppression of *Schistosoma mansoni* infected mice, with particular reference to survival and hepatotoxicity after thymectomy and treatment with antithymocyte serum, and treatment with hydrocortisone acetate. *Transactions of the Royal Society of Tropical Medicine and Hygiene* **74**, 633–643.

LURIE, M. B. & DANNENBERG, A. M. (1965). Macrophage function in infectious disease with inbred rabbits. *Bacteriological Reviews* **29**, 466–476.

METWALI, A., ELLIOT, D., BLUM, A. M., JIE, L., SANDOR, M., LYNCH, R., NOBEN-TRAUTH, N. & WEINSTOCK, J. V. (1996). The granulomatous response in murine schistosomiasis does not switch to Th1 in IL-4 deficient mice. *Journal of Immunology* **157**, 4546–4533.

MONTESANO, M. A., FREEMAN, G. L., GAZZINELLI, G. & COLLEY, D. G. (1990). Immune responses during human *Schistosoma mansoni*. XVII. Recognition by monoclonal anti-idiotypic antibodies of several idiotypes on a monoclonal anti-soluble schistosomal egg antigen antibody and anti-soluble schistosomal egg antibodies from patients with different clinical forms of infection. *Journal of Immunology* **145**, 3095–3099.

MONTESANO, M. A., FREEMAN, G. L., SECOR, W. E. & COLLEY, D. G. (1997). Immunoregulatory idiotypes stimulate T helper 1 cytokine responses in experimental *Schistosoma mansoni* infections. *Journal of Immunology* **158**, 3800–3804.

MOORE, D. L., GROVE, D. I. & WARREN, K. S. (1977). The *Schistosoma mansoni* egg granuloma: quantitation of cell populations. *Journal of Pathology* **121**, 41–50.

MURARE, H. M., AGNEW, A. M., BALTZ, M. N., LUCAS, S. B. & DOENHOFF, M. J. (1987). The response to *Schistosoma bovis* in normal and T-cell deprived mice. *Parasitology* **95**, 517–530.

MURARE, H. M. & DOENHOFF, M. J. (1987). Parasitological observations of *Schistosoma bovis* in normal and T-cell deprived mice. *Parasitology* **95**, 507–516.

MURARE, H. M., DUNNE, D. W., BAIN, J. & DOENHOFF, M. J. (1992). *Schistosoma mansoni*: control of hepatotoxicity and egg excretion by immune serum in infected immunosuppressed mice is schistosome species-specific, but not *S. mansoni* strain-specific. *Experimental Parasitology* **75**, 329–339.

MURRAY, J. F. (1990). Cursed duet – HIV-infection and tuberculosis. *Respiration* **57**, 210–220.

NGAIZA, J. R. & DOENHOFF, M. J. (1990). Blood platelets and schistosome egg excretion. *Proceedings of the Society for Experimental Biology and Medicine* **193**, 73–79.

NGAIZA, J. R., DOENHOFF, M. J. & JAFFE, E. A. (1993). *Schistosoma mansoni* egg attachment to cultured human umbilical vein endothelial cells: an *in vitro* model of an early step of parasite egg excretion. *Journal of Infectious Diseases* **168**, 1576–1580.

PEARCE, E. J., VASCONCELOS, J. P., BRUNET, L. R. & SABIN, E. A. (1996*a*). IL-4 in schistosomiasis. *Experimental Parasitology* **84**, 295–299.

PEARCE, E. J., CHEEVER, A., LEONARD, S., COVALESKY, M., FERNANDEZ-BOTRAN, R., KOHLER, G. & KOPF, M. (1996*b*). Schistosomiasis in IL-4 deficient mice. *International Immunology* **8**, 435–444.

PITCHENIK, A. E. & RUBINSON, H. A. (1985). The radiographic appearance of tuberculosis in patients with the acquired immune-deficiency syndrome (AIDS) and pre-AIDS. *American Review of Respiratory Disease* **131**, 393–396.

POZNIAK, A. L., MACLEOD, G. A., NDLOVU, D., ROSS, E., MAHARI, M. & WEINBERG, J. (1995). Clinical and chest radiographic features of tuberculosis associated with human immunodefficiency virus in Zimbabwe. *American Journal of Respiratory and Critical Care Medicine* **152**, 1558–1561.

RAGHEB, S. & BOROS, D. L. (1989). Characterization of granuloma T lymphocyte function from *Schistosoma mansoni*-infected mice. *Journal of Immunology* **142**, 3239–3246.

ROOK, G. A. W. & BLOOM, B. R. (1994). Mechanisms of pathogenesis in tuberculosis. In *Tuberculosis: Pathogenesis, Protection and Control* (ed. Bloom, B. R.) American Society for Microbiology, Washington DC, USA.

SHER, A., COFFMAN, R. L., HIENY, S., SCOTT, P. & CHEEVER, A. W. (1990). Interleukin 5 is required for the blood and tissue eosinophilia but not granuloma formation induced by infection with *Schistosoma mansoni*. *Proceedings of the National Academy of Sciences, USA* **87**, 61–65.

SMITH, M. D. (1977). The ultrastructural development of the schistosome egg granuloma in mice. *Parasitology* **75**, 119–123.

SMITH, R. L., YEW, K., BERKOWITZ, K. A. & ARANDA, C. P. (1994). Factors affecting the yield of acid-fast sputum smears in patients with HIV and tuberculosis. *Chest* **106**, 684–686.

SMITHERS, S. R. & TERRY, R. J. (1965). The infection of laboratory hosts with cercariae of *Schistosoma mansoni*. *Parasitology* **55**, 695–700.

STENGER, R. J., WARREN, K. S. & JOHNSON, E. A. (1967). An ultrastructural study of hepatic granulomas and schistosome egg shells in murine hepatosplenic schistosomiasis mansoni. *Experimental Molecular Pathology* **7**, 116–132.

VON LICHTENBERG, F. (1962). Host response to eggs of *S. mansoni*. I. Granuloma formation in the unsensitized laboratory mouse. *American Journal of Pathology* **41**, 711–731.

VON LICHTENBERG, F. (1964). Studies on granuloma formation. III. Antigen sequestration and destruction in the schistosome pseudotubercle. *American Journal of Pathology* **45**, 75–93.

VON LICHTENBERG, F., ERICKSON, D. G. & SADUN, E. H. (1973). Comparative histopathology of schistosome granulomas in the hamster. *American Journal of Pathology* **72**, 149–168.

WARREN, K. S. (1973). The pathology of schistosome infections. *Helminthological Abstracts Series A* **42**, 590–633.

WARREN, K. S. (1975). Hepatosplenic schistosomiasis mansoni: an immunologic disease. *Bulletin of the New York Academy of Medicine* **51**, 545–550.

WARREN, K. S. & DOMINGO, E. O. (1970). *Schistosoma mansoni*: stage-specificity of granuloma formation around eggs after exposure to irradiated cercariae, unisexual infections, or dead worms. *Experimental Parasitology* **27**, 60–66.

WARREN, K. S., DOMINGO, E. O. & COWAN, R. B. T. (1964). Granuloma formation around schistosome eggs as a manifestation of delayed hypersensitivity. *American Journal of Pathology* **51**, 735–756.

WEINSTOCK, J. L. & BOROS, D. L. (1983). Organ-dependent difference in composition and function observed in hepatic and intestinal granulomas isolated from mice

with schistosomiasis mansoni. *Journal of Immunology* **130**, 418–422.

WYNN, T. A. & CHEEVER, A. W. (1995). Cytokine regulation of granuloma formation in schistosomiasis. *Current Opinion in Immunology* **7**, 505–511.

WYNN, T. A., CHEEVER, A. W., JANKOVIS, D., POINDEXTER, R. W., CASPAR, P., LEWIS, F. A. & SHER, A. (1995*a*). An IL-12 based vaccination method for preventing fibrosis induced by schistosome infection. *Nature* **376**, 594–596.

WYNN, T. A., ELTOUM, I., OSWALD, I. P., CHEEVER, A. W. & SHER, A. (1994). Endogenous interleukin-12 (IL-12) regulates granuloma formation induced by eggs of *Schistosoma mansoni* and endogenous IL-12 both inhibits and prophylactically immunizes against egg pathology. *Journal of Experimental Medicine* **179**, 1551–1561.

WYNN, T. A., JANKOVIC, D., HIENY, S., ZIONCHECK, K., JARDIEU, P., CHEEVER, A. W. & SHER, A. (1995*b*). IL-12 exacerbates rather than suppressed T helper 2-dependent pathology in the absence of endogenous IFN-γ. *Journal of Immunology* **154**, 3999–4009.

YAMAMURA, Y., OGAWA, Y., MAEDA, H. & YAMAMURA, Y. (1974). Prevention of tuberculous cavity formation by desensitization with tuberculin-active peptide. *American Review of Respiratory Disease* **109**, 594–601.

ZHU, Y., LUKACS, N. W. & BOROS, D. L. (1994). Cloning of Th0 and Th2-type helper lymphocytes from liver granulomas of *Schistosoma mansoni*-infected mice. *Infection and Immunity* **62**, 994–999.

Enhancing antibodies in HIV infection

G. FÜST

*3rd Department of Medicine, Semmelweis Medical University and Research Group of Membrane Biology and
Immunopathology, Hungarian Academy of Science, Budapest, Eötvös út 12, H-1121, Hungary*

SUMMARY

The author has summarized the history of discovery, the mechanism and the clinical significance of antibody-dependent
enhancement (ADE) of HIV infection. ADE has two major forms: (*a*) complement-mediated antibody-dependent
enhancement (C-ADE) and (*b*) complement-independent Fc receptor-dependent ADE (FcR-ADE). The most important
epitope responsible for the development of C-ADE-mediating antibodies is present in the immunodominant region of
gp41 while antibodies mediating FcR-ADE react mainly with V3 loop of gp120. There are at least three fundamentally
different hypotheses for the explanation of ADE *in vitro*: (*a*) increased adhesion of HIV-antibody-(complement) complexes
to FcR or complement receptor carrying cells; (*b*) facilitation of HIV-target cell fusion by complement fragment deposited
on the HIV-virions and (*c*) complement activation products may have a non-specific stimulatory effect on target cells
resulting in enhanced virus production. FcR-ADE and C-ADE have been measured *in vitro* mostly by using FcR-carrying
and complement receptor-carrying cell lines, respectively; no efforts have been made to standardize these methods. Several
data support the possible clinical significance of FcR-ADE and C-ADE: (*a*) Cross-sectional and longitudinal studies
indicate a correlation between the amounts of FcR-ADE and C-ADE-mediating antibodies and clinical, immunological
and virological progression of the HIV-disease; (*b*) ADE may facilitate maternal–infant HIV-1 transmission; (*c*) According
to experiments in animal models, ADE are present and may modify the course of SIV (simian immunodeficiency) infection
as well. The author raises a new hypothesis on the mechanism of the *in vivo* effect of C-ADE. According to the hypothesis,
C-ADE-mediating antibodies exert their effect through enhancement of HIV propagation and consequent facilitation of
the progression of HIV disease. Finally, according to observations from animal experiments and human clinical trials it
cannot be excluded that ADE-mediating antibodies may develop, diminish the beneficial effect or may be harmful in
volunteers vaccinated with HIV-1 candidate vaccines.

Key words: AIDS, HIV, HIV infection, SIV, enhancement, C-ADE, complement, neutralization vaccines.

INTRODUCTION

Antibody-dependent enhancement (ADE) of infection with parasites is a phenomenon by which uptake
of the complexes of microorganism–antibody (and
complement proteins or fragments) is facilitated by
interactions of antibodies with the Fc receptors
(FcR) and/or of complement proteins or fragments
with the complement receptors (CR) on the target
cells. The first type of antibody-dependent enhancement is designated as complement-independent or FcR-dependent. In the following, the term
FcR-ADE will be used for the effect of that type of
infection-enhancing antibodies. The effect of the
second type of antibodies is mediated by the
activation of the complement system and it is
dependent on the presence of CR on the target cells.
The term C-ADE (complement-mediated, antibody-dependent enhancement) will be applied for
this type of infection enhancement.

A significant, 4 log-fold or even greater increase in
the *in vitro* production of several viruses (alpha, pox,
bunya, reo and herpes) was observed following
exposure to low-affinity antibodies (Porterfield,

1986). ADE may be clinically important especially in
the case of flaviviruses. For example, dengue haemorrhagic fever is usually more severe in individuals
or experimentally infected monkeys with low levels
of anti-dengue antibodies than in seronegative
individuals (Halstead, Shortwell & Casals, 1979;
Halstead, 1988). The presence of infection-enhancing antibodies was found to be a significant (relative
risk = 6·2) risk factor for severe dengue haemorrhagic fever among children in Bangkok, a dengue
haemorrhagic fever endemic region (Kliks *et al.*
1989).

Antibodies that may enhance HIV infection *in
vitro* were described shortly after the first isolation of
HIV. Their presence in the blood of HIV-infected
patients as well as in HIV- or SIV-infected experimental animals was confirmed by several groups.
The data on the mechanism and clinical significance
of FcR-ADE and C-ADE, are however, highly
controversial.

The purpose of the present article is to review the
literature on the antibody-mediated enhancement in
HIV infection. The paper is based partly on the
critical analysis of literature data and partly on the

results of our own group. This group consists of researchers from different Hungarian, Austrian, French, Swiss and German institutions and started to study the mechanism and clinical significance of C-ADE in 1990; our first paper was published in 1991 (Tóth *et al.* 1991).

DIFFERENT FORMS OF ANTIBODY-DEPENDENT ENHANCEMENT IN HIV INFECTION

Two types of enhancing antibodies were described approximately at the same time in the late eighties. Robinson and colleagues (Robinson, Montefiori & Mitchell, 1987) found that sera from HIV-infected individuals enhance *in vitro* HIV infection of the CR2 (complement receptor type2)-bearing T lymphoblastoid cell line MT2. The same authors demonstrated this enhancement to be dependent on antibodies and mediated by complement (Robinson, Montefiori & Mitchell, 1988). Gras, Strub & Dormont (1988) reproduced the same phenomenon with peripheral blood lymphocytes. According to further studies (Robinson *et al.* 1989*a*) C-ADE of HIV-1 infection is characterized by $> 1 \log_{10}$ increases in infectious virus release.

FcR-ADE, that is the ability of heat-inactivated sera from HIV-seropositive patients or IgG purified from such sera to accelerate and/or enhance production of HIV by cells infected with mixtures containing these antibodies, was first described by Homsy, Tateno & Levy (1988) and by Takeda, Tuazon & Ennis (1988). Shortly thereafter, FcR-ADE of HIV-1 infection was observed *in vitro* with many different HIV-1 isolates of both T cell- and monocyte-tropic phenotypes, using different target cells (Homsy *et al.* 1989; Jouault *et al.* 1989; Laurence *et al.* 1990; Perno *et al.* 1990; Matsuda *et al.* 1989; Takeda, Sweet & Ennis, 1990). The extent of *in vitro* enhancement of HIV infection by C-ADE is one or two orders of magnitude greater than that by FcR-ADE (Mascola *et al.* 1993).

It has been clear since the first description of FcR-ADE and C-ADE that different types of FcR and CR, respectively, are indispensable for these phenomena to occur. The contribution of the CD4 receptors in antibody-dependent HIV-1 infection enhancement especially in FcR-ADE has been, however, highly controversial until now. In principle, both FcR-ADE and C-ADE may have two forms, a CD4-dependent and a CD4-independent one. Probably the co-receptors (chemokine receptors), like CCR5 an essential cofactor for the infection of CD4-carrying cells by the macrophage-tropic HIV-strains (Alkhatib *et al.* 1996; Deng *et al.* 1996) or CXCR4 an essential cofactor for the T-cell adapted HIV-1 strains (Feng *et al.* 1996) are also involved in the HIV-entry These forms of ADE are summarized schematically in Fig. 1.

CD4-dependent and CD4-independent forms of FcR-ADE

According to the experiments of Takeda, Sweet & Ennis (1990) two receptors, FcγR and CD4 are required for antibody-dependent enhancement of HIV-1 infection. Pretreatment of human macrophages and monocytic cell lines with anti-CD4 or soluble recombinant CD4 was found to markedly inhibit FcR-ADE. Similar results were obtained by Perno *et al.* (1990) by using human peripheral blood monocytes/macrophages, and by Zeira, Byrn & Groopman (1990) in the case of the infection of U937 monocytoid cells with the HTLV$_{IIIB}$ HIV-1 strain. Connor *et al.* (1991) demonstrated that at lower levels of antibody opsonization, there are enough interactions with FcγR to stabilize the virus at the cell surface allowing antibody-dependent enhancement of HIV-1 infection of monocytes and monocyte-derived macrophages through high-affinity CD4 interaction. Nottet *et al.* (1992) were also able to abrogate antibody-dependent enhancement of HIV-1 infection of an EBV-transformed B cell line with a monoclonal antibody against CD4. By contrast, other authors like Homsy *et al.* (1989) or more recently Trischmann, Davis & Lachmann (1995) failed to inhibit FcR-ADE of human macrophages by anti-CD4 antibodies. Use of different strains, experimental conditions or target cells may be responsible for these discrepancies.

CD4-dependency of C-ADE, non-antibody-dependent complement-mediated enhancement of HIV infection

It was demonstrated very early that C-ADE of HIV-1 infection of MT-2 cells requires CD4 and CR type 2 (CR2) (Robinson, Montefiori & Mitchell, 1990). Similar findings were obtained by Tremblay *et al.* (1990) and Gras & Dormont (1991) with the CR2-dependent C-ADE of Epstein–Barr virus-transformed B-lymphocytic cell lines. The CD4 dependency of C-ADE was confirmed more recently by Lund *et al.* (1995) and by our group (Prohászka *et al.* 1997) as well.

A special form of the HIV-1 infection enhancement is the complement-mediated non-antibody-dependent one. Although HIV-1 can bind and activate purified complement components (Dierich *et al.* 1993), non-antibody dependent complement-mediated HIV infection is usually studied by mixing HIV-1 stocks with HIV-seronegative serum samples. Under these conditions enhancement may (Gras & Dormont, 1991) or may not (Boyer *et al.* 1992) require the CD4 receptors to occur. According to the most recent studies of Saarloos, Lint & Spear (1995) 'antibody-independent' mechanisms of complement-activation by HIV-infected cells are at least partly due to a cross-reactive IgM type antibody

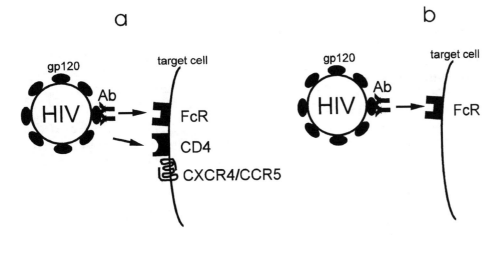

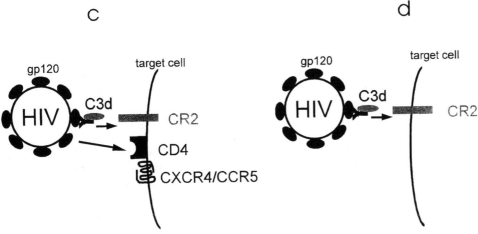

Fig. 1. Four possible forms of the antibody-dependent enhancement of HIV-1 infection (ADE). (*a*) FcR- and CD4-dependent ADE. Out of FcRs, FcγRIII seems to be the most important. Probably the co-receptors (chemokine receptors), like CCR5 for the macrophage-tropic or CXCR4 for the T-cell adapted HIV-1 strains are also involved in the HIV-entry. (*b*) ADE dependent only on FcR. (*c*) Complement-mediated, CD4-dependent ADE: virus-antibody complexes activate complement, complement activation products, like, C4b, C3b, C3bi, C3dg covalently bind to the complexes. Out of complement receptors, CR2 seems to be the most important. Interaction between CR2 and its ligand, C3dg facilitates ADE. Probably the co-receptors (chemokine receptors), like CCR5 for the macrophage-tropic or CXCR4 for the T-cell adapted HIV-1 strains are also involved in the HIV entry. (*d*) Complement-mediated, non-CD4-dependent ADE. Existence of this form of ADE is doubtful.

present in normal human serum. It cannot be excluded that similar mechanisms are responsible for the complement-mediated non-antibody dependent HIV-1 infection enhancement, too.

MAPPING OF EPITOPES ON HIV ENVELOPE PROTEINS WHICH BIND ENHANCING ANTIBODIES

According to the studies of Robinson and his colleagues, the most important epitope responsible for the development of C-ADE-mediating antibodies seems to be the immunodominant epitope (amino acids 586 to 620) on gp41. Three of 16

human monoclonal antibodies (mAbs) were found to enhance HIV-1 infection of the MT-2 target cells. Using analyses by radioimmunoprecipitation, Western blot and ELISA, it was demonstrated that antigen specificity of all the 3 mAbs can be localized to the *N*-terminal end of gp41 containing the so-called immunodominant domain of the protein (Robinson *et al.* 1990*a*, 1990*b*, 1991). This epitope on gp41 is recognized by sera from essentially all HIV-infected subjects. Similar results were obtained with 2 other mAbs. The same group characterized further the antigen specificity of C-ADE-mediating antibodies. They found 4 different anti-gp41 mAbs to mediate C-ADE. These mAbs were mapped to 2 distinct domains on the gp41 by using synthetic peptides. The first domain (amino acids 579 to 613) was recognized by 3/4 of the mAbs, while the 4th one bound to a 2nd gp41 domain (amino acids 644 to 663). C-ADE of MT-2 cells mediated by sera from HIV patients was blocked only by the mAbs to the first domain. We have found 3 mAbs against this domain to mediate C-ADE of another CR2-carrying cell line, MT-4 (Füst *et al.* 1994*b*). Interestingly enough one of these mAbs, 181-D, was previously found to be non-enhancing by Robinson *et al.* (1991).

The 2 domains on gp41 that can bind enhancing antibodies are conserved between HIV-1 isolates as well as between HIV-2 and SIV isolates. These data suggest that 2 conserved regions within the HIV-1 gp41 are most important in the development of C-ADE-mediating antibodies during HIV infection. In addition to anti-gp41 antibodies, however, antibodies against the V3 loop of gp120 can also mediate C-ADE at least according to the experiments of Jiang, Lin & Neurath (1991) performed with heterologous (rabbit) antisera.

In contrast to the findings on C-ADE, it seems that the antibodies mediating FcR-ADE react mainly with the gp120, although Eaton *et al.* (1994) described an anti-gp41 human monoclonal antibody which enhanced HIV-1 infection the absence of complement. Murine mAbs directed against a domain (amino acids 724 to 752) on gp41 were also found to enhance HIV-1 infection *in vitro*. No antibodies reactive with this region were, however, found in the sera of 100 HIV-infected individuals, indicating that this domain is not immunogenic in humans (Niedrig *et al.* 1992). One human anti-gp120 mAb which was able markedly to enhance FcR mediated infection on monocytic cells was found to bind to a conformational site of non-variable sequences in the C-terminal part (amino acids 272 to 509) of gp120 (Takeda *et al.* 1992). Interestingly, deletions in both the N and C terminus of gp120 significantly decreased the binding of this antibody to the gp120 (Lee *et al.* 1994). Kliks *et al.* (1993) studied the neutralizing *vs.* FcR-ADE-mediating effect of different human mAbs against the V3 loop of gp120. These mAbs show different effects on 3 distinct HIV-1 strains: neutralization, enhancement or no effect. Only one amino acid in the mAb-binding epitopes proximal to the crown of V3 loop was different among these 3 strains. According to the recent study of Trischmann *et al.* (1995), the extent of enhancing effect markedly depends on the antigen structure of the virus strain: while rat antisera against the 5 variable regions (V1–V5) of gp120 and conserved parts of gp120 and gp41 facilitated infection of primary human macrophages with the homologous virus HIV-SF2mc, infection of the same cell with a heterologous strain, HTLVIIIB, was enhanced only by antisera to V4 and V5. Recently Auewarakul *et al.* (1996) also found FcR-

ADE activity in individuals infected with subtype B and E viruses in Thailand to be mostly isolate-specific and independent of genetic subtypes.

POSSIBLE MECHANISMS OF *IN VITRO* EFFECT OF ENHANCING ANTIBODIES

According to Lund *et al.* (1995) there are at least three fundamentally different hypotheses for the explanation of ADE *in vitro*: (*a*) increased adhesion of HIV-antibody-(complement) complexes to FcR or CR carrying cells; (*b*) facilitation of HIV-target cell fusion by complement fragment deposited on the HIV-virions and (*c*) complement activation products may have non-specific stimulatory effect on target cells resulting in enhanced virus production. Recent experiments of Mouhoub *et al.* (1996) demonstrated that ligation of complement receptors type 1 (CR1) on CD4[+] T lymphocytes enhances viral replication in HIV-infected cells. This observation lends credence to the third hypothesis. The most probable explanation for the mechanism of the FcR-ADE and C-ADE, however, is the increase of adhesion of virus particles to the target cells. This may facilitate entry of the virus particle via the usual route. Alternatively, binding can be followed by a receptor-mediated endocytosis of the FcR or CR-bound HIV which may lead to the fusion of viral envelope with the membrane of an endosomal vesicle then allowing the insertion of viral core to the cytoplasm before it is degraded in a lysosome (Kozlowski *et al.* 1995).

The increased adhesion is mediated by the interaction to the virus-bound antibody and FcR, and the interaction of the virus or virus-antibody complex-bound complement fragment to the CR. It seems that although all 3 types of FcγR can mediate FcR-ADE, the FcγRIII (CD16) is the most important in the FcR-ADE of primary human macrophages (Trischmann *et al.* 1995) and U937 cells (Laurence *et al.* 1990). According to the studies of Zoellner, Feucht & Laufs (1992), proteases produced by macrophages may significantly increase FcR-ADE.

As for C-ADE, June *et al.* demonstrated first in 1991 that it is mediated by an increased virus binding to the MT-2 target cells. This increased adhesion strongly correlated to higher level of provirus formation 8–28 h after infection. Bakker *et al.* (1992) found that antibody and complement enhance binding and uptake of HIV-1 by human monocytes.

In recent experiments (Prohászka *et al.* 1997), using 3 different anti-gp41 mAbs, we have demonstrated the existence of an alternative route of the C-ADE. Addition of purified C1q to mixtures of mAbs and HTLV$_{IIIB}$ used for infection of MT-4 target cells significantly increased HIV production of the infected cells as compared to the cell cultures

mAb 246-D

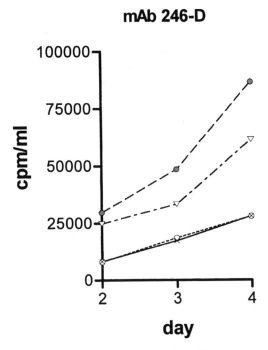

Fig. 2. C-ADE with purified C1q. Human monoclonal anti-gp41 antibody 246-D (alone: ○) was mixed with normal human serum (●), purified C1q (▽) or with buffer (×) and thereafter with a stock of HIV-1$_{\text{IIIB}}$. MT-4 cells were infected with the mixtures and virus production was measured on days 2, 3, and 4 by the reverse transcriptase assay. For methodical details, see text. (Adapted from Prohászka *et al.* 1997.)

infected with the mAb-virus mixtures (Fig. 2). We assume that this enhancement is due to an additional bridge between the virus particles and target cells though interaction of C1q bound to the virus-antibody complexes and the C1q (collectin) receptors on the target cells. Experiments studying this hypothesis are in progress in our laboratories.

SIGNIFICANCE OF THE ADE *IN VIVO*

Data on the presence of ADE-mediating antibodies in vivo

Measurement of FcR-ADE and C-ADE in the serum and other bodily fluids of humans and experimental animals. The extent of enhancement and the ratio of neutralization/enhancement measured in a given sample depend in several factors specific for the assay system employed. Variables such as target cells, the density of FcRs or CRs on the target cells, serum dilutions tested and the strains of HIV-1 used can markedly influence the final outcome of the measurement. Usually each laboratory applies a different method for the measurement of the titre of enhancing antibodies, and to my best knowledge no efforts have been made for standardization of the procedures. For the measurement of FcR-ADE, primary human monocytes/macrophages or the

U937 cell line are used while lymphoblastoid cell lines MT-2 or MT-4 are mostly applied for the measurement of C-ADE. Although in some studies primary isolates were used, laboratory strains – most frequently HTLVIIIB – are applied in the clinical studies.

ADE in the blood of HIV patients and infected experimental animals. Using sensitive methods C-ADE were detected from 95% of HIV-infected individuals by Robinson *et al.* (1989*a*) but they found that the titres of such antibodies varied greatly between patients. FcR-ADE was found in different frequencies in the sera of HIV patients: while Takeda *et al.* (1988) found FcR-ADE in almost all serum samples tested, Laurence *et al.* (1990) and Homsy, Meyer & Levy (1990) found FcR-ADE-mediating antibodies in only 5/16 and 7/16 patients, respectively. ADE could be detected also in sera of HIV-infected chimpanzees and SIV-infected rhesus macaques (Robinson *et al.* 1989*b*, Montefiori *et al.* 1990).

Cross-sectional and longitudinal studies of FcR-ADE and C-ADE-mediating and related antibodies in HIV patients. Results concerning the clinical relevance of the enhancing antibodies are contradictory. Homsy *et al.* (1990) were the first to demonstrate that the appearance of the Fc-receptor-dependent enhancing type antibodies are associated with the progression of HIV disease. FcR-ADE with homotypic isolates was primarily observed in AIDS patients and in longitudinal studies clinical progression was found to be associated with the appearance of FcR-ADE-mediating antibodies.

Our group demonstrated first a similar association between C-ADE and the clinical progression of HIV-disease. In our first study (Tóth *et al.* 1991) we found C-ADE only in 4/20 (20%) of asymptomatic HIV-patients but in 12/19 (63%) of patients in the symptomatic (ARC or AIDS) stage of the disease. In a more recent study (Füst *et al.* 1994*a*) these findings were confirmed. At longitudinal measurements a strong negative correlation ($R = -0.458$, $P = 0.00044$) was found between the extent of C-ADE and the CD4$^+$ cell counts. Moreover, high titre C-ADE predicted rapid decline of the CD4$^+$ cell counts and an increased probability of the AIDS development. In a 5-year follow up period AIDS developed in 33 and 83% of patients with missing or low titre C-ADE and in those with high titre C-ADE, respectively.

Very recently (F. D. Tóth *et al.* unpublished) similar results were obtained at a relatively short-term longitudinal study performed in 125 serum and plasma samples from 25 HIV patients untreated or treated according to different antiviral protocols.

The most important results of this study can be summarized as follows. (1) Addition of complement abrogated neutralization in a part of the serum samples tested. (2) High extent of C-ADE predicted immunological progression and, to a lesser extent, clinical progression in the patients tested. (3) A significant positive correlation was found between the levels of C-ADE and the amounts of plasma HIV-1 RNA in samples obtained from the patients at the same time. These findings, which confirm and extend our previous results, indicate that C-ADE measured in serum samples from HIV-1-infected individuals, correlate with immunosuppression and disease not only in the early and intermediate but in advanced-stage patients as well. Besides these observations, some indirect evidence seems to support the proposition that C-ADE-mediating antibodies are associated with immunosuppression and disease in HIV infection. In a collaborative study (Tóth *et al.* 1994) we have found a strong positive correlation (R = 0·408, *P* = 0·003) between C-ADE and the IgA type anti-Fab antibodies measured in the same serum samples. Anti-Fab autoantibodies are known to be strongly associated with the progression of HIV disease (Süsal, Daniel & Opelz, 1996). It is most probable (see above) that C-ADE mediating antibodies are mainly directed against the immunodominant epitope of gp41. In this context, it is interesting that Monell *et al.* (1993) found the antibody response of HIV-infected individuals for the immunodominant (amino acids 598 to 609) sequence to continue to increase over time in HIV patients. Similarly, Radkowski *et al.* (1993) and Thomas *et al.* (1996) observed increased avidity of the anti-gp41 antibodies in the late symptomatic stage of HIV disease. According to the studies of Zwart *et al.* (1994) a high level of antibodies against the gp41 immunodominant peptide correlated with rapid progression of disease. Lallemant *et al.* (1994) also noted a correlation between higher levels of antibodies to this peptide and increased risk of maternal–fetal transmission.

In contrast to these findings, Montefiori *et al.* (1991) did not find any correlation between the clinical course of the HIV disease and the C-ADE. More recently they were unable to detect any difference in the amounts of C-ADE-mediating antibodies between long-term non-progressors and progressor patients (Montefiori *et al.* 1996). Two main methodological differences exist between the procedures used for the measurement of C-ADE between Montefiori's group and ours. While we use a near-physiological (25 %) NHS concentration as complement source, the other group always applied a strongly (1:20–1:27) diluted NHS. It seems reasonable to suppose that the closer the complement concentration is in the mixtures used for infecting the MT-4 cells to that in the blood, the higher the physiological relevance of the measurement. The mathematical model of Lund *et al.* (1995) also suggests that the enhancing effect of complement is dose-dependent. On the other hand, when we compared the main titres in HIV-seropositive sera reported by Montefiori *et al.* (1995) and our group (Füst *et al.* 1994 *a*) the assay of the other group seems to be about 10 times more sensitive, which might decrease the discriminative power of the measurement.

We are aware of the fact, however, that the test system we applied for the measurement of neutralization/enhancement is far from the physiological conditions. The target cell, MT-4, like MT-2 used in other studies, an HTLV-1 infected cell-line and HIV1$_{IIIB}$ is a laboratory strain passaged many times through cell lines. In spite of these draw-backs, the measurement seems to have clinical relevance. This apparently paradoxical observation can be best explained by assuming that complement-dependent enhancement is due to broadly reacting antibodies which can bind and activate complement not only with HIV-I$_{IIIB}$ but most HIV-1 strains present in the body of infected persons (Berman, Eastman & Wilkes, 1994; Nkengasong *et al.* 1994).

Possible role of ADE in the maternal–infant HIV-1 transmission. As mentioned previously, Kliks *et al.* (1993) demonstrated that anti-V3 loop mAbs that neutralize one HIV strain enhance infection of another and have no effect on a third strain. In a group of North American mothers they have observed that transmission mainly occurs to infants infected with viruses that are either resistant to neutralization by the mother's serum or enhanced by this serum sample (Kliks & Levy, 1994). Similar findings were reported by Lallemant *et al.* (1994) and Markham *et al.* (1994). Both groups observed that mothers who transmitted infection to their offspring had a significantly higher mean concentration of IgG1 antibodies to the V3 loop of the gp120 than the non-transmitters. In the studies of Markham *et al.* (1994), the geometric mean of IgG was 16·8, and 9·4 μg/ml in the 23 transmitter and 103 non-transmitter Haitian mothers, respectively. Moreover, the experiments of Lallemant *et al.* (1994) demonstrated that higher antibody titres to the immunodominant domain of gp41 were correlated to a higher risk of perinatal transmission. Finally, Tóth *et al.* (1994 *b*) demonstrated that antibodies mediating both FcR-ADE and C-ADE are able to increase the HIV infection of human syncytiotrophoblast cells. This could be one of the mechanisms of the facilitating effect of ADE on the spread of HIV-1 from mother to the fetus.

Data from an animal model. According to the studies of Montefiori *et al.* (1990) SIV infection enhancing

Table 1. *Possible mechanisms of the antibody-dependent enhancement of HIV-1 infection in vivo*

1. Abrogation of the effect of neutralizing antibodies
2. Enhancement of the HIV production by different cells *in vivo*
   ● monocytes/macrophages (mainly FcR-ADE)
   ● CR2$^+$ CD4$^+$ T lymphocytes (mainly C-ADE)
   ● short-living CD4$^+$ lymphocytes responsible for > 99 % of HIV production in the body of infected persons (see hypothesis in Fig. 3)
3. Effect of the trapping of virus-(antibody)-complement complexes on FDC in germinal centres of lymph nodes

complement-dependent antibodies appeared in rhesus macaques soon (28 days) after experimental inoculation of SIV. C-ADE titre increased over time and peaked just prior to the death of macaques from opportunistic infections or lymphoma. According to the experiments of Gardner *et al.* (1994, 1995), passive immunization of rhesus macaques with pooled plasma or purified immunoglobulin from healthy SIVmac251-infected animals prior to or after challenge with the same virus did not confer protection or even enhanced infection and facilitated the development of disease, although plasma and immunoglobulin preparations contained a high level of SIV-binding and neutralizing antibodies. Another finding that seems to be important was reported by Montefiori *et al.* (1995). These authors found that C-ADE activity in sera during acute primary infection in macaques inoculated with the SIVmac251 strain appeared before neutralizing antibodies and coincided with the initial peak and decline of plasma antigenaemia. As the neutralizing antibodies developed and their titre increased in the infected animals, the extent of C-ADE activity gradually decreased.

Possible mechanisms of ADE in vivo

In principle, enhancing antibodies may exert their effect through different routes (Table 1).

Abrogation of the effect of neutralizing antibodies. Sera of HIV-infected subjects may contain different types of antibodies in parallel. When several human monoclonal antibodies derived from lymphocytes of HIV patients were studied, 10/20 mAbs were found to mediate exclusively neutralization, 5 exclusively mediated antibody-dependent cellular cytotoxicity (ADCC) while 2 mAbs had both ADCC-mediating and C-ADE effect (Forthal *et al.* 1995). Kostrikis *et al.* (1996), using quantitative analysis of serum neutralization of HIV-1 from different subtypes, found enhancement in more than one-quarter (28 %) of the 1213 combinations of sera and HIV-1 isolates.

The extent of neutralization and enhancement was comparable. Therefore, it is conceivable that during natural HIV-1 infection the balance of the 'good', that is ADCC-mediating or neutralizing, and the 'bad', that is FcR-ADE- or C-ADE-mediating-antibodies, determines the effect of antibodies on *in vitro* HIV production. In addition, complement may have a dual effect on the fate of HIV virions. According to the studies of Spear *et al.* (1990), HIV-1 can be neutralized by viral lysis with subneutralizing doses of antibody from HIV-infected persons in the presence of complement. This finding indicates that complement may stimulate neutralizing capacity of specific antibodies. Recently Sullivan *et al.* (1996) proved the role of complement-mediated lysis of the antibody-bound HIV-1 plasma virions in the clearance of virus *in vivo*. On the other hand, according to the early study of Robinson *et al.* (1988), addition of complement of some sera may decrease or fully abrogate the neutralizing effect of the human antibodies or even may induce neutralizing antibodies to mediate C-ADE. Similar results were obtained with rabbit antibodies against the V3 loop of gp120 by Jiang *et al.* (1991): antisera to V3 of 21 distinct HIV-1 isolates were tested for their neutralizing and enhancing effect in the presence or absence of human complement. In the absence of complement, each antibody was virus-neutralizing whereas addition of complement enhanced virus infection by 10 strains. In our most recent study (F. D. Tóth *et al.* unpublished), we also found that addition of complement to heat-treated serum samples of some HIV-infected patients did not affect neutralization while in the case of other serum samples neutralizing sera became strongly enhancing. These observation, taken together, also suggest that a mixture of various antibodies is present in the sera of HIV-infected individuals. Sera of some subjects contain, while those of others do not contain C-ADE-mediating antibodies. Therefore when neutralization is measured in the presence of complement and by using a CR-carrying target cell, antibodies in the serum samples tested may neutralize, enhance or do not affect *in vitro* virus production. A similar situation may exist *in vivo* (see above).

Another factor which may result in ADE could be the emergence of strains in the HIV-infected patients in which growth is enhanced *in vitro* and possibly *in vivo* by the antibodies which effectively neutralize T-cell adapted laboratory isolates. For example, in the recent studies of Sullivan *et al.* (1995), an antibody reacting with the CD4-binding site of gp120 was found markedly to enhance infection of PBMC by two primary HIV-1 isolates.

Enhancement of HIV-1 production by different cells in vivo. According to *in vitro* studies, HIV production by cells of the monocyte–macrophage lineage may be increased by FcR-ADE *in vivo*. These cells carry FcγRIII and therefore complexes of HIV and FcR-ADE-mediating antibodies may attach to these cells and increase HIV production by them. Although Schadduck *et al.* (1991) failed to detect enhancement of HIV-1 infection of human blood monocytes and peritoneal macrophages, most papers published both before (reviewed in Mascola *et al.* (1993)) and after publication of this paper (Trischmann *et al.* 1995; Kostrikis *et al.* 1996) demonstrated the ability of the sera of HIV patients to mediate FcR-ADE. As for the C-ADE, most *in vitro* studies have demonstrated the ability of this type of ADE to enhance HIV-1 production of B lymphocytes and lymphoblastoid T cell lines. B cells are usually not infected in the patients, therefore the increase of their infection may not be pathologically important. Recently it was demonstrated (Fischer, Delibrias & Kazatchkine, 1991; June *et al.* 1992) that a large proportion, at least 30%, of the peripheral T cells both CD4+ and CD8+ subsets carry CR2, the receptor which is used for the C-ADE of the T lymphoblastoid cell lines. According to June *et al.* (1992) the percentage of CD4+ cells carrying CR2 is significantly lower in the blood of both asymptomatic and symptomatic HIV patients compared with the uninfected control donors. This finding indicates that the CD4+CR2+ cells are eliminated early from the blood of HIV-infected individuals. Therefore it is probable that if C-ADE has a biological significance *in vivo*, its effect is exerted at least partly by the increase of HIV-1 production and facilitation of destruction of CD4+ T cells.

Very recent findings of Sullivan *et al.* (1996) demonstrated that the effect of antibodies and complement on HIV-1 replication *in vivo* could be more complex than was claimed so far. The authors isolated HIV-1 virions from the plasma of HIV-infected individuals and demonstrated that exogenously added complement caused lysis of a fraction of the virions whereas another fraction of the virions was opsonized by complement without being lysed. They also proved that isolated virions already had bound specific antibodies *in vivo*, complement activation leading to lysis and opsonization was initiated by the virus–antibody complexes via the classical pathway. It is conceivable that lysis observed by the authors *in vitro* occurs *in vivo* as well and has a significant role in the rapid clearance of the virus (according to Perelson *et al.* (1996), the half-life of the virus is 6 h). In principle, ADE-mediating antibodies may exert their effect to enhance virus production *in vivo* in 2 different ways (Fig. 3). (*a*) They can inhibit complement-mediated lysis of plasma viruses induced by non-ADE mediating antibodies. (*b*) The non-lysed virus particles that are coated by antibody and complement fragments can be eliminated and killed by the phagocytic cells. On the other hand, it cannot be excluded that a part of the antibody on the non-lysed particles can mediate FcR-ADE or C-ADE. If it is really the case, ADE triggered by these antibodies may result in an increased rate of infection of the short-living CD4+ cells which produce more than 99% of the virus load in the body of the HIV-infected individuals (Perelson *et al.* 1996). Viral load measured by the concentration of HIV-1 RNA in plasma is the most important predictive marker of HIV disease. The higher the plasma HIV-RNA concentration the quicker progression can be expected (Mellors *et al.* 1996; O'Brien *et al.* 1996).

If our hypothesis, summarized in Fig. 3, is correct, a correlation should exist between the HIV-RNA concentration and the amount of ADE-mediating antibodies. To the best of our knowledge, no data on a parallel measurement of viral load and FcR-ADE or C-ADE have been reported until now. Findings of Montefiori *et al.* (1995) on the very high C-ADE activity in acute primary infection of SIV-inoculated macaques and the gradual but finally highly significant decrease of C-ADE in parallel to the drop of p24 antigenaemia at the switch from the acute primary infection to the asymptomatic phase seems to support our assumption. Similarly, it is well known that viral load increases during the symptomatic stage of HIV disease as compared to the asymptomatic stage. Our findings on the more frequent occurrence of C-ADE in the ARC/AIDS patients as compared to the asymptomatic ones (Tóth *et al.* 1991), and on a gradual increase in the C-ADE titres with the progression of HIV disease (Füst *et al.* 1994*a*; Tóth *et al.* unpublished) also seem to support the hypothesis summarized in Fig. 3. This hypothesis can be experimentally tested as soon as the HIV-propagating short-lived CD4+ cell population is better characterized.

Effect on the trapping of virus-antibody-complement complexes on follicular dendritic cells (FDC) in germinal centres of lymph nodes. Lymphoid germinal centres are most important reservoirs of human immunodeficiency virus type 1 particles (Tenner-Racz *et al.* 1988; Fox *et al.* 1991). Trapping of HIV-1 on the surface of the cells may be one of the main protective mechanisms against increase of the virus load. On the other hand, CD4 cells can be infected by HIV-1 present on the FDC. According to the *in vitro* studies of Jolling *et al.* (1993), binding of human immunodeficiency virus type 1 particles to FDC is mediated by complement. It is tempting to speculate that *in vivo* the enhancing antibodies may influence the bonds between virus-complement complexes and FDC during the early phase of lymph node infection, and contribute by this way the sharp increase of viral load during the symptomatic stages of HIV disease.

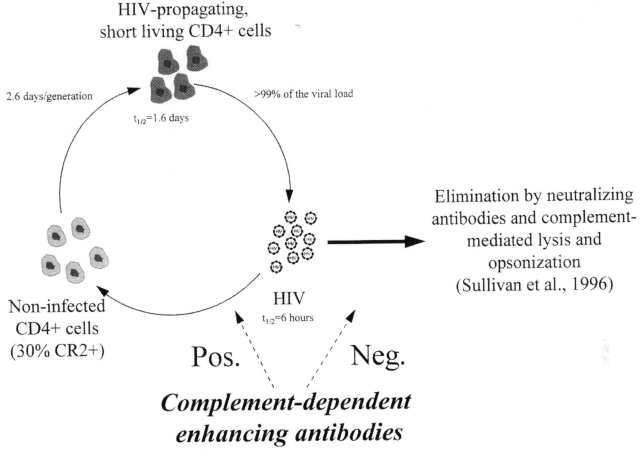

Fig. 3. A new hypothesis on the mechanism of the effect of complement-mediated enhancing antibodies on the facilitation of HIV propagation and consequently progression of HIV disease. According to Sullivan *et al.* (1996), antibody-dependent, complement-mediated lysis and opsonization have a major role in the rapid elimination of plasma virions. Enhancing antibodies may down-regulate elimination by inhibiting lysis of plasma virions or by facilitating the HIV-1 infection of not infected CD4[+] cells and their entry in the rapid-turnover cycle which is responsible for the production of more than 99 % of HIV-1 virions in the body (Perelson *et al.* 1996). Since high levels of plasma HIV-1 predicts a quick progression of HIV-disease (Mellors *et al.* 1996; O'Brien *et al.* 1996) any factor interfering with the elimination of plasma virions may contribute to the progression of HIV disease.

POTENTIAL RISKS OF ADE IN HUMAN HIV VACCINE TRIALS

A report on a workshop of this topic, held in December, 1992 was given by Mascola *et al.* (1993). Based on early data from SIV model (Montefiori *et al.* 1990) and volunteers vaccinated with candidate subunit HIV vaccines (Bernard *et al.* 1990; Dolin *et al.* 1991; Haubrich *et al.* 1992), authors of the report concluded that although ADE can be detected in a part of vaccinated people, no evidence is available which indicates that ADE occurs in these subjects when they are infected with HIV-1. They stressed that determination of *in vivo* correlates of protection/enhancement as well as development of an HIV animal model is necessary to evaluate the ADE risk accurately. Unfortunately, controversial results were obtained in recent studies performed in SIV infected macaques, too. Montefiori *et al.* (1995) failed to find any correlation between the extent of C-ADE measured on the day of challenge with wild virus and the protection by vaccination in macaques immun-

ized with live attenuated SIVmac239/nef deletion strain or primed with recombinant SIVmne gp120 vaccinia virus and boosted with SIVmne rgp160. Some protected animals had a C-ADE activity similar to that measured in unprotected macaques. According to the authors, however, 'it remains to be determined whether C-ADE-mediating antibodies interfere with vaccine efficacy after heterologous virus challenge or when animals are challenged with homologous or heterologous virus after vaccine-induced immune responses are waned'. In another study (Mitchell *et al.* 1995), it was found that the beneficial effect of a recombinant SIV gp160 vaccine was markedly diminished when the animals were boosted with a synthetic peptide corresponding to the immunodominant gp41 epitope of SIV gp41 (amino acids 603–622) and challenged finally with SIVmac251. Animals immunized with this protocol had detectable p27 antigen longer and died of simian AIDS before the animals of the control groups immunized with the gp160 vaccine but not boosted at all or boosted with an irrelevant peptide. Enhancement of infection in animals vaccinated with recombinant subunit vaccines after challenge with the homologous virus was observed in two other animal retrovirus models: in feline immunodeficiency virus (FIV) infection (Hosie *et al.* 1992; Siebelink *et al.* 1995) and in equine infectious anaemia virus (EIAV) infection (Wang *et al.* 1994).

In a human vaccine trial performed with a gp160 candidate vaccine (VaxSyn) Keefer *et al.* (1994) reported on the detection of C-ADE in 11/19 subjects immunized with the higher dose of 640 µg.

CONCLUSIONS

ADE was first described 10 years ago. Since that time several important details of the 2 types (FcR-dependent and complement-mediated) ADE and mechanisms of *in vitro* effects of the ADE-mediating antibodies were revealed. FcR-ADE and C-ADE were detected in HIV-infected patients and in experimental animals infected with HIV and SIV, several findings seem to demonstrate the role of ADE in the progression of HIV disease; these observations, however, need further support. Standardization of the methods used for the measurement of FcR-ADE and C-ADE is crucial for comparative data from different labs. Development of C-ADE was observed in macaques immunized with SIV vaccines and in phase I/II trials of human subunit gp160 vaccines. Nevertheless, no studies on the *in vivo* relevance of ADE in vaccinated people have been published. It would be most interesting to study if 'breakthrough' infections which sometimes occur in volunteers immunized with the currently used candidate subunit vaccines were associated with the development of ADE-mediating antibodies when HIV-1 infection occurred. Independently of the result of these studies, however, efforts to eliminate the epitopes which are responsible for induction of ADE-mediating antibodies (Lee *et al.* 1994) such as epitopes corresponding to the immunodominant epitopes of gp41 or the V3 loop of gp120 in the development of new vaccine candidates seems to be reasonable.

ACKNOWLEDGEMENT

I am highly indebted to Dr Zoltán Prohászka for his kind help in preparing the Figures.

REFERENCES

ALKHATIB, G., COMBADIERE, C., BRODER, C. C., FENG, Y., KENNEDY, P. E., MURPHY, P. M. & BERGER, E. A. (1996). CC CKR5: A RANTES, MIP1α, MIP1β receptor as a fision cofactor for macrophage-tropic HIV-1. *Science* **272**, 1955–1958.

AUEWARAKUL, P., LOUISIRIROTCHANAKUL, S., SUTTHENT, R. TAECHOWISAN, T., KANOKSINSOMBAT, C. & WASI, C. (1996). Analysis of neutralizing and enhancing antibodies to human immunodeficiency virus type 1 primary isolates in plasma of individuals infected with env genetic subtype B and E viruses in Thailand. *Viral Immunology* **9**, 175–185.

BAKKER, L. J., NOTTET, H. S. L. M., DE VOS, N. M., DE GRAAF, L., VAN STRIJP, J. A. G., VISSER, M. R. & VERHOEF, J. (1992). Antibody and complement enhance binding and uptake of HIV-1 by human monocytes. *AIDS* **6**, 35–41.

BERMAN, P. W., EASTMAN, D. J. & WILKES, D. M. (1994) Comparison of the immune response to recombinant gp120 in human and chimpanzees. *AIDS* **8**, 591–601.

BERNARD, J., REVEIL, B., NAJMAN, I., LIAUTAUD-ROGER, F., FOUCHRAD, M., PICARD, O., CATTAN, A., MABONDZO, A., LAVERNE, S., GALLO, R. C. & ZAGURY, D. (1990). Discrimination between protective and enhancing HIV antibodies. *AIDS Research and Human Retroviruses* **6**, 243–249.

BOYER, V., DELIBRIAS, C., NORAZ, N., FISCHER, E., KAZATCHKINE, M. D. & DESGRANGES, C. (1992). Complement receptor type 2 mediates infection of the human DC4-negative Raji B-cell line with opsonized HIV. *Scandinavian Journal of Immunology* **36**, 879–883.

CONNOR, H. I., DINCES, N. B., HOWELL, A. L., ROMET-LEMONNE, J. L., PASQUALI, J. L. & FANGER, M. W. (1991). Fc receptors for IgG (Fc gamma Rs) on human monocytes and macrophages are not infectivity receptors for human immunodeficiency virus type 1: studies using bispecific antibodies to target HIV-1 to various myeloid cell surface molecules, including the Fc gamma R. *Proceedings of the National Academy of Sciences, USA* **88**, 9593–9597.

DENG, H., LIU, R., ELLMEIER, W., CHOE, S., UNUTMAZ, D., BURKHART, M., DI MARZIO, P., MARMON, S., SUTTON, R. E., HILL, C. M., DAVIS, C. B., PEIPER, S. C., SCHALL, T. J., LITTMAN, D. R. & LANDAU, N. R. (1996). Identification of a major co-receptor for primary isolates of HIV-1. *Nature* **381**, 661–666.

DIERICH, M. P., EBENBICHLER, C., MARSCHANG, P., FÜST, G., THIELENS, N. & ARLAUD, G. (1993). HIV and human complement: Mechanisms of interaction and biological implications. *Immunology Today* **14**, 435–440.

DOLIN, R., GRAHAM, B. S., GREENBERG, S. B., TACKET, C. O., BELSCHE, R. B., MIDTHUN, K., CLEMENTS, N. L., GORSE, G. J., HORGAN, B. W. & ALTMAR, R. L. (1991). The safety and immunogenicity of human immunodeficiency virus type 1 (HIV-1) recombinant gp160 vaccine in humans. *Annals of Internal Medicine* **114**, 119–127.

EATON, A. M., UGEN, K. E., WEINER, D. B., WILDE, S. T. & LEVE, J. A. (1994). An anti-gp41 human monoclonal antibody that enhances HIV-1 infection in the absence of complement. *AIDS Research and Human Retroviruses* **10**, 13–18.

FENG, Y., BRODER, C. C., KENNEDY, P. E. & BERGER, E. A. (1996). HIV-1 entry cofactor: functional cDNA cloning of a seven-transmembrane, G-protein-coupled receptor. *Science* **272**, 872–877.

FISCHER, E., DELIBRIAS, C. & KAZATHCHKINE, M. D. (1991). Expression of CR2 The C3dg/EBV receptor, CD21) on normal human peripheral blood lymphocytes. *Journal of Immunology* **146**, 865–869.

FORTHAL, D. N., LANDUCCI, G., GORNY, M. K., ZOOLA-PAZNER, S. & ROBINSON, E. W. JR. (1995). Functional activities of 20 human immunodeficiency virus type 1 (HIV-1)-specific human monoclonal antibodies. *AIDS Research and Human Retroviruses* **11**, 1095–1099.

FOX, C. H., TENNER-RACZ, K., RACZ, P., FIRPO, A., PIZZO, P. A. & FAUCI, A. S. (1991). Lymphoid germinal centers are reservoirs of human immunodeficiency virus type 1 RNA. *The Journal of Infectious Diseases* **164**, 1051–1057.

FÜST, G., PROHÁSZKA, Z., HIDVÉGI, T., THIELENS, N., ARLAUD, G., TÓTH, F. D., KISS, J., DIERICH, M. P. & UJHELYI, E. (1994*b*). Competition of complement proteins end specific antibodies for binding to HIV-1 envelope antigens. *Acta Microbiologica et Immunologica Hungarica*, **41**, S27–S31.

FÜST, G., TÓTH, F. D., KISS, J., UJHELYI, E., NAGY, I. & BÁNHEGYI, D. (1994*a*). Neutralizing and enhancing antibodies measured in complement-restored serum samples from HIV-1 individuals correlate with immunosuppression and disease. *AIDS* **8**, 603–609.

GARDNER, M., ROSENTHAL, A., JENNINGS, M., YEE, J. A., ANTIPA, L. & MACKENZIE, M. (1994). Passive immunization of macaques against SIV infection. *Journal of Medical Primatology* **23**, 164–174.

GARDNER, M., ROSENTHAL, A., JENNINGS, M., YEE, J., ANTIPA, L. & ROBINSON, E. JR (1995). PASSIVE IMMUNIZATION OF RHESUS MACAQUES AGAINST SIV INFECTION AND DISEASE. *AIDS Research and Human Retoviruses* **11**, 843–854.

GRAS, G. S. & DORMONT, D. (1991). Antibody-dependent and antibody-independent complement-mediated enhancement of human immunodeficiency virus type 1 infection in a human, Epstein–Barr virus-transformed B-lymphocytic cell line. *Journal of Virology* **65**, 541–545.

GRAS, G., STRUB, T. & DORMONT, D. (1988). Antibody-dependent enhancement of HIV infection. *The Lancet* **1**, 1285.

HALSTEAD, S. B. (1988). Pathogenesis of dengue: challenges to molecular biology. *Science* **239**, 476–481.

HALSTEAD, S. B., SHORTWELL, B. H. & CASALS, J. (1979). *In vivo* enhancement of dengue virus infection in rhesus monkeys by passively transferred antibody. *Journal of Infectious Diseases* **140**, 527–535.

HAUBRICH, R. H., TAKEDA, O., KOFF, W., SMITH, G. & ENNIS, F. A. (1992). Studies on antibody-dependent enhancement of human immunodeficiency virus (HIV) type 1 infection mediate by Fc receptors using sera from recipients of a recombinant gp160 experimental HIV-1 vaccine. *Journal of Infectious Diseases* **165**, 545–548.

HOMSY, J., MEYER, M. & LEVY, J. A. (1990). Serum enhancement of human immunodeficiency virus (HIV) infection correlates with disease in HIV-infected individuals. *Journal of Virology* **64**, 1437–1440.

HOMSY, J., MEYER, M., TATENO, S., CLARKSON, S. & LEVY, J. A. (1989). The Fc and not CD4 receptor mediates antibody enhancement of HIV infection in human cells. *Science* **244**, 1357–1360.

HOMSY, J., TATENO, M. & LEVY, J. A. (1988). Antibody-dependent enhancement of HIV infection. *The Lancet* **1**, 1285–1286.

HOSIE, M. J., OSBORNE, R., REID, G., NEIL, J. C. & JARRETT, O. (1992). Enhancement after feline immunodeficiency virus vaccination. *Veterinary Immunology and Immunopathology* **35**, 191–197.

JIANG, S., LIN, K. & NEURATH, A. R. (1991). Enhancement of human immunodeficiency virus type 1 infection by antisera to peptides from the envelope glycoprotein gp120/gp41. *Journal of Experimental Medicine* **174**, 1557–1563.

JOLLING, P., BAKKER, L. J., VAN STRIJP, J. A. G., MEERLOO, T., DE GRAAF, L. DEKKER, M. E. M., GOUDSMIT, J., VEHOEF, J. & SCHUURMAN, H.-J. (1993). Binding of human immunodeficiency virus type 1 to follicular dendritic cells *in vitro* is complement dependent. *Journal of Immunology* **150**, 1065–1073.

JOUAULT, T., CHAPUIS, F., OLIVIER, R. PARRAVICINI, C., BAHRAOUI, E. & GLUCKMAN, J.-C. (1989). HIV infection of monocytic cells: role of antibody-mediated virus binding to Fc-gamma receptors. *AIDS* **3**, 125–133.

JUNE, R. A., LANDAY, A. L., STEFANIK, K., LINT, T. F. & SPEAR, G. T. (1992). Phenotypic analysis of complement receptor 2^+ T lymphocytes: reduced expression by $CD4^+$ cells in HIV-infected persons. *Immunology* **75**, 59–65.

JUNE, R. A., SCHADE, S. Z., BANKOWSKI, M. J., KUHNS, M., McNAMARA, A., LINT, T. G., LANDAY, A. L. & SPEAR, G. T. (1991). Complement and antibody mediate enhancement of HIV infection by increasing virus binding and provirus formation. *AIDS* **5**, 269–274.

KEEFER, M. C., GRAHAM, B. S., BELSCHE, R. B., SCHWARTZ, D., COREY, L., BOLOGNESI, D. P., STABLEIN, D. M., MONTEFIORI, D. C., McELRATH, M. J. & CLENETS, M. L. (1994). Studies of high doses of a human immunodeficiency virus type 1 recombinant glycoprotein 160 candidate vaccine in HIV-1 type 1-seronegative humans. *AIDS Research and Human Retroviruses* **10**, 1713–1723.

KLIKS, S. C. & LEVY, J. A. (1994). Maternal antibody response and maternal–infant HIV-1 transmission. *The Lancet* **343**, 1364.

KLIKS, S. C., NISALAK, A., BRANDT, W. E., WAHL, L. & BURKE, D. S. (1989). Antibody-dependent enhancement of dengue virus growth in human monocytes as a risk factor for dengue hemorrhagic fever. *American Journal of Tropical Medicine and Hygiene* **40**, 444–451.

KLIKS, S. C., SHIODA, T., HAIGWOOD, N. L. & LEVY, J. A. (1993). V3 variability can influence the ability of an antibody to neutralize or enhance infection by diverse strains of human immunodeficiency virus type 1. *Proceedings of the National Academy of Sciences, USA* **90**, 11518–11522.

KOSTRIKIS, L. G., CAO, Y., NGAI, H., MOORE, J. P. & HO, D. H. (1996). Quantitative analysis of serum neutralization of human immunodeficiency virus type 1 from subtypes, A, B, C, D, E, F, and I: lack of direct correlation between neutralization serotypes and genetic subtypes and evidence for prevalent serum-dependent infectivity enhancement. *Journal of Virology* **70**, 445–458.

KOZLOWSKI, P. A., BLACK, K. P., SHEN, L. & JACKSON, S. (1995). High prevalence of serum IgA HIV-1 infection-enhancing antibodies in HIV-infected persons. Masking by IgG. *Journal of Immunology* **154**, 6163–6173.

LALLEMANT, M., BAILLOU, A., LALLEMANT-LE COEUR, S., NZINGOULA, S., MAMPAKA, M., M'PELÉ, F., BARIN, F. & ESSEX, M. (1994). Maternal antibody response at delivery and perinatal transmission of human immunodeficiency virus type 1 in African women. *The Lancet* **343**, 1001–1005.

LAURENCE, J., SAUNDERS, A., EARLY, E. & SALMON, J. E. (1990). Human immunodeficiency virus infection of monocytes: relationship to Fc-gamma receptors and antibody-dependent viral enhancement. *Immunology* **70**, 338–343.

LEE, C. N., ROBINSON, J., CHENG, Y. L., ESSEX, M. & LEE, T. H. (1994). Influence in deletion in N or C terminus of HIV-1 glycoprotein 120 on binding of infectivity enhancing antibody. *AIDS Research and Human Retroviruses* **10**, 1065–1069.

LUND, O., HANSEN, J., SORENSEN, A. M., MOSEKILDE, E., NIELSEN, J. O. & HANSEN, J.-E. S. (1995). Increased adhesion as a mechanism of antibody-dependent and antibody-independent complement-mediated enhancement of human immunodeficiency virus infection. *Journal of Virology* **69**, 2393–2400.

MARKHAM, R. B., COBERLY, J., RUFF, A. J., HOOVER, D., GOMEZ, J., HOLT, E., DESORMEAUX, J., BOULOS, R., QUINN, T. C. & HALSEY, N. A. (1994). Maternal IgG1 and IgA antibody to V3 loop consensus sequence and maternal-infant HIV-1 transmission. *The Lancet* **343**, 390–391.

MASCOLA, J. R., MATHIESON, B. J., ZACK, P. M., WALKER, M. C., HALSTEAD, S. B. & BURKE, D. S. (1993). Summary report: workshop on the potential risks of antibody-dependent enhancement in human HIV vaccine trials. *AIDS Research and Human Retroviruses* **12**, 1175–1184.

MATSUDA, S., GIDLUND, M., CHIODI, A., CAFARO, A., NYGREN, A., MOREIN, B., NILSSON, K., FENYO, E.-M. & WIGZELL, H. (1989). Enhancement of human immunodeficiency virus replication in human monocytes by low titres of anti-HIV antibodies *in vitro*. *Scandinavian Journal of Immunology* **30**, 425–434.

MELLORS, J. W., RINALDO, C. R., GUPTA, P., WHITE, R. M., TODD, S. A. & KINGSLEY, R. A. (1996). Prognosis of HIV-1 infection predicted by the quantity of virus in plasma. *Science* **272**, 1167–1170.

MITCHELL, W. M., TORRES, J., JOHNSON, P. R., HIRSCH, V., YILMA, T., GARDNER, M. R. & ROBINSON, W. E. JR. (1995). Antibodies to the putative SIV infection-enhancing domain diminish beneficial effects of an SIV gp160 vaccine in rhesus macaques. *AIDS* **9**, 27–34.

MONELL, C. R., HOOVER, D. R., ODAKA, N., HE, X., SAAH, A. J. & STRAND, M. (1993). Assessment of antibody response to the immunosuppressive/immunodominant region of HIV gp41 in a 5-year longitudinal study. *Journal of Medical Virology* **39**, 125–130.

MONTEFIORI, D. C., LEFKOWITZ, L. B. JR, KELLER, R. E., HOLMBERG, V., SANDSTROM, E., PHAIR, J. P. & MULTICENTER AIDS COHORT STUDY. (1991). Absence of a clinical correlation for complement-mediated infection-enhancing antibodies in plasma or sera from HIV-1-infected individuals. *AIDS* **5**, 513–517.

MONTEFIORI, D. C., PANTALEO, G., FIND, L. M., ZHOU, J. T., ZHOU, J. Y., BILSKA, M., MIRALLES, G. D. & FAUCI, A. S. (1996). Neutralizing and infection-enhancing antibody responses to human immunodeficiency virus type 1 in long-term non-progressors. *Journal of Infectious Diseases* **173**, 60–67.

MONTEFIORI, D. C., REIMANN, K. A., LETVIN, N. L., ZHOU, J. & HU, S.-L. (1995). Studies of complement-activating antibodies in the SIV/macaque model of acute primary infection and vaccine protection. *AIDS Research and Human Retroviruses* **11**, 963–970.

MONTEFIORI, D. C., ROBINSON, W. E., JR, HIRSCH, V. M., MODLISZEWSKI, A., MITCHELL, W. M. & JOHNSON, P. R. (1990). Antibody-dependent enhancement of simian immundeficiency virus (SIV) infection *in vitro* by plasma from SIV-infected rhesus macaques. *Journal of Virology* **64**, 113–119.

MOUHOUB, A., DELIBRIAS, C. C., FISCHER, E., BOYER, V. & KAZATCHKINE, M. D. (1996). Ligation of CR1 (C3b receptor, CD35) on CD4$^+$ T lymphocytes enhances viral replication in HIV-infected cells. *Clinical Immunology and Immunopathology* **106**, 297–303.

NIEDRIG, M., BROKER, M., WALTER, G., STUBER, W., HARTHUS, H. P., MEHDI, S., GELDERBLOM, H. R. & PAULI, G. (1992). Murine monoclonal antibodies directed against the transmembrane protein gp41 of human immunodeficiency virus type 1 enhance its infectivity. *Journal of General Virology* **73**, 951–954.

NKENGASONG, J. N., PEETERS, M., NDUMBE, P., JANSSENS, V., EILLEMS, B., FRANSEN, K., NGOLDE, M., PIOT, P. & VAN DER GROEN, G. (1994). Cross-neutralizing antibodies to HIV-1ANT70 and HIV-1$_{IIIB}$ in sera of African and Belgian HIV-1-infected individuals. *AIDS* **8**, 1089–1096.

NOTTET, H. S., JANSE, I., DE VOS, N. M., BAKKER, J. L., VISSER, M. R. & VERHOEF, J. (1992). Antibody mediated enhancement of HIV-1 infection of an EBV transformed B cell line is CD4 dependent. *European Journal of Clinical Investigations* **22**, 670–675.

O'BRIEN, T. B., BLATTNER, W. A., WATERS, D., EYSTER, E., HILGARTNER, M. W., COHEN, A. R., LUBAN, N., HATZAKIS, A., ALADORT, L. M., ROSENBERG, P. S., MILEY, W. J., KRONER, B. L., GOEDERT, J. J. & THE MULTICENTER HEMOPHILIA COHORT STUDY. (1996). Serum HIV-1 RNA levels and time to development of AIDS in the Multicenter Hemophilia Cohort Study. *Journal of the American Medical Association* **276**, 105–110.

PERELSON, A. S., NEUMANN, A. U., MARKOWITZ, M., LEONARD, J. M. & HO, D. D. (1996). HIV-1 dynamics *in vivo*: virion clearance rate, infected cell life-span, and viral generation time. *Science* **271**, 1582–1586.

PERNO, C.-F., BASELER, M. W., BRODER, S. & YARCHOAN, R. (1990). Infection of monocytes by human immunodeficiency virus type 1 is blocked by inhibitors of CD4-gp120 binding, even in the presence of enhancing antibodies. *Journal of Experimental Medicine* **141**, 1043–1056.

PORTERFIELD, J. S. (1986). Antibody-dependent enhancement of viral infectivity. *Advances in Virus Research* **31**, 335–355.

PROHÁSZKA, Z., NEMES, J., HIDVÉGI, T., TÓTH, F. D., SZABÓ, J., UJHELYI, J., THIELENS, N., DIERICH, M. P., SPATH, P., HAMPL, H., KISS, J., ARLAUD, G. & FÜST, G. (1997). Two parallel routes of the complement-mediated antibody-dependent enhancement of the human immunodeficiency virus type 1 infection. *AIDS*, **11**, 949–958.

RADKOWSKI, M., LASKUS, I., GOCH, A. & SLUSARCHIK, J. (1993). Affinity of anti-gp41 antibody in patients infected with human immunodeficiency virus type 1. *European Journal of Clinical Investigations* **25**, 455–458.

ROBINSON, W. E. JR, KAWAMURA, T., GORNY, M. K., LAKE, D., XU'J.-Y., MATSUMUTO, Y., SUGAO, T., MASUHO, Y., MITCHELL, W. M., HERSCH, E. & ZOLLA-PAZNER, S. (1990a). Human monoclonal antibodies to the human immunodeficiency virus type 1 (HIV-1) transmembrane glycoprotein gp41 enhance HIV-1 infection *in vitro*. *Proceedings of the National Academy of Sciences, USA* **87**, 3185–3189.

ROBINSON, W. E. JR, KAWAMURA, T., LAKE, D., MASUHO, Y., MITCHELL, W. M. & HERSCH, E. M. (1990b). Antibodies to the primary immunodominant domain of human immunodeficiency virus type 1 (HIV-1) glycoprotein enhance HIV-1 infection *in vitro*. *Journal of Virology* **64**, 5301–5305.

ROBINSON, W. E. JR, GORNY, M. K., XU, J.-Y., MITCHELL, W. M. & ZOLLA-PANER, S., (1991). Two immunodominant domains of gp41 bind antibodies which enhance human immunodeficiency virus type 1 infection *in vitro*. *Journal of Virology* **65**, 4169–4176.

ROBINSON, W. E. JR, MONTEFIORI, D. C., GILLESPIE, D. H. & MITCHELL, W. M. (1989a). Complement-mediated, antibody-dependent enhancement of HIV-1 infection *in vitro* is characterized by increased protein and RNA syntheses and infectious virus release. *Journal of Acquired Immune Deficiency Syndromes* **2**, 33–42.

ROBINSON, W. E. JR, MONTEFIORI, D. C. & MITCHELL, W. M. (1987). A human immunodeficiency virus type 1 (HIV-1) infection-enhancing factor in seropositive sera. *Biochemical and Biophysical Research Communications* **149**, 693–699.

ROBINSON, W. E. JR, MONTEFIORI, D. C. & MITCHELL, W. M. (1988). Antibody-dependent enhancement of human immunodeficiency virus type 1 infection. *The Lancet* **1**, 830–831.

ROBINSON, W. E. JR, MONTEFIORI, D. C. & MITCHELL, W. M. (1990). Complement-mediated antibody-dependent enhancement of HIV-1 infection requires CD4 and complement receptors. *Virology* **175**, 600–604.

ROBINSON, W. E. JR, MONTEFIORI, D. C., MITCHELL, W. M., PRINCE, A. M., ALTER, H. J., DREESMEN, G. R. & EICHBERG, J. W. (1989b). Antibody-dependent enhancement of human immunodeficiency virus type 1 (HIV-1) infection *in vitro* by serum from HIV-1-infected and passively immunized chimpanzees. *Proceedings of the National Academy of Sciences, USA* **86**, 4710–4714.

SAARLOOS, M. N., LINT, T. F. & SPEAR, G. T. (1995). Efficacy of HIV-specific and 'antibody-independent' mechanisms for complement activation by HIV-infected cells. *Clinical and Experimental Immunology* **99**, 189–195.

SCHADDUCK, P. P., WEINBERG, J. B., HANEY, A. F., BARTLETT, J. A., LANGLOIS, A., BOLOGNESI, D. & MATTHEWS, T. J. (1991). Lack of enhancing effect of human anti-human immunodeficiency virus type 1 (HIV-1) antibody on HIV-1 infection of human blood monocytes and macrophages. *Journal of Virology* **65**, 4309–4316.

SIEBELINK, K. H., TIJHAAR, E., HUISMAN, R. C., HUISMAN, W., DE RONDE, A., DARBY, I. H., FRANCIS, M. J., RIMMELZWAAN, F. F. & OSTERHAUS, A. D. (1995). Enhancement of feline immunodeficiency virus infection after immunization with envelope glycoprotein subunit vaccines. *Journal of Virology* **69**, 3704–3711.

SPEAR, G. T., SULLIVAN, B. L., LANDAY, A. L. & LINT, T. F. (1990). Neutralization of human immunodeficiency virus type 1 by complement occurs by viral lysis. *Journal of Virology* **64**, 5869–5873.

SULLIVAN, B. L., KNOPOFF, E. J., SAIFUDDIN, M., TAKEFMAN, D. M., SAARLOOS, M-L., SHA, B. E. & SPEAR, G. T. (1996). Susceptibility of HIV-1 plasma virus to complement-mediated lysis. Evidence for a role in clearance of virus *in vivo*. *Journal of Immunology* **157**, 1791–1798.

SULLIVAN, N., SUN, Y., LI, J., HOFMANN, W. & SODROSKI, J. (1995). Replicative function and neutralization sensitivity of envelope glycoproteins from primary and T-cell passaged human immunodeficiency virus type 1 isolates. *Journal of Virology* **69**, 4413–4422.

SÜSAL, C., DANIEL, V. & OPELZ, G. (1996). Does AIDS emerge from a disequilibrium between two complementary groups of molecules that mimic the MHC? *Immunology Today* **17**, 114–119.

TAKEDA, A., ROBINSON, J. E., HO, D. D., DEBOUCK, C., HAIGWOOD, N. L. & ENNIS, F. A. (1992). Distinction of human immunodeficiency virus type 1 neutralization and infection enhancement by human monoclonal antibodies to glycoprotein 120. *Journal of Clinical Investigations* **89**, 1952–1997.

TAKEDA, A., SWEET, R. W. & ENNIS, F. A. (1990). Two receptors are required for antibody-dependent enhancement of human immunodeficiency virus type 1 infection: CD4 and FcγR. *Journal of Virology* **64**, 5605–5610.

TAKEDA, A., TUAZON, C. & ENNIS, F. (1988). Antibody-enhanced infection of HIV-1 via Fc receptor-mediated entry. *Science* **242**, 580–583.

TENNER-RACZ, K., RACZ, P., SCHMIDT, H., DIETRICH, M., LOUIE, A., GARTNER, S. J. & POPOVIC, M. (1988). Immunohistochemical, electron microscopic and *in situ* hybridization evidence for the involvement of lymphatics in the spread of HIV. *AIDS* **2**, 299–309.

THOMAS, H. I. J., WILSON, S., O'TOOLE, C. M., LISTER, C. M., SAEED, A. M., WATKINS, R. P. F. & MORGAN-CAPNER, P. (1996). Differential maturation of avidity of IgG antibodies to gp41, p24 and p17 following infection with HIV-1. *Clinical and Experimental Immunology* **103**, 185–191.

TÓTH, F. D., MOSBORG-PETERSEN, P., KISS, J., ABOAGYE-MATHIESEN, G., ZDRAKOVIC, M., HAGER, H., ARANYOSI, J., LAMPÉ, L. & EBBESEN, P. (1994b). Antibody-dependent enhancement of HIV-1 infection in human term syncytiotrophoblast cells cultured *in vitro*. *Clinical and Experimental Immunology* **96**, 89–394.

TÓTH, F. D., SÜSAL, C., UJHELYI, E., BÁNHEGYI, D., KISS, J., DANIEL, V., NAGY, I., OPELZ, G. & FÜST, G. (1994). Comparative studies of antibodies that are associated with disease progression in HIV disease. *Immunology Letters* **41**, 33–36.

TÓTH, F. D., SZABÓ, B., UJHELYI, E., PÁLÓCZI, K., FÜST, G., KISS, J., BÁNHEGYI, D. & HOLLÁN, S. R. (1991).

Neutralizing and complement-dependent enhancing antibodies in different stages of HIV infection. *AIDS* **5**, 263–268.

TREMBLAY, M., MELOCHE, S., SEKALY, R.-P. & WINBERG, M. A. (1990). Complement receptor 2 mediates enhancement of human immunodeficiency virus type 1 infection in Epstein–Barr virus carrying B cells. *Journal of Experimental Medicine* **171**, 1791–1976.

TRISCHMANN, H., DAVIS, M. D. & LACHMANN, P. J. (1995). Lymphocytotropic strains of HIV type 1 when complexed with enhancing antibodies can infect macrophages via Fc gamma RIII, independently of CD4. *AIDS Research and Human Retroviruses* **11**, 343–352.

WANG, S. Z., RUSHLOW, K. E., ISSEL, C. J., COOK, R. F., RAABE, M. L., CHONG, Y. H., COSTA, L. & MONTELARO, R. C. (1994). Enhancement of EIAV replication and disease by immunization with a baculovirus-expressed recombinant envelope surface glycoprotein. *Virology* **199**, 247–251.

ZEIRA, M., BYRN, R. A. & GROOPMAN, J. E. (1990). Inhibition of serum-enhanced HIV-1 infection of U937 monocytoid cells by recombinant soluble CD4 and anti-CD4 monoclonal antibody. *AIDS Research and Human Retroviruses* **6**, 629–639.

ZOELLNER, B., FEUCHT, H. H. & LAUFS, R. (1992). Role of proteases as cofactors for antibody-dependent enhancement of HIV. *AIDS* **8**, 887–888.

ZWART, G., VEN DER HOEK, L., VALK, M., CORNELISSEN, M. T., BEAN, E., DEKKER, J., KOOT, M., KULKEN, Z. L. & GOUDSMIT, J. (1994). Antibody responses to HIV-1 envelope and gag epitopes in HIV-1 seroconverters with rapid versus slow disease progression. *Virology* **201**, 285–293.

Immunological enhancement of breast cancer

T. H. M. STEWART[1] *and* G. H. HEPPNER[2]*

[1] *Department of Medicine, Ottawa General Hospital, 501 Smyth Road, Ottawa, Ontario, Canada K1H 8L6* and
[2] *Breast Cancer Program, Karmanos Cancer Institute, 110 E Warren Avenue, Detroit MI 48201, USA*

SUMMARY

Breast cancer is a complex disease. Its aetiology is multifactorial, its period of development can span decades, and its clinical course is highly variable. Evaluation of the role of the immune response in either the development or control of breast cancer is also complex. Nevertheless, there is substantial information that in this disease, the immune response is not a host defence reaction and may even serve to facilitate cancer development. This evidence comes from a variety of sources including clinical–pathological investigations in women that show a correlation between the intensity of lymphocytic infiltration into the tumour mass with poor prognosis, studies in breast cancer patients that demonstrate a similar correlation between delayed hypersensitivity reactivity or *in vitro* assays of immune reactivity to tumour cell membranes or non-specific antigens and poor prognosis, and analyses of cancer incidence in chronically immunosuppressed, kidney transplant recipients who develop an unexpectedly low incidence of breast cancer. The overall conclusions from these human studies are corroborated by observations in mouse mammary tumour models that also demonstrate immune enhancement of breast cell proliferation *in vitro* and of breast cancer development *in vivo*. Potential mechanisms for these effects include production, by inflammatory cell infiltrates, of direct or indirect modulators of breast cell growth, e.g. cytokines, peptide or steroid hormones, enzymes involved in steroid metabolism, as well as of antibodies to growth factors or their receptors. These immune facilitatory mechanisms must be overcome if immune-based therapies are to be applied successfully in breast cancer.

Key words: breast cancer, immune facilitation, cell-mediated immunity, NK cells, lymphocytic infiltration, immune competence.

INTRODUCTION

In this chapter we discuss the possible involvement of the immune response in breast cancer development. First, we present the literature from pathologists and surgeons on evidence of immune recognition in breast cancer tissue and its relationship to prognosis. We next review studies on the linkage between specific or non-specific immune reactivity and breast cancer prognosis, as well as studies on breast cancer incidence or outcome in situations of immune suppression. From the experimental literature, principally in the mouse, we review evidence for immune system influence in breast cancer development. Our focus is on the natural immunobiology of breast cancer development. We will not discuss the perturbations of the immune system in advanced disease, nor, in the case of the mouse, on the response to transplantable tumours or lines.

Breast cancer is a complex disease. Its aetiology is multifactorial, its clinical manifestations are diverse and its outcome is variable. The natural history of breast cancer is prolonged; from epidemiological studies we know that circumstances early in life, in pre-adolescence or early adulthood, can influence

whether or not cancer develops many decades later (Henderson *et al.* 1981). This prolonged natural history is carried out against a background of profound physiological change, particularly in regard to reproductive hormone status, as well as during the gradual unfolding of the processes of maturation and ageing. The breast itself is a complex and dynamic organ. For most of life the epithelial components are relatively inconspicuous, but, when called upon, they are capable of marked hyperplasia, structural differentiation and function, followed in due time by rapid involution and quiescence. The breast is also an immunological organ. It is the source of immunoglobulins and T cells, passed from mother to nursing child (Eglinton, Roberton & Cummings, 1994), and therefore the breast must be hospitable to cells of the immune system.

IMMUNE RECOGNITION IN BREAST CANCER PATIENTS

We have recently co-authored reviews on the topic of the immune system in breast cancer and the reader is referred to them for more detailed documentation of the sources of information discussed here (Stewart & Tsai, 1993; Wei & Heppner, 1996). Overall, breast cancer patients are not grossly abnormal immuno-

* Corresponding author.

Parasitology (1997), **115**, S141–S153. © 1997 Cambridge University Press

logically, at least until the time of terminal disease. T and B cell counts and functional assays are well within the admittedly broad range of normal. There appears to be no strong correlation among various parameters of immune competence and accepted clinical indices of breast cancer prognosis, although it is usually possible to tease out some association if the patient numbers are large enough and the statistician persistent (Head, Elliott & McCoy, 1993).

A potential indicator of the involvement of the immune response in breast cancer development is the extent of inflammatory cell infiltration into the tumour mass. The literature is not consistent on this point. In a 1993 review, Stewart & Tsai found that 43 articles could be divided into 3 groups. (1) Nine reported that an intense infiltrate was associated with good prognosis; 6 of these articles were descriptions of medullary cancer. (2) Two articles found no relationship, of any sort, between prognosis and infiltration. (3) Twenty-three other papers reported poor prognosis when the stromal infiltrate was intense; 11 of these concerned inflammatory breast cancer. Six of the remaining 12 publications were particularly impressive: Champion, Wallace & Prescott (1972), Fisher, Saffer & Fisher (1974), Meyer & Hixon (1979), Fisher *et al.* (1983), Parkes *et al.* (1988) and Rosen & Groshen (1990). Overall, it appears that the immune system may down-regulate growth of medullary cancer, whereas it may stimulate other types of breast cancers. A particularly provocative study was provided by Kurtz *et al.* (1990). This group studied 18 factors in 496 stage I–II ductal cancers treated by conservative surgery and radiotherapy. Multivariate analysis showed that a major lymphocytic stromal reaction was the factor most strongly correlated with recurrence in younger women. The authors concluded that the intensity of the cellular reaction may reflect a cancer–host response that favours, rather than impedes, cancer growth.

In a retrospective study (length of follow-up was 19 years), Pupa *et al.* (1992) reported that over-expression of the c-erbB2 oncogene was associated both with the presence of lymphoplasmacytic infiltration and a bad prognosis in breast cancer patients in whom cancer cells could be demonstrated in the draining lymph nodes. Similarly, Tang *et al.* (1990) found a very strong association between amplification of the oncogenes c-erbB2, int-2 or c-myc and dense lymphocyte infiltration of the tumour in a series of 106 primary breast cancers. The authors suggested that cytokine production may be associated with paracrine immunological phenomena which are themselves associated with poor prognosis.

A very recent publication (Leek *et al.* 1996) focused specifically on the relationship among infiltrating macrophages, degree of tumour angiogenesis (vascularity) and survival of breast cancer patients.

There was a strong correlation between the first 2 parameters and between both with reduced survival. The authors suggested that macrophages promote angiogenesis by secreting a variety of endothelial cell cytokines or extracellular matrix-degrading enzymes. In turn, angiogenesis promotes metastatic spread, establishment of systemic disease, and ultimately, poor survival. A similar association between tumour-associated macrophage infiltration and propensity for metastasis has been seen in a mouse mammary tumour model (Mahoney, Fulton & Heppner, 1983).

Interpretation of studies on the significance of inflammatory infiltrates in breast cancer specimens is complicated by the heterogeneity of diseases lumped under the term 'breast cancer'. Many of the studies may be skewed by having significant numbers of medullary carcinomas in their mix, a disease with a good prognosis, or inflammatory carcinoma, a type with a very bad prognosis overall. In addition, the subclinical period of breast cancer development is quite prolonged, so that inflammatory cell infiltration with cancer may reflect earlier events, prior to clinical evidence of disease. In this context, the early work of Black (1972), who focused on morphological and functional indicators of immune reactivity in patients with 'precancerous' lesions or carcinoma-in-situ (CIS), may be instructive. Black found that a variety of immune-associated parameters signified a good prognosis in early breast cancer. Black's work remains unique in its emphasis on immune events in very early disease.

Histological detection of inflammatory cell infiltration cannot address the functional status. A number of investigators have isolated lymphocytes from primary breast cancers and tested their functionality in *in vitro* assays (Vose & Moore, 1979; Eremin, Coombs & Ashby, 1981; Whiteside *et al.* 1986). In general, T cells tend to predominate in isolated infiltrates, with CD8 out-numbering CD4 cells. NK cells are relatively infrequent. Functionally, the cells tend to be deficient, or even suppressive in lymphocyte proliferation or cytotoxicity assays (Vose & Moore, 1979; Eremin *et al.* 1981). However, in a detailed clonal analysis of T cells isolated from breast cancers, Whiteside *et al.* (1986), showed the presence of potential cytolytic function, demonstrable in expanded cell cultures *in vitro*. The relevance of these observations to cell function *in situ* is not clear.

DELAYED HYPERSENSITIVITY IN BREAST CANCER PATIENTS

Hope that the immune response can be used to interfere with cancer growth and development has undergone dramatic shifts over the past 30–40 years. Some of the first excitement developed out of the area of autoimmune disease. In the late 1950s and early 60s evidence was rapidly accumulating that autoimmune disease was a reality. Grave's disease was found to be the result of stimulation of thyroid function by antibodies (reviewed by McKenzie & Zakerija, 1986) and Hashimoto's thyroiditis resulted in destruction of the thyroid by cellular immune mechanisms (reviewed by Volpe, 1986). An occasional patient shows features of each disease, with a period of stimulation of function in an otherwise dying gland.

Animal models were developed to study the mechanisms underlying the initiation of autoimmune diseases. In several models a specific disease was induced by inoculation with the appropriate organ antigen, autologous or allogeneic in origin, homogenized in Freund's complete adjuvant (FCA). As reviewed by Waksman (1959, 1962), the severity of the induced disease correlated very well with the intensity of the delayed hypersensitivity reaction (DHR) to the organ antigen in the animal in which the disease was produced. Severe disease was accompanied by an intense DHR, mild disease with a weak DHR. Brent, Brown & Medawar (1958) showed that the tempo of graft rejection was reflected in the intensity expressed by recipient animals of DHR toward donor histocompatibility antigens; intense DHR signified rapid rejection. Similarly, in autoimmune disease, the severity and tempo of the immune reaction correlated well with the intensity of the DHR to the appropriate antigen. In both rapid graft rejection and severe autoimmune disease, the 'round cell' stromal infiltrate of the organ was found to be marked.

In 1964, Hughes & Lytton published a paper showing that 27 % of 50 patients with cancer gave a positive DHR when skin tested with acellular extracts of their own tumours. This experiment was repeated in a series of 144 patients and the earlier test results were confirmed (Stewart, 1969 *a*, *b*). The strongest reactions were induced by cell membranes; and there was a low incidence of positive reactions without lymphocytic or round cell infiltration of the tumour stroma. Only 6 % of such patients were positive, compared with a much greater frequency of positive reactions when the infiltrate was marked, 44 %, or very marked, 100 %. These findings suggested that not only was there an autoimmune reaction to autologous tumour in at least some patients, but that it might be of benefit to patients to strengthen this reaction, using the method that successfully induced autoimmune disease in animals, namely vaccination with membrane antigen homogenized in FCA. However, although the prospect of using cell membrane tumour antigens, homogenized with FCA, to vaccinate patients intradermally held promise, it also had potential dangers. If normal tissue antigens were present there was danger of inducing severe autoimmune disease, as had been shown in men in whom severe acute orchitis was

induced following autovaccination with an homogenate of normal testicular tissue in FCA, prior to completing orchidectomy as treatment for prostate cancer (Mancini *et al.* 1965). Thus, before embarking on a potentially dangerous clinical experiment, it was thought prudent to examine the survival experience of the largest cohort of tumour patients that had been skin-tested earlier with extracts of their own tumours, namely women with breast cancer (Stewart, 1969 *a*, *b*). There were 56 such women of whom 12 had shown a DHR to extracts of their tumours. The results of this study were published in 1971 (Stewart & Orizaga, 1971) and were shocking in their implication. There were 3 major findings; (1) Of 52 patients tested and followed for at least 2·5 years, 40 had a negative DHR to their own tumour; of these 31 (77·5 %) were alive, and 9 (22·5 %) had died of the cancer. Of 12 patients showing a positive DHR to their tumours, 5 (41·5 %) were alive and 7 (58·5 %) were dead. The difference between the 2 groups was statistically significant. (2) A significant association was found between a positive DHR and the degree of nuclear differentiation; DHR was seen only in 2 of 21 patients with moderate to highly differentiated tumours but in 10 of 35 patients with anaplastic tumours. (3) Antigens from one patient gave a positive reaction in another patient with breast cancer, suggesting common shared antigens, which from animal studies were indicative of a viral aetiology. Two conclusions were suggested; that if anything, the immune response might facilitate tumour growth, and that a viral influence might be present in the origin of breast cancer (see below).

IMMUNE FACILITATION OF HUMAN BREAST CANCER

Cell-mediated reactivity and prognosis

The hypothesis of the possible immune facilitation of breast cancer is disturbing and counterintuitive. However, it is not without support. In 1993, Stewart & Tsai reviewed the experience of others in regard to DHR and related assays of cell-mediated immunity in breast cancer patients. The highlights of this review include: (1) Non-specific stimulation of breast cancer patients using the method of Bacillus Calmette–Guérin inoculation or scarification as an adjuvant therapy significantly worsened prognosis by 20 % (Early Breast Cancer Trialist Collaborative Group, 1992). (2) In a study of 134 patient with breast cancer and 63 patients with benign breast lesions, low lymphocyte reactivity *in vitro* to various mitogens and antigens correlated with low overall risk compared to that of patients with high or intermediate risk disease (Wanebo *et al.* 1976). (3) In a series of 77 patients with inflammatory breast cancer treated with radiotherapy and chemotherapy, half of whom were randomized to receive BCG, patients who were tuberculin negative on skin testing survived significantly better than those who were tuberculin positive (Pouillart *et al.* 1981). (4) Tunisian women, who have a high frequency of inflammatory breast cancer, had a DHR to soluble breast cancer antigen with a frequency 3 times greater than that in women with other forms of breast cancer (Mourali *et al.* 1978). Tunisian adults were shown to have increased immune reactivity when compared to a comparable cohort of adult Americans (Levine *et al.* 1981). As mentioned before, inflammatory breast carcinoma is extremely aggressive. (5) The thymidine labeling index (TLI) was studied in primary invasive breast carcinomas of 133 patients (Meyer & Hixon, 1979). Operable patients with TLIs above the median had a significantly higher rate of occurrence than those with indices below. There was a significant linear increase of log TLI with increasing degrees of inflammatory cell reaction at the margin of the breast cancer. (6) Cannon, Dean & Herberman (1981) studied the lymphoproliferative responses of peripheral blood mononuclear cells from 95 stage I and II breast cancer patients in the early postoperative period following mastectomy. The test used was one-way mixed leucocyte culture (MLC) against a pool of mitomycin-C-treated lymphocytes from allogeneic donors. Depressed lymphoproliferative responses were associated with significantly longer disease-free intervals.

Breast cancer in immunosuppressed patients

Reinforcing the suspicion of immune facilitation of tumour growth in a substantial subset of women with breast cancer is the fact that all non-surgical treatments that have given positive results in breast cancer have one thing in common; they are immunosuppressive to a greater or lesser degree. And this topic has been reviewed in detail (Stewart *et al.* 1994) and only salient points will be emphasized here: (1) Median response rates of 30%, lasting 3–14 months following treatment with corticosteroids, have been reported in patients with metastatic breast cancer. Additive effects are seen when corticosteroids are combined with endocrine therapy. The immunosuppressive effects of corticosteroids are quite clear cut: a clear dose response exists for suppression of the production of lymphokines, including IL-1, IL-2, IL-6 and TNF (Stewart *et al.* 1994). (2) Adjuvant chemotherapy with CMF is immunosuppressive and causes prolonged impairment of certain aspects of B, T, and NK cell function. Zielinski *et al.* (1990) reviewed their studies showing a depression of antibody production following vaccination, prolonged impairment of mitogen-induced soluble IL-2 receptor production, and a decrease in proliferation of peripheral blood mononuclear cells following phytohemagglutinin stimulation. The depressed

activity was seen up to 3 years following cessation of CMF therapy. (3) Escalating doses of chemotherapy followed by allogenic or autologous bone marrow transplantation causes marked and prolonged immunosuppression lasting up to 2 years (Welte, *et al.* 1984; Olsen *et al.* 1988). Defective production of IL-2 is seen up to 18 months following high-dose cyclophosphamide and whole body irradiation. (4) Locoregional radiotherapy following mastectomy for breast cancer has been found to confer a systemic benefit for survival (Stewart, 1994). One explanation for improved survival is that there is a significant T cell lymphopenia and decreases in T cell responses that can persist for as long as 11 years in women so treated for breast cancer. (5) Robinson *et al.* (1993) found that in women treated for bilateral breast cancer and without evidence of disease, NK cell activity was higher than in normal controls, but that this activity declined on long-term tamoxifen therapy. Tamoxifen also reduced CD4 cells and the CD4/CD8 ratio in such patients.

If successful treatment regimens for breast cancer are immunosuppressive, one might ask whether women who are chronically immunosuppressed show a low population incidence of *de novo* breast cancers. To investigate this hypothesis the incidence of breast cancer was assessed in female transplant recipients. Results based on data provided since 1983 to the Collaborative Transplant Study in Heidelberg were published in 1995 (Stewart *et al.* 1995). Overall incidence of breast cancer was found to be significantly lower among 25 914 transplant recipients than would be expected from background rates. During a follow-up period of 1–11 years, 86 cases were observed compared with 113·8 expected ($\chi_1^2 = 6·75$, $P = 0·009$). Incidence was particularly low in the first year following transplant with a relative risk of 0·49 (95% Cl 0·64–1·03). A subset of 13 003 women received a combination of cyclosporin, azathioprine and steroids (CSA), the remainder only one or two of these drugs. There were 30 cases of breast cancer in the 13 003 patients against 53·8 expected, and 56 cases in 12 911 patients treated otherwise against 60 expected (RR 0·58, 95% Cl 0·36–0·93, $P = 0·011$). The CSA group had a very low rate in the years following the first year, with 24 observed cases rather than 39·8 expected, giving an SIR of 0·60. The low breast cancer rate applied only to kidney recipients, particularly the 8166 North American women where the incidence was halved. (The number of heart recipients was small, 2185, and the results are less reliable.) *All other major cancers* had higher than expected incidence, in some cases substantially, with marginal increases in others.

IMMUNOBIOLOGY OF MOUSE MAMMARY TUMOURS

About the time that immunization of breast cancer patients with membrane preparations was being considered, a new era of experimental tumour immunology with animal models was beginning. A number of important observations on the immunobiology of mouse mammary cancers were made in the laboratories of Weiss (Weiss *et al.* 1966). First, it was found possible to immunize mice, by a number of ways, to transplants of either autografts or isografts of mammary tumours. However, the specificity of the response depended upon the type of host being tested. The tumours used were ones associated with the mouse Mammary Tumour Virus (MMTV). In hosts that had been neonatally infected (through their mother's milk) with MMTV, the immune response only extended to the tumour used for immunization. In syngeneic, but MMTV-free hosts, immune responsiveness included other tumours of the same origin as the immunizing tumour. The interpretation of these results was that neonatal infection resulted in a state of effective 'tolerance' to MMTV and associated antigens, so that only non-infected mice could respond to the common, viral-associated antigens expressed by tumour cells. The fact that infected mice could, however, mount an immune response to the particular tumour used for immunization meant that there were other, non-MMTV associated immunogens in the system. The direction of the immune response was also interesting. Although immunized mice could respond to a challenge by inhibiting tumour growth, the opposite was also seen, that is stimulation of tumour growth in previously sensitized hosts.

Much of this work, as well as the vast majority of studies in experimental tumour immunology, was done with transplantable tumours. However, Weiss and associates (Weiss *et al.* 1964, 1966) also studied the immune reactions to autochthonous mammary tumours, that is to tumours as they developed, in their original host. The overall results of these studies were sobering: about 1/3 of the cancers produced no detectable response in the host of origin, whereas about 1/3 grew better than when transplanted into naïve, syngeneic mice. Only 1/3 appeared to induce a protective response. Although these studies were logistically difficult and statistically problematic, it was clear that, even in inbred strains of mice, the immunological relationship between mammary cancers and their own hosts is not necessarily protective.

Further evidence along these lines came from a different approach. A number of investigators had shown that immunosuppression of mice by neonatal thymectomy greatly increased susceptibility to the growth of various kinds of non-mammary tumours (Miller, Grant & Roe, 1963; Law, 1966). When this technique was applied to MTV-infected mice, however, the results were in the opposite direction in that the spontaneous development of mammary tumours, as well as of their morphological precursors, hyperplastic alveolar nodules (HANs), was inhibited or delayed by neonatal thymectomy (Martinez, 1964; Heppner, Wood & Weiss, 1968; Squartini, Olivi & Bolis, 1970; Belyaev & Gruntenko, 1972). Other means of immunosuppression, for example, by anti-lymphocyte serum, had the same effects (Lappe & Blair, 1970). The mechanisms behind these unexpected findings were complex and included hormonal as well as immunological and viral factors. However, the 'bottom line' result did not suggest that the immune response was protective.

By the late 1960s the overall excitement in the field of tumour immunology had led to attempts to measure anti-tumour reactivity in patients. A variety of *in vitro* techniques, including the colony inhibition and microtoxicity assays, were developed in which lymphocytes from patients or animal hosts were tested for their ability to kill or to inhibit the proliferation of cancer cells. Application of these assays to lymphocytes from mice that had been transplanted with syngeneic mammary tumours, resulted in the reproduction of observations from the *in vivo* transplantation experiments in Weiss's laboratory (Heppner & Pierce, 1969). Again depending upon the MMTV status of the host, immunity was demonstrable either only to the immunizing tumour or extended to other tumours of similar origin. With this assay, colony inhibition was fairly easy to demonstrate with lymphocytes that had been isolated from their host. However, the addition of serum from the same host often abrogated the inhibitory activity of their lymphocytes, a circumstance that was attributed to the presence of 'blocking antibody' or antigen–antibody complexes (Heppner, 1969). These observations were not unique to mammary tumours, but were seen with a variety of other tumour types, both animal and human (Heppner, 1972).

'IMMUNE STIMULATION' OF MAMMARY TUMOUR CELL PROLIFERATION

The generally optimistic view of tumour immunology during the 1960s and 1970s was not without challenge. In particular, Prehn & Lappe (1971) suggested that a weak immune response, of the sort most likely to be seen against naturally occurring (not-transplanted) tumours, that gradually develop, without the involvement of 'strong' carcinogens, and become established without alerting their host in a threatening way, might be stimulatory, rather than inhibitory (see Prehn, 1994 for a recent expansion of this concept). Evidence for this concept was obtained in another mouse mammary tumour model, namely tumours that arose spontaneously in mice implanted with a series of HAN lines that Medina (1973) had established and characterized. With these transplantable HAN lines, particularly the lines known as D2 and C4, it was possible to focus on the immune events during the critical time of progression from the preneoplastic to neoplastic phenotype. When colony inhibition or microcytotoxicity assays were applied to D2 or C4 HANs and tumours, stimulation, rather than inhibition, of HAN or tumour cell growth was often the result of *in vitro* exposure to host lymphocytes (Medina & Heppner, 1973; Heppner, Kopp & Medina, 1976). However, as reported in 1973 (Medina & Heppner) and more extensively in 1976 (Heppner *et al.*), stimulation, rather than inhibition, of HAN or tumour cell growth was often the result of *in vitro* exposure to host lymphocytes. Stimulation was not a random, or 'feeder' effect, but instead was specific to particular combinations of lymphocytes and tumours.

Fidler, Brodey & Beck-Nielsen (1974) reported stimulation of tumour growth *in vitro* by low ratios of sensitized lymphocytes to tumour cells. Two of the 13 dogs used in their studies had breast cancer. The addition of autologous serum to the lymphocyte–tumour cultures potentiated the stimulation of growth. These authors also concluded that their results supported the hypothesis of Prehn.

More recently Ögmundsdottir, Petersdottir & Gudmunsdottir (1995) have provided *in vitro* evidence for immune stimulation of growth of human breast cancers. These authors were successful in establishing co-cultures of fresh samples of breast carcinoma and autologous peripheral lymphocytes in 20 cases. Cancer cell growth significantly above control levels was seen with lymphocytes in 11 such cultures. Optimal growth was also seen in 5 of 17 cocultures of uninvolved breast tissue and lymphocytes. Growth stimulation in response to lymphocytes was significantly associated with the expression of MHC class I by the tumour cells, suggesting that it depended upon immune recognition. Growth stimulation was not seen with a series of breast cell lines, only with primary cultures.

IMMUNE FACILITATION OF MOUSE MAMMARY TUMOUR PROGRESSION

As mentioned, the HAN model of Medina offers a good opportunity to investigate whether the immune response affects preneoplastic progression to mammary cancer. The HAN lines are transplanted into 'cleared' (epithelium-free) mammary fat pads of syngeneic, strain BALB/c mice; the HAN implants grow out to fill the pads; and eventually, after periods of many months, focal tumours arise within the HAN tissue. Thus, tumour development is spontaneous and within the mammary site, although

it occurs in transplanted HAN tissue. Using the C4 HAN model, a variety of approaches has been tried to detect an immune response to mammary tumour development, with intriguing results: the host response appears to stimulate the development of the cancers by a mechanism that involves the NK cells. Thus, activated NK cells were found to be present in higher numbers in HANs (as compared with normal mammary epithelium) (Wei & Heppner, 1987) and suppression of NK activity was accompanied by a lengthening of the latency period and reduction in the frequency of tumour formation (Wei *et al.* 1989). In contrast, NK stimulation correlated with shortening of the latency period and increased tumour formation (Wei *et al.* 1989; Tsai, Loeffler & Heppner, 1992). The idea that NK activity may be a stimulatory factor in mammary tumour growth is also supported by results in another animal model, the androgen-responsive Shionogi's mouse mammary tumour, in which Rowse, Weinberg & Emerman (1995) have demonstrated a positive correlation between tumour size and NK activity as a function of stress.

As stated above, NK activity has been reported to be low in human breast cancers and breast cell infiltrates have been found capable of suppressing NK activity (Vose & Moore, 1979; Eremin *et al.* 1981). Similarly, although the numbers of NK cells were elevated in both C4 HAN and tumours, in comparison with normal glands, unlike in the HAN, tumour NK activity was depressed, probably due to the presence of suppressor cells. These C4 results resemble those of Pross, Sterns & MacGillis (1984), who found that a subset of women, characterized as being of high risk for developing breast cancer due to a mammographic pattern defined as 'benign breast syndrome', had a history of abnormally high, peripheral NK activity. Pross and associated speculated that prolactin (PRL), reported to be at high levels in patients with proliferative breast disease (Franks *et al.* 1974; Cole *et al.* 1977) might contribute to NK stimulation. Interestingly, inhibition of pituitary PRL production by bromocriptine simultaneously decreases both progression of C4 HAN to tumours and the HAN-infiltrating NK activity (Tsai *et al.* 1992). Prolactin also stimulated NK activity *in vitro* (Tsai & Heppner, 1994). Taken together, these results suggest a possible influence at the level of the interface of the immune and endocrine systems during the early progression of breast cancer.

One reason that the C4 system was chosen for studies of immune activity during neoplastic progression was the belief that there was no viral agent, i.e. MMTV, present. However, this belief has since been shown to be false. Wei and associates (Wei *et al.* 1991) have described a new, milk-transmitted virus, MMTV (C4), which is produced by C4 HAN. This virus encodes a superantigen that causes thymic deletion of T cells with the VB2 segment in their T cell receptor. MMTV (C4) has been shown to be responsible for at least some of the NK activation in this system (Gill *et al.* 1994).

MECHANISMS OF IMMUNE FACILITATION OF BREAST CANCER DEVELOPMENT

The idea that immune responses may facilitate cancer development is initially counterintuitive, given our basic assumption that immunity is a 'host-defence' reaction. The immune system does protect us from external threats, ones which are sufficiently foreign to be perceived as not-self. There are many examples from parasitology, however, in which the foreign invader avoids immune destruction by becoming 'like' the host. It should not be surprising that cancer cells, which differ from their hosts in only the most minute ways, can also escape immune destruction. Nevertheless, escaping the immune response is not the same as using it to promote growth. What mechanisms might underlie immune facilitation?

In addressing this question, it is necessary to consider the difference between a cause–effect relationship, versus an association of events which are themselves quite independent. It has become recognized in recent years that the immune system is not an 'island' with its own, unique, mechanisms of modulation and control, but rather is connected, via a variety of shared growth factors and hormones, with the other organ systems of the body. Many, if not all, of the same factors that control the differentiation or proliferation of lymphoid cells also control other cell types (Paneri, 1993; Weigent & Blalock, 1995). For example, in the C4 HAN system described earlier, PRL appears to act as a mitogen for infiltrating lymphocytes (Tsai & Heppner, 1994), along with its well-known stimulatory effects on mammary epithelial cells (Welsch & Nagasawa, 1977). Thus, although suppression of pituitary PRL suppresses HAN-infiltrating NK activity and simultaneously, interferes with C4 HAN progression, this correlation is not in itself evidence that NK activity facilitates HAN progression. Quite simply, both cell types may be sensitive to the same control factor and yet be independent of each other.

There is evidence, however, that tumour-associated inflammatory cells are themselves the source of many factors that can act, in a paracrine fashion, as direct growth factors for cancer cells, as well as to influence cancer growth indirectly by, for example, enhancing angiogenesis. Again in reference to the C4 HAN example, activated lymphocytes have been reported to produce a molecule with PRL activity (Montgomery *et al.* 1990), so that C4 HAN epithelial cells may be stimulated by PRL produced *in situ* by infiltrating lymphocytes, as well as to that produced by the pituitary.

Human breast tumour infiltrating lymphocytes (TIL) have been extensively studied for their capacity to produce cytokines (Rubbert *et al.* 1991; Schwartzentruber *et al.* 1992; Vitolo *et al.* 1992). Two recent reviews (Mantovani *et al.* 1992; Michiel & Oppenheim, 1992) show clearly that cancer infiltrating mononuclear cells produce cytokines that can either down-regulate *or* stimulate cancer growth. Rubbert *et al.* (1991) found that TIL could produce tumour necrosis factor (TNF) when exposed to autologous tumour cells *in vitro*. Mitogen-stimulated TIL also produced TNF, as well as IL-2 and IFNα. Vitolo *et al.* (1992) studied the TIL expression of cytokine mRNAs. In the stroma of ductal breast cancers which contained intracellular or intraductal mucous, up to 30% of lymphoid cells expressed IL-2, TNFα, IFNα, and IL-2R mRNA. Schwartzentruber *et al.* (1992) showed that TIL stimulated by autologous breast cancers secreted TNFα, GM-CSF and IFNα. Peoples *et al.* (1995) have shown that TIL produce heparin-binding epidermal growth factor-like growth factor (HB-EGF) and basic fibroblast growth factor (bFGF) *in vitro*, as well as *in vivo*. HB-EGF and bFGF derived from TILs directly stimulated breast cancer cell proliferation *in vitro*, and also stimulated vascular smooth muscle cells, whereas bFGF displayed angiogenic properties. Another potent angiogenic factor, vascular endothelial growth factor (VEGF), is produced by TIL *in situ* at bioactive concentrations in human prostate and bladder cancers (Freeman *et al.* 1995). Recent data (Lu & Brodie, 1996) shows that both hormone-dependent and -independent human breast cancer cell lines are stimulated to grow by VEGF. Thus, a wealth of cytokines are produced by infiltrating inflammatory cells. The next question is which of these are most likely to be responsible for immune facilitation of breast cancer development.

TNF is a leading candidate for an inflammatory cell-produced cytokine that stimulates breast cancer development. Using a model of rat mammary epithelial cells growing in serum-free medium in a reconstituted basement membrane, IP, Shoemaker & Darcy (1992) found that TNFα stimulates epithelial cell proliferation at optimal growth conditions and markedly stimulates morphogenesis in suboptimal or EGF-deficient medium. Tsai *et al.* (1992) have found TNFα to be a positive growth factor for normal mouse mammary cells, as well as for preneoplastic C4 HAN and C4 tumour cells. They have also demonstrated TNFα mRNA expression in both HAN infiltrating lymphocytes (HILs) and HAN epithelial cells. In the milieu of HAN tissues, HILs are in close contact with HAN cells and may be constitutively secreting TNFα in response to autologous HAN stimulation. C4 HAN epithelial cells also express mRNA for both the p55–60 receptor and the p75-80 TNFα receptors. Most importantly, *in vivo* treatment of C4 HAN bearers with TNFα decreases the latency period and enhances the frequency of HAN progression to tumour (Heppner & Miller, 1996). These results indicate that stimulation of epithelial cell proliferation by HIL-produced TNFα is one mechanism responsible for the immune facilitation of neoplastic progression in the HAN model. However, growth stimulation may not be the only mechanism by which TNFα affects progression: TNFα may promote tumour progression indirectly by its ability to stimulate angiogenesis (Vukanovic & Isaacs, 1995), to enhance epithelial cell motility (Rosen et al. 1991) or through induction of increased expression of adhesion molecules (Spriggs et al. 1987), among other possibilities.

Other candidate inflammatory cell-produced, breast cancer stimulating cytokines include IL-6 and colony-stimulating factor (CSF). Although IL-6 can inhibit the proliferation of several duct carcinomas of breast in vitro, it also enhances motility of breast cancer cells and causes increased cell–cell separation with decreased adherens type function formation (Sehgal & Tamm, 1991). Furthermore, human mammary epithelial cells transfected with the int-2 gene are stimulated to grow by IL-6 (Basolo et al. 1993). CSF-1 has been suggested as a key mediator of breast cancer invasion and metastasis (Scholl et al. 1993). Autocrine growth stimulation due to the combined expression of CSF-1 and its receptor in breast cancer cells is possible.

In addition to cytokines, infiltrating lymphocytes may be the source of other factors of specific relevance to breast cancer. Prolactin has already been mentioned. A number of investigators are exploring the possibility that lymphocytes either produce enzymes involved in oestrogen synthesis (Berstein et al. 1993a), or that inflammatory cell-produced cytokines affect the activity of such enzymes (Reed et al. 1995). Indeed, Singh et al. (1997) have recently shown that exposure of fibroblasts from normal breast tissue to conditioned medium from cultures of lipopolysaccharide-stimulated monocytes and lymphocytes obtained from an immunosuppressed kidney transplant recipient receiving cyclosporin. A therapy resulted in a significant reduction in aromatase activity stimulation, compared with the marked stimulation seen using conditioned medium derived from monocytes and lymphocytes taken from a woman with breast cancer. The authors postulate that reduction in cytokine-induced oestrogen synthesis in breast tissue of immunosuppressed patients may play an important role in lessening the incidence of breast cancer in such patients (Stewart et al. 1995). So far, there appear to be no studies of oestrogen synthesis modulation by TIL, although Berstein et al. (1993b) have reported that peripheral lymphocytes from breast cancer patients have a higher capacity to convert androstenedione to oestrogen than do peripheral lymphocytes of age-matched control women.

Antibodies are also a possible mechanism of breast cancer stimulation. Parkes et al. (1988) analysed material taken from 34 patients who had been operated for primary breast cancer in 1976 and followed for 11 years. Plasma cells were found in cancers from 84% of women who had relapsed and died (19 patients) whereas they were detected in only one tumour of 15 women who had survived (6%), a case of medullary carcinoma, free of disease. The authors concluded that the presence of plasma cells in infiltrating duct carcinoma and mixed infiltrating duct and lobular carcinoma is associated with a poor prognosis.

Using the Prausnitz–Kustner reaction, Grace & Dao (1958) found that a patient's antibodies may be responsible for inflammation in inflammatory breast cancer. HER-2/neu protein is amplified and overexpressed in inflammatory breast cancer and non-inflammatory breast cancer with positive axillary nodes (Guerin et al. 1989). Antibody responses have been seen in breast cancer patients having this oncoprotein, in 647 cases which also had a lympho-plasma cell infiltrate (Pupa et al. 1993), and in 11 of 20 premenopausal breast cancer patients (Disis et al. 1994). One patient also showed a significant proliferative T cell response to the HER-2/neu protein and peptides. Antibodies to growth factors or growth factor receptors may facilitate cancer cell growth by contributing to the stimulus for signal transduction through that receptor. For example, in a recent study of a newly isolated human breast cancer cell line it was shown that a monoclonal antibody that binds to the extracellular domain of erbB2 is a potent growth factor in vitro for these cells (Ethier et al. 1996). Complementing this finding is an earlier study by Stancovski et al. (1991), showing that monoclonal antibodies to the extracellular portion of the erbB2 protein of human breast cancer cell lines could either inhibit or strongly stimulate the growth of these cancers in athymic mice. They caution that antibody therapy could accelerate tumour growth.

CONCLUSIONS

The hope that the immune response may be harnessed to control cancer development and growth continues to burn brightly and, indeed, has received new fuel in the form of recent advances in gene and molecular therapy. It is likely that the treatment of many types of cancers will be impacted favourably by these advances. Will breast cancer be one of them? Immune responses of various types to breast cancer can be demonstrated, and in some types of laboratory settings shown to be growth inhibitory. However, in human breast cancer, as well as in experimental models of mammary cancer development (as opposed to growth of transplantable lines), the inflammatory/immune response is often associated with facilitation of cancer growth and progression. Potential mechanisms of facilitation include inflammatory cell-produced cytokines, modulators of steroid hormone metabolism, and antibodies, working either directly on the cancer cells themselves or indirectly through effects on angiogenesis or stromal cell behaviour. Optimistically, one may argue that any type of immune response to cancer may be better than no response at all, in the sense that it indicates that the host does recognize the presence of 'an invader' and, therefore, that there may be a basis for engineering that recognition to achieve a clinically favourable outcome. To do so, it will be necessary to select those immunogens and conditions that are able either to overcome or to deflect the immune-faciliatory responses which appear to play a significant role in the natural history of breast cancer development.

ACKNOWLEDGEMENTS

Our acknowledgements to all our colleagues and co-workers whose work has contributed to our views, including Drs Bonnie Miller, Fred Miller, S.-C. Jane Tsai, and Wei-Zen Wei. We thank Dr Jacob Karsh for editing the manuscript and Liz Lacasse, McKenzie Duke and Elaine Weber for their care in typing it. We also acknowledge the support of USPHS Grant CA61217 and of the Concern and Concern II Foundation.

REFERENCES

BASOLO, F., CALVO, S., FIORE, L., CONALDI, P. G., FALCON, V. & TONIOLO, A. (1993). Growth-stimulating activity of interleukin-6 on human mammary epithelial cells transfected with the *int-2* gene. *Cancer Research* **53**, 2957–2960.

BELYAEV, D. K. & GRUNTENKO, E. V. (1972). The influence of the thymus on the development of transplanted mammary tumors in mice. *International Journal of Cancer* **9**, 1–7.

BERSTEIN, L. M., LARIONOV, A. A., KRJUKOVA, O. G. & SEMIGLAZOV, V. F. (1993*a*). Conversion of androstenedione in blood lymphocytes of breast carcinoma patients. *Journal of Endocrinology* **139**, 17.

BERSTEIN, L. M., SANTNER, S. J., BRODIE, A. M., KOOS, R. D., NAFTOLIN, F. & SANTEN, R. J. (1993*b*). Pseudoaromatase in circulating lymphocytes. *Journal of Steroid Biochemistry and Molecular Biology* **44**, 647–649.

BLACK, M. M. (1972). Cellular and biologic manifestations of immunogenicity in precancerous mastopathy. *National Cancer Institute Monographs* **35**, 73–82.

BRENT, L., BROWN, J. & MEDAWAR, P. B. (1958). Skin transplantation immunity in relation to hypersensitivity. *Lancet* **34**, 561–563.

CANNON, G. B., BARSKY, S. H., ALFORD, T. C., JEROME, L. F., TINLEY, V., McCOY, J. L. & DEAN, J. H. (1982). Cell-mediated immunity to mouse mammary virus antigens by patients with hyperplastic benign breast disease. *Journal of the National Cancer Institute* **68**, 935–943.

CANNON, G. B., DEAN, J. H. & HERBERMAN, R. B. (1981). Lymphoproliferative responses to alloantigen by postoperative patients with stage I lung and breast cancer and their application to prognosis. In *Biological Relevance of Immune Suppression* (ed. Dean, J. H. & Padarathsingh, M.), pp. 98–118. Van Nostrand Reinhold, New York.

CHAMPION, H. R., WALLACE, I. W. J. & PRESCOTT, R. J. (1972). Histology in breast cancer prognosis. *British Journal of Cancer* **26**, 129–138.

COLE, E. N., SELLWOOD, R. A., ENGLAND, P. C. & GRIFFITHS, K. (1977). Serum prolactin concentrations in benign disease throughout the menstrual cycle. *European Journal of Cancer* **13**, 597–603.

DISIS, M. L., CALENOFF, E., McLAUGHLIN, G., MURPHY, M. E., CHEN, W., GRONER, B., TESCHKE, M., LYDON, L., McGLYNN, E., LIVINGSTON, R. B., MOE, R. & CHEEVER, M. A. (1994). Existent T-cell and antibody immunity to HER-2/neu protein in patients with breast cancer. *Cancer Research* **54**, 16–20.

EARLY BREAST CANCER TRIALISTS COLLABORATIVE GROUP (1992). Systemic treatment of early breast cancer by hormonal, cytotoxic or immune therapy. *Lancet* **339**, 1–15.

EGLINTON, B. A., ROBERTON, D. M. & CUMMINGS, A. G. (1994). Phenotype of T-cells, their soluble receptor levels, and cytokine profile of human breast milk. *Immunology and Cell Biology* **72**, 306–313.

EREMIN, O., COOMBS, R. R. A. & ASHBY, J. (1981). Lymphocytes infiltrating human breast cancers lack K-cell activity and show low levels of NK-cell activity. *British Journal of Cancer* **4**, 166–176.

ETHIER, S. P., KOKENY, K. E., RIDINGS, J. W. & DILTS, C. A. (1996). ErbB family receptor expression and growth regulation in a newly isolated human breast cancer cell line. *Cancer Research* **56**, 899–907.

FIDLER, I. J., BRODEY, R. S. & BECK-NIELSEN, S. (1974). *In vitro* immune stimulation – inhibition to spontaneous canine tumors of various histologic type. *Journal of Immunology* **112**, 1051–1060.

FISHER, B., SAFFER, E. A. & FISHER, E. R. (1974). Studies concerning the regional lymph nodes in cancers. VII. Thymidine uptake by cells from nodes of breast cancer patients relative to axillary location and histopathologic discrimination. *Cancer* **33**, 271–279.

FISHER, E. R., KOTWAL, N., HERMANN, C. & FISHER, B. (1983). Types of tumor lymphoid response and sinus histiocytosis. *Archives of Pathology and Laboratory Medicine* **107**, 222–227.

FRANKS, S., RALPHS, D. N. L., SEAGROATT, V. & JACOBS, H. S. (1974). Prolactin concentrations in patients with breast cancer. *British Medical Journal* **2**, 320–321.

FREEMAN, M. R., SCHNECK, F. X., GAGNON, M. L., CORLESS, C., SOKER, S., NIKNEJAD, K., PEOPLES, G. E. & KLAGSBRUN, M. (1995). Peripheral blood T lymphocytes and lymphocytes infiltrating human cancer express vascular endothelial growth factor: a potential role for T cells in angiogenesis. *Cancer Research* **55**, 4140–4145.

GILL, R., WANG, H., BLUETHMANN, H., INGLESIAS, A. & WEI, W.-z. (1994). Activation of natural killer cells by mouse mammary tumor virus C4 in BALB/c mice. *Cancer Research* **54**, 1529–1535.

GRACE, J. T. & DAO, T. L. (1958). Etiology of inflammatory reactions in breast cancer. *Surgical Forum* **9**, 611–614.

GUÉRIN, M., GABILLOT, M., MATHIEU, M. C., TRAVALGI, J. P., SPIELMAN, M., ANDRIEU, N. & RIOU, G. (1989). Structure and expression of C-erb B-2 and EGF receptor genes in inflammatory and non-inflammatory breast cancer: prognostic significance. *International Journal of Cancer* **43**, 201–208.

HEAD, J. F., ELLIOTT, R. L. & McCOY, J. L. (1993). Evaluation of lymphocyte immunity in breast cancer patients. *Breast Cancer Research and Treatment* **26**, 77–88.

HENDERSON, B. E., PIKE, M. C. & GRAY, G. E. (1981). The epidemiology of breast cancer. In *Breast Cancer* (ed. Hoogstraten, B. & McDivitt, R. W.), pp. 1–25. CRC Press, Boca Raton.

HEPPNER, G. H. (1969). Studies on serum-mediated inhibition of cellular immunity to spontaneous mouse mammary tumors. *International Journal of Cancer* **4**, 608–615.

HEPPNER, G. H. (1972). Blocking antibodies and enhancement. *Series Haematologica* **4**, 41–66.

HEPPNER, G. H., KOPP, J. S. & MEDINA, D. (1976). Microcytotoxicity assay of immune responses to non-MTV-induced, preneoplastic and neoplastic mammary lesions in strain BALB/c mice. *Cancer Research* **38**, 753–758.

HEPPNER, G. H. & MILLER, B. E. (1996). Enhanced tumor progression following TNFα or interferon treatment of mice bearing preneoplastic lesions. *Proceedings of the American Association of Cancer Research* **37**, 157.

HEPPNER, G. H. & PIERCE, G. E. (1969). *In vitro* demonstration of tumor specific antigens in spontaneous mammary tumors of mice. *International Journal of Cancer* **4**, 212–218.

HEPPNER, G. H., WOOD, P. C. & WEISS, D. W. (1968). Studies on the role of thymus in viral tumorigenesis and effect of thymectomy on induction of hyperplastic alveolar nodules and mammary tumors in Balb/cf C$_3$H mice. *Israel Journal of Medical Science* **4**, 1195–1203.

HUGHES, L. E. & LYTTON, B. (1964). Antigenic properties of human tumors: delayed cutaneous hypersensitivity reactions. *British Medical Journal* **1**, 209–212.

IP, M. M., SHOEMAKER, S. F. & DARCY, K. M. (1992). Regulation of rat mammary epithelial cell proliferation and differentiation by tumor necrosis factor-α. *Endocrinology* **130**, 2833–2840.

KURTZ, J. M., JACQUEMIER, J., AMALRIC, R., BRANDONE, H., AYME, Y., HANS, D., BRESSAC, C. & SPITALIER, J. M. (1990). Why are local recurrences after breast-conserving therapy more frequent in younger patients? *Journal of Clinical Oncology* **8**, 591–598.

LAPPE, M. A. & BLAIR, P. B. (1970). Interference with mammary tumorigenesis by antilymphocyte serum. *Proceedings of the American Association of Cancer Research* **11**, 47.

LAW, L. W. (1966). Studies on thymic function with emphasis on the role of the thymus in oncogenesis. *Cancer Research* **26**, 551–574.

LEEK, R. D., LEWIS, C. E., WHITEHOUSE, R., GREENALL, M., CLARKE, J. & HARRIS, A. L. (1996). Association of macrophage infiltration with angiogenesis and prognosis in invasive breast carcinoma. *Cancer Research* **56**, 4625–4629.

LEVINE, P. H., MOURALI, N., TABANNE, F., LOON, J., TERASAKI, P., TSANG, P. & BEKESI, J. G. (1981). Studies on the role of cellular immunity and genetics in the etiology of rapidly progressing breast cancer in Tunisia. *International Journal of Cancer* **27**, 611–615.

LU, Q. & BRODIE, A. (1996). Stimulation of the growth of MCF-7 and MDA-MB-468 breast cancer cells by vascular endothelial growth factor. *Proceedings of the American Association of Cancer Research* **37**, 220.

MAHONEY, K. H., FULTON, A. M. & HEPPNER, G. H. (1983). Tumor-associated macrophages of mouse mammary tumors. II. Differential distribution of macrophages from metastatic and non-metastatic tumors. *Journal of Immunology* **131**, 2079–2085.

MANCINI, R. E., ANCHADA, J. A., SARACENI, A. E., BACHMAN, J. C., LAVIERI, J. C. & REMIROUSKY, M. J. (1965). Immunological and testicular response in man sensitized with human testicular homogenate. *Journal of Clinical Endocrinology and Metabolism* **25**, 859–875.

MANTOVANI, A., BOTAZZI, B., COLOTTA, F., SOZZANI, S. & RUCO, L. (1992). The origin and function of tumor-associated macrophages. *Immunology Today* **13**, 265–270.

MARTINEZ, C. (1964). Effect of early thymectomy on development of mammary tumors in mice. *Nature* **203**, 1188.

McKENZIE, J. M. & ZAKARIJA, M. (1986). Assays of thyroid-stimulating antibodies (TSAb) of Grave's disease. In *Werner's The Thyroid, a fundamental and clinical text*, 5th edn (ed. Ingbar, S. H. & Braverman, L. E.), pp. 559–575. J. P. Lippincott Company, Philadelphia.

MEDINA, D. (1973). Preneoplastic lesions in mouse mammary tumorigenesis. *Methods in Cancer Research* **7**, 3–53.

MEDINA, D. & HEPPNER, G. H. (1973). Cell-mediated immunostimulation induced by mammary tumor virus-free BALB/c mammary tumors. *Nature* **242**, 329–330.

MEYER, J. S. & HIXON, B. (1979). Advanced stage and early relapse of breast carcinomas associated with high thymidine labeling indices. *Cancer Research* **39**, 4042–4047.

MICHIEL, D. F. & OPPENHEIM, J. J. (1992). Cytokines as positive and negative regulators of tumor promotion and progression. *Cancer Biology* **3**, 3–15.

MILLER, J. F. A. P., GRANT, G. H. & ROE, F. J. C. (1963). Effect of thymectomy on the induction of skin tumors by 3, 4-benzopyrene. *Nature* **199**, 920–922.

MONTGOMERY, D. W., LE FEVRE, J. A., ULRICH, E. D., ADAMSON, C. R. & ZUKO, C. F. (1990). Identification of prolactin-like proteins synthesized by normal murine lymphocytes. *Endocrinology* **127**, 2601–2605.

MOURALI, N., LEVINE, P. H., TABANNE, F., BELHASSEN, S., BAKI, J., BENNACEUR, M. & HERBERMAN, R. (1978). Rapidly progressing breast cancer (pousée évolutive) in Tunisia: studies on delayed hypersensitivity. *International Journal of Cancer* **22**, 1–3.

ÖGMUNDSDOTTIR, H. M., PETERSDOTTIR, I. & GUDMUNSDOTTIR, I. (1995). Interactions between the immune system and breast cancer. *Acta Oncologica* **34**, 647–650.

OLSEN, G. A., GOCKERMAN, J. P., BASR, R. C., BOROWITZ, M. & PETERS, W. P. (1988). Altered immunologic reconstitution after standard-dose chemotherapy or high-dose chemotherapy with autologous bone marrow support. *Transplantation* **46**, 57–60.

PANERAI, A. E. (1993). Lymphocytes as a source of hormones and peptides. *Journal of Endocrinology Investigations* **16**, 549–557.

PARKES, H., COLLIS, C., BAILDAM, A., RALPHS, D., LYONS, B., HOWELL, A. & CRAIG, R. (1988). *In situ* hybridization and S_1 mapping show that the presence of plasma cells is associated with poor prognosis in breast cancer. *British Journal of Cancer* **58**, 715–722.

PEOPLES, G. E., BLOTNIK, S., TAKAHASHI, K., FREEMAN, M. R., KLAGSBRUN, M. & EBERLEN, T. J. (1995). T lymphocytes that infiltrate tumors and atherosclerotic plaques produce heparin-binding epidermal growth factor-like growth factor and basic fibroblast growth factor: a potential pathologic role. *Proceedings of the National Academy of Sciences, USA* **92**, 6547–6551.

PETERSEN, O. W. & VAN DEURS, B. (1987). Preservation of defined phenotypic traits in short-term cultured human breast carcinoma derived epithelial cells. *Cancer Research* **47**, 850–866.

POUILLART, P., PALANGIE, T., JOUVE, M., GARCÍA-GIRALT, E., VILCOQ, J. R., BATAINI, J. P., CALLE, R., FENTON, J., MATHIEU, S., ROUSSEAU, J. & ASSELAIN, B. (1981). Cancer inflammatoire du sein traité par une association de chimio-thérapie et d'irradiation, résultat d'un essai randomisé étudiant le rôle d'une immunothérapie par le BCq. *Bulletin du Cancer* **68**, 171–186.

PREHN, R. T. (1994). Stimulatory effects of immune reactions upon the growth of untransplanted tumors. *Cancer Research* **54**, 908–914.

PREHN, R. T. & LAPPE, M. A. (1971). An immunostimulation theory of tumor development. *Transplant Review* **7**, 26–54.

PREHN, R. T. & MAIN, J. M. (1957). Immunity to methylcholanthrene-induced sarcomas. *Journal of the National Cancer Institute* **18**, 769–778.

PROSS, H. F., STERNS, E. & MacGILLIS, D. R. (1984). Natural killer cell activity in women at 'high risk' for breast cancer, with or without benign breast syndrome. *International Journal of Cancer* **34**, 303–308.

PUPA, S. M., MENARD, S., ANDREOLA, S., RILKE, F., CASCINELLI, N. & COLNAGHI, M. I. (1992). Prognostic significance for C-erb B-2 oncoprotein overexpression in breast carcinoma patients and its immunological role. *Proceedings of the American Association of Cancer Research* **31**, 312.

PUPA, S. M., MENARD, S., ANDREOLA, S. & COLNAGHI, M. I. (1993). Antibody response against the c-erb B-2 oncoprotein in breast carcinoma patients. *Cancer Research* **53**, 5864–5866.

REED, M. J., PUROBIT, A., DUNCAN, L. J., SINGH, A., ROBERTS, C. J., WILLIAMS, G. J. & POTTER, B. V. L. (1995). The role of cytokines and sulphatase inhibitors in regulating estrogen synthesis in breast tumors. *Journal of Steroid Biochemistry and Molecular Biology* **53**, 413–420.

ROBINSON, E., RUBIN, D., MEKORI, T., SEGAL, R. & POLLACK, S. (1993). *In vivo* modulation of natural killer cell activity by tamoxifen in patients with bilateral primary breast cancer. *Cancer Immunology and Immunotherapy* **37**, 209–212.

ROSEN, E. M., GOLDBERG, I. D., LIU, D., SETTER, E., DONOVAN, M. A., BHARGAVA, M., REISS, M. & KACINKSI, B. M. (1991). Tumor necrosis factor stimulates epithelial tumor cell motility. *Cancer Research* **51**, 5315–5321.

ROSEN, P. P. & GROSHEN, S. (1990). Factors influencing survival and prognosis in early breast carcinoma: assessment of 644 patients with median follow-up of 18 years. *Surgery Clinics of North America* **70**, 937–962.

ROWSE, G., WEINBERG, J. & EMERMAN, J. (1995). Role of natural killer cells in psychosocial stressor-induced changes in mouse mammary tumor growth. *Cancer Research* **55**, 617–622.

RUBBERT, A., MANGER, B., LANG, N., KALDEN, J. R. & PLATZER, E. (1991). Functional characterization of tumor-infiltrating lymphocytes, lymph-node lymphocytes and peripheral blood lymphocytes from patients with breast cancer. *International Journal of Cancer* **49**, 25–31.

SCHOLL, S. M., CROCKER, P., TANG, R., POUILLART, P. & POLLARD, J. W. (1993). Is colony-stimulating factor-1 a key mediator of breast cancer invasion and metastasis? *Molecular Carcinogenesis* **7**, 207–211.

SCHWARTZENTRUBER, D. J., SOLOMON, D., ROSENBERG, S. A. & TOPALIAN, S. L. (1992). Characterization of lymphocytes infiltrating human breast cancer: specific immune reactivity detected by measuring cytokine secretion. *Journal of Immunotherapy* **12**, 1–12.

SEGHAL, P. B. & TAMM, I. (1991). Interleukin-6 enhances motility of breast carcinoma cells. In *Cell Motility Factors* (ed. Goldberg, I. D.), pp. 178–193. Basel, Birkhäuser Verlag.

SINGH, A., PUROHIT, A., DUNCAN, L. J., MOKBEL, K., GHILCHIK, M. W. & REED, M. J. (1977). Control of aromatase activity in breast tumors: the role of the immune system. *Journal of Steroid Biochemistry and Molecular Biology, in press.*

SPRIGGS, D., INAMURA, K., RODRIGUEZ, C., HORIGUCHI, J. & KUFE, D. W. (1987). Induction of tumor necrosis factor expression and resistance in a human breast tumor cell line. *Proceedings of the National Academy of Sciences, USA* **84**, 6563–6566.

SQUARTINI, F., OLIVI, M. & BOLIS, G. B. (1970). Mouse strain and breeding stimulation as factors influencing the effect of thymectomy on mammary tumorigenesis. *Cancer Research* **30**, 2069–2072.

STANCOVSKI, L., HURWITZ, E., LEITNER, O., ULLRICH, A., YARDEN, Y. & SELA, M. (1991). Mechanistic aspects of the opposing effects of monoclonal antibodies to the ERBB-2 receptor on tumor growth. *Proceedings of the National Academy of Sciences, USA* **88**, 8691–8695.

STEWART, T. H. M. (1969a). The presence of delayed hypersensitivity reactions in patients toward cellular extracts of their own tumors. 1. The role of tissue antigen, non-specific reactions of nuclear material, and bacterial antigen as a cause for this phenomenon. *Cancer* **23**, 1368–1379.

STEWART, T. H. M. (1969b). The presence of delayed hypersensitivity reactions in patients toward cellular extracts of their own tumors. 2. A correlation between the histologic picture of lymphocytic infiltrate of the tumor stroma, the presence of such a reaction, and a discussion of the significance of this phenomenon. *Cancer* **23**, 1380–87.

STEWART, T. H. M. (1994). Post-mastectomy radiotherapy: more than a local matter. *Lancet* **344**, 351.

STEWART, T. H. M. & ORIZAGA, M. (1971). The presence of delayed hypersensitivity reactions in patients toward cellular extracts of their malignant tumors. 3. The frequency, duration and cross reactivity of this phenomenon in patients with breast cancer and its correlation with survival. *Cancer* **28**, 1472–1478

STEWART, T. H. M. & RAMAN, S. (1992). Adjuvant treatment of non-small-cell lung cancer: a meta-analysis. In *Cellular Immune Mechanisms and Tumor Dormancy* (ed. Stewart, T. H. M. & Wheelock, E. F.), pp. 213–238. Boca Raton, CRC Press.

STEWART, T. H. M., RETSKY, M. W., TSAI, S. C. J. & VERMA, S. (1994). Dose response in the treatment of breast cancer. *Lancet* **343**, 402–404.

STEWART, T. H. M. & TSAI, S.-C. (1993). The possible role of stromal cell stimulation in worsening the prognosis of a subset of patients with breast cancer. *Clinical and Experimental Metastasis* **11**, 295–305.

STEWART, T. H. M., TSAI, S. C. J., GRAYSON, H., HENDERSON, R. & OPELZ, G. (1995). Incidence of *de-novo* breast cancer in women chronically immunosuppressed after organ transplantation. *Lancet* **346**, 796–798.

TANG, R., KACINSKI, B., VALIDIRE, P., BEURON, F., SASTRE, X., BENOIT, P., DE LA ROCHEFORDIÈRE, MOSSERI V., POUILLART, P. & SCHOLL, S. (1990). Oncogene amplification correlates with dense lymphocytic infiltration in human breast cancers: a role for hematopoietic growth factor release by tumor cells? *Journal of Cell Biochemistry* **44**, 189–198.

TSAI, S.-J. & HEPPNER, G. H. (1994). Immunoendocrine mechanisms in mammary tumor progression direct prolactin modulation of peripheral and preneoplastic hyperplastic-alveolar-nodule infiltrating lymphocytes. *Cancer Immunology and Immunotherapy* **39**, 291–298.

TSAI, S.-J., LOEFFLER, D. A. & HEPPNER, G. H. (1992). Assorted effects of bromocriptine on neoplastic progression of mouse mammary preneoplastic hyperplastic alveolar nodule (HAN) line C4 and on HAN-infiltrating and splenic lymphocyte function. *Cancer Research* **52**, 2209–2215.

TSAI, S.-J., MILLER, F., WEI, W.-Z. & HEPPNER, G. H. (1992). Tumor necrosis factor (TNF) modulates growth of normal and neoplastic murine mammary cells. *Proceedings of the American Association of Cancer Research* **33**, 269.

VITOLO, D., ZERBE, T., KANBOUR, A., DAHL, C., HEBERMAN, R. B. & WHITESIDE, T. L. (1992). Expression of mRNA for cytokines in tumor-infiltrating mononuclear cells in ovarian adenocarcinoma and invasive breast cancer. *International Journal of Cancer* **51**, 573–580.

VOLPÉ, R. (1986). Pathogenesis of autoimmune thyroid disease. In *Werner's The Thyroid, a fundamental and clinical text*, 5th edn (ed. Ingbar, S. H. & Braverman, L. E.), pp. 747–767. J. P. Lippincott Company, Philadelphia.

VOSE, B. M. & MOORE, M. (1979). Suppressor cell activity of lymphocytes infiltrating human lung and breast tumors. *International Journal of Cancer* **24**, 579–585.

VUKANOVIC, J. & ISAACS, J. T. (1995). Linomide inhibits angiogenesis, growth, metastasis, and macrophage infiltration within rat prostatic cancers. *Cancer Research* **55**, 1499–1504.

WAKSMAN, B. H. (1959). Experimental allergic encephalomyelitis and the 'autoallergic' diseases. *International Archives of Allergy, Supplement* **14**, 1–87.

WAKSMAN, B. H. (1962). Autoimmunization and the lesions of autoimmunity. *Medicine (Balt)* **14**, 93–141.

WANEBO, H. J., ROSEN, P. P., THALER, T., URBAN, J. A. & OETTGEN, H. F. (1976). Immunobiology of operable breast cancer: an assessment of biologic risk by immunoparameters. *Annals of Surgery* **184**, 258–267.

WANG, Y., HOLLAND, J. F., BLEIWEISS, I. J., MELANA, S., LIU, X., PELISSON, I., CANTARELLA, A., STELLRECHT, K., MANI, S. & POGO, B. G. T. (1995). Detection of mammary tumor virus ENV gene-like sequences in human breast cancer. *Cancer Research* **55**, 5173–5179.

WEI, W.-Z., FICSOR-JACOBS, R., TSAI, S. J. & PAULEY, R. (1991). Elimination of Vβ2 bearing T cells in BALB/c mice implanted with syngeneic preneoplastic and neoplastic mammary lesions. *Cancer Research* **51**, 3331–3333.

WEI, W.-Z., FULTON, A., WINKELHAKE, J. & HEPPNER, G. (1989). Correlation of natural killer activity with tumorigenesis of a preneoplastic mouse mammary lesion. *Cancer Research* **49**, 2709–2715.

WEI, W.-Z., GILL, R. F. & WANG, H. (1993). Mouse mammary tumor virus associated antigens and superantigens: immuno-molecular correlates of neoplastic progression. *Seminars in Cancer Biology* **4**, 205–213.

WEI, W.-Z. & HEPPNER, G. H. (1987). Natural killer problem of lymphocytic infiltrates in mouse mammary lesions. *British Journal of Cancer* **55**, 589–594.

WEI, W.-Z. & HEPPNER, G. H. (1996). Breast cancer immunology. In *Mammary Tumor Cell Cycle, Differentiation and Metastasis* (ed. Dickson, R. & Lippman, M.), pp. 395–410. Boston, Kluwer Academic Publishers.

WEI, W.-Z., MILLER, F. R., BLAZAR, B. A., MEDINA, D. & HEPPNER, G. (1979). Opposing effects of cryostat sections of preneoplastic and neoplastic mouse mammary lesions on *in vitro* migration of peritoneal exudate cells. *Journal of Immunology* **122**, 2059–2067.

WEIGENT, D. A. & BLALOCK, J. E. (1995). Associations between the neuroendocrine and immune systems. *Journal of Leukocyte Biology* **58**, 137–150.

WEISS, D. W., FAULKIN, L. J. & DE OME, K. B. (1964). Acquisition of heightened resistance and susceptibility to spontaneous mouse mammary carcinomas in the original host. *Cancer Research* **24**, 732–741.

WEISS, D. W., LAVRIN, D. H., DEZFULIAN, M., VAAGE, J. & BLAIR, P. B. (1966). Studies on the immunology of spontaneous mammary carcinoma of mice. In *Viruses Inducing Cancer* (ed. Burdette, W. J.), pp. 138–168. Salt Lake City, University of Utah Press.

WELSH, C. W. & NAGASAWA, H. (1977). Prolactin and murine mammary tumorigenesis: a review. *Cancer Research* **37**, 951–963.

WELTE, K., CIOBAÑU, N., MOORE, M. A. S., GULATI, S., O'REILLY, R. J. & MERTELSMAN, R. (1984). Defective interleukin 2 production in patients after bone marrow transplantation and *in vitro* restoration of defective T lymphocyte proliferation by highly purified interleukin 2. *Blood* **64**, 380–385.

WHITESIDE, T. L., MIESCHER, S., HURLIMANN, J., MORETTA, L. & VON FLIEDNER, V. (1986). Clonal analysis and *in situ* characterization of lymphocytes infiltrating human breast carcinomas. *Cancer Immunology and Immunotherapy* **23**, 169–178.

ZIELINSKI, C. C., MÜLLER, C., KUBISTA, E., STAFFEN, A. & EIBL, M. M. (1990). Effects of adjuvant chemotherapy on specific and nonspecific immune mechanisms. *Acta Medica Austriaca (Wien)* **17**, 11–14.

Mathematical models of parasite responses to host immune defences

R. ANTIA *and* M. LIPSITCH

Department of Biology, Emory University, Atlanta, GA 30322, USA

SUMMARY

We examine the evolution of microparasites in response to the immune system of vertebrate hosts. We first describe a simple model for an acute infection. This model suggests that the within-host dynamics of the microparasite will be a 'race' between parasite multiplication and a clonally expanding response by the host immune system, resulting either in immune-mediated clearance or host death. In this very simple model, in which there is only a single parasite and host genotype, maximum transmission is obtained by parasites with intermediate rates of growth (and virulence). We examine how these predictions depend on key assumptions about the parasite and the host, and consider how this model may be expanded to incorporate the effect of additional complexities such as host–parasite co-evolution, host polymorphism, and multiple infections.

Key words: parasite–host evolution.

INTRODUCTION

The immune response of vertebrate hosts has evolved to protect the host against attack by parasites.[1] Parasites have, in turn, evolved a wide array of mechanisms to counter the immune response. In this paper we restrict ourselves to consideration of microparasites (viruses, bacteria and protozoa), and their interaction with the adaptive or antigen-specific immune response of vertebrate hosts. The constraints the immune system imposes on microparasites are particularly severe as they must undergo considerable replication within the host in order to reach sufficiently high densities to be transmitted to new hosts.

In general (i.e. for many parasites) we would expect both the rate of transmission of the parasite from a host and its deleterious effects on the host will be proportional to its within-host density (Topley, 1919; Lipsitch & Moxon, 1997). Consequently we would expect selection on the parasite population will favour parasites which attain high densities for as long a duration as possible, while selection on the host population will favour hosts which reduce the density of parasites as well as the duration of infection. This is likely to result in a co-evolutionary 'arms race' between the parasites and host. The

asymmetrical antagonistic relationship which characterizes such an 'arms race' is likely to be unstable and may involve rapid, erratic co-evolution (Haldane, 1949; Seger, 1992).

Microparasites usually have relatively short generation times and large population sizes, while vertebrate hosts have comparatively long generation times and small population sizes. Thus we may expect parasites to be able to evolve much more rapidly than their vertebrate hosts. Vertebrate hosts counter this assault by potentially rapidly evolving parasites in several ways. First, they have sexual reproduction. Sexual reproduction permits more rapid evolution by allowing for the generation of new genotypic variants by recombination. This also allows for the generation and maintenance of a high degree of polymorphism (Ebert & Hamilton, 1996). The diversity in the MHC molecules of the host is probably generated and maintained in this manner (Hill, 1996; Parham & Ohta, 1996). Second, the immune system of vertebrates works by clonal selection in which only useful (i.e. parasite-specific) responses are selected from an enormous repertoire of possible responses. Thus the rate at which the host is able to generate novel parasite-specific responses equals the rate of generation of specific immune responses rather than the rate of evolution of the host.

Many studies of parasite evolution have considered the transmission of the parasite at the epidemiological level. In their simplest form these models predict that parasites will evolve to maximize their total transmission over time (i.e. their basic reproductive rate) and examine where maximal transmission obtains for different tradeoffs among transmissibility, the duration of infection and parasite-induced mortality (Anderson & May, 1982;

[1] An alternative hypothesis proposed by Sir McFarlane Burnett is the 'immune surveillance' hypothesis which postulates that the immune system evolved to protect the host from cancers and other insurrections arising from within the host rather than against the invasion of parasites from outside the host (Burnet, 1970). The observation that immunocompromised hosts do not have a greatly elevated incidence of cancer, but do have higher rates of mortality following infection with pathogens, suggests that immune surveillance primarily protects the host from invasion by parasites rather than cancer.

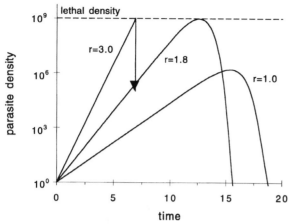

Fig. 1. Within-host dynamics of parasites with different growth rates r. The parasite density is plotted (on a logarithmic scale) against the time after infection. The dynamics of parasite and immunity are calculated by numerical simulations of eqns (1) and (2) for parasites with growth rates $r = 1\cdot0$, $1\cdot8$ and $3\cdot0$ as indicated and with parameters: $s = 1$; $d = 0\cdot1$; $k = 10^{-3}$; $h = 10^3$; $D = 10^9$; and initial densities of parasite and immunity set to unity i.e. $P(0) = X(0) = 1\cdot0$. Parasites with a slow growth rate are cleared by the host's immune response before they reach a high density, parasites with intermediate growth rates reach higher densities, and parasites with very fast growth rates reach the lethal density and kill the host before the immune response can control the infection.

May & Anderson, 1983; Lenski & May, 1994; Anderson, 1995). These models have been modified to take into account other factors such as vertical transmission (Lipsitch *et al.* 1996), and multiple infection (Levin & Pimentel, 1981; Nowak & May, 1994; May & Nowak, 1995). While these models help examine the consequences of a particular tradeoff between transmissibility, duration of infection and parasite-induced mortality, the tradeoff needs to be specified extrinsic to the model.

A few recent studies of parasite evolution (Sasaki & Iwasa, 1991; Antia, Levin & May, 1994; Bonhoeffer & Nowak, 1994*a*, *b*) have taken into account the process of infection within the host. These studies describe the types of tradeoffs among transmissibility, duration of infection and virulence which might be expected to arise as a consequence of factors such as the immune response of vertebrate hosts or competition between multiple strains of the parasite. These models thus attempt to provide the biological explanations for the types of tradeoffs that were used in the epidemiological models mentioned earlier.

In this paper we consider the evolution of rapidly replicating microparasites in response to the immune system. Towards this end we describe a simple model for the interaction between a rapidly replicating microparasite and the immune response of the host, and use this model to evaluate how evolutionary forces may act on particular characteristics of the organism (its growth rate, immunogenicity, susceptibility to immune attack etc.). This approach allows us to consider the following aspects of the evolution of microparasites. First, we determine the rate of replication of the microparasite at which transmission would be maximized, and consider the implications for the evolution of virulence of the parasite. Second, we consider the effects of polymorphism in the host population on this model, focusing on how it alters the tradeoff between rate of transmission, duration of infection and virulence. We then consider parasite–host evolution in terms of changes in the parameters of this model. Finally, we consider the limitations of the model and mechanisms of parasite evolution which are not encompassed by this simple model.

The role of the relatively simple models described in this paper is primarily heuristic; the models allow us rigorously to analyse the outcomes arising from a set of well-defined assumptions. Models are required because the complex (in general nonlinear) nature of the interactions between parasite and immune response can make for surprising, and sometimes counterintuitive, results, which would not be understood in the absence of mathematical treatment.

MODEL OF AN ACUTE INFECTION

Basic model

We present a simple model to describe the interaction between a rapidly replicating microparasite and the immune response of the host, and use this model to consider several aspects of the evolution of microparasites. This model of parasite evolution (in the absence of host polymorphism) is very similar to that described in a previous paper (Antia *et al.* 1994), and the reader is referred to this paper for further details. In this model the rate of change in the density of a parasite population, P, within a host depends on the rate, r at which it replicates in the host, and the rate at which it is cleared by the host's defences. The immune responses of the host may be divided into non-specific and specific components. For simplicity, we let the non-specific defences and the other causes of parasite loss be time- and density-independent and let them be reflected by a decrease in the rate of proliferation of the parasite r (Antia & Koella, 1994). We assume that the magnitude of the specific immune response, X, is proportional to the density of immune cells (cytolytic and helper T lymphocytes and antibody-producing B lymphocytes) specific to the microparasite. The rate of change in the density of these cells is described by 2 terms: the first term describes their proliferation, and the second their death. The proliferation of antigen-specific immune cells occurs by clonal expansion at a rate proportional to the parasite density at low parasite densities, and saturates at high parasite densities. The death of immune cells is

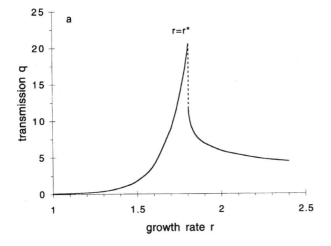

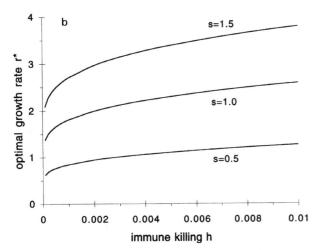

Fig. 2. Transmission of parasite from infected hosts over the course of infection. In Fig. 2*a* we plot the total transmission of parasite from an infected host over the course of infection (*q*) for parasites with different growth rates. The total transmission *q* is obtained from integration of the rate of transmission over the course of the infection as described in eqn 3, with $u = 10^{-8}$. The optimal transmission occurs at an intermediate growth rate $r = r*$ which corresponds to the situation in which the maximum parasite density just falls short of the lethal density before being controlled by the immune response. In Fig. 2*b* we plot how the optimal growth rate $r*$ changes when the rate of generation of immune responses or the rate of clearance of parasite by immunity is altered. Increasing the rate of generation of immune responses or the rate at which they can clear the parasite results in an increase in the growth rate of the parasite at which maximum transmission occurs, suggesting that more effective immune responses will select for faster growing parasites. Parameters same as in Fig. 1.

assumed to occur at a parasite-independent death rate *d*. We assume there is a threshold density *D* of parasites, above which the host dies and the parasite density is set to zero, and also that if the parasite density falls below unity the parasite is eliminated and its density is set to zero. We let *h*, *s* and *k* represent, respectively, the rate constant for elim-

ination of the parasite population by the immune response, the maximum rate of proliferation of immune cells, and the parasite density at which the immune cells proliferate at half the maximum rate. We also set the parasite density and the immune response so that their initial values equal one, and adjust the other constants accordingly. With these definitions and assumptions, the rates of changes in the densities of parasite and immunity are given by,

$$\frac{dP}{dt} = rP - hPX \quad \text{if } P < D \tag{1}$$

and

$$P \to 0 \quad \text{if } P > D \quad \text{or} \quad P < 1$$

$$\frac{dX}{dT} = sX\left(\frac{P}{P+k}\right) - dX. \tag{2}$$

We now consider the relative magnitudes of the various parameters. First we note that as an acute infection is of relatively short duration the death rate of immune cells during this time is small and can be set to zero without much quantitative change in the outcome (Antia, Nowak & Anderson, 1996). We also expect that the initial parasite inoculum is much smaller than the parasite density *k* at which the immune response grows at half its maximum rate, which is, in turn, much smaller than the density *D* at which the parasite kills the host. Finally, since parasite-specific immune cells must attain a high density in order to control the parasite, we expect that the initial rate of clearance by the specific immune response (*h*) is much less than the initial growth rate of the parasite, *r*. Thus,

$$h, d \ll r \quad \text{and} \quad s, P(0), X(0) \ll k \ll D. \tag{3}$$

Within-host parasite dynamics and the optimal transmission of the parasite

We illustrate the within-host dynamics of the parasite population with different rates of replication *r*, and the parameters *h*, *s* and *k* held constant (Fig. 1). At the slowest rate of replication the parasite is cleared by the immune system well before it reaches a lethal density, *D*. At the intermediate rate of replication the parasite is cleared by the immune system just as its density reaches *D*, and at the fastest rate of replication the parasite soon exceeds *D*, and the host dies before the parasite can be cleared.

If we know the relationship between the within-host parasite density and transmissibility then we can calculate the total number of parasites released by an infected host *q* as the integral over the duration of infection (τ) of the rate of release of transmissible parasites per unit time, i.e.

$$q = \int_0^\tau u[P(t)]\,dt, \tag{4}$$

where $u[P]$ is the rate of transmission when the within-host parasite density equals P. In the following discussion we let the rate of transmission be directly proportional to the within-host density, i.e. $u[P] = uP$. We note that qualitatively similar results will obtain if we choose a transmission function which saturates at a density greater than the density k at which the immune response is stimulated (see Antia et al. 1994).

In Fig. 2a, we plot the total number of parasites released from an infected host q, as a function of the rate of replication of the parasite r. We find that parasites with intermediate rates of replication release the greatest number of free parasites, and this transmission is maximal at that value of r where the immune response is just able to control the parasite population before it reaches its lethal density D. The growth rate of the parasite r^* at which maximum transmission is obtained is given approximately by the solution of the equation:

$$\frac{D^{s-d}}{k^s} \approx \left(\frac{r}{he}\right)^r. \tag{5}$$

The model makes a simple prediction, namely, that as a consequence of the interaction between a proliferating microparasite population and the host's immune defences, selection can favour parasites with intermediate rather than high or low rates of growth. Slowly growing parasites are cleared by the immune response before they reach high within-host densities, and are therefore released at lower rates than faster-growing parasites. Very fast growing parasites, however, kill the host before releasing as many transmissible forms as parasites of somewhat lower growth rates that persist longer in the living host.

The optimal growth rate depends on the values of the parameters describing the interaction of the parasite with the immune response. In general, parasites eliciting rapid, highly effective immune responses will have higher optimal growth rates, since without rapid growth they will be cleared quickly, before reaching high densities. By contrast, parasites eliciting slower, less effective immune responses will have slower optimal growth rates because rapid growth would result in early death of the host, thus curtailing transmission. Fig. 2b shows how the optimal growth rate increases as a function of the parameters s and h.

Consequences for parasite virulence

The model predicts that optimal transmission occurs when the parasite is controlled by the immune response just as its density reaches the lethal density, and this obtains at intermediate rates of growth of the parasite. The maximum density attained by the parasite increases with increasing growth rate of the parasite; therefore, provided virulence is positively

correlated with the parasite density within the host, our results suggest that optimal transmission obtains at intermediate levels of virulence.

A more quantitative relationship between growth rate and virulence can be obtained if we use a quantitative measure of parasite virulence. There are several measures of parasite virulence including: (i) the LD_{50} or the inoculum which on average kills half of infected hosts; and (ii) the mortality rate due to infection, which can be expressed in various ways including as the rate of parasite-induced mortality or the fraction of hosts which die following infection (i.e. the percentage mortality). Since our model is deterministic, a given dose either kills or does not kill the host. Under these conditions the LD_{50} simply corresponds to the lethal dose (LD). For the model presented, the $\log(LD)$ is linearly related to the rate of growth of the parasite. In the absence of any heterogeneity in parasite and host populations and stochastic effects (in, for example the inoculum size) our model predicts that the parasite optimum is obtained by the parasite whose maximum density within the host is infinitesimally short of the lethal density: i.e. host mortality is zero. The consequences of introducing heterogeneity are discussed in the following section.

Consequences of host polymorphism

As mentioned in the introduction, sexual reproduction and recombination can generate a considerable amount of polymorphism in the host population. This polymorphism is particularly evident at the MHC loci which play an important role in immune recognition. In this section we examine the consequences of this polymorphism for the results of our model. Polymorphism in the MHC loci can be incorporated by changes in 2 of the parameters of the model. First, polymorphism in both the class I and the class II MHC loci can be expressed by variability in the immunogenicity of parasite (and consequently the rate at which the immune response to the parasite is generated). This can be expressed as changes in the parameter k of the model. Additionally, polymorphism in the class I MHC loci can result in changes in the rate at which CTL kill infected cells. This can be expressed by changes in the parameter h of the model.

We incorporated polymorphism in the host population by including a stochastic term in the choice of the parameters k and h in our model. For each growth rate we ran a number of simulations in which the parameters k and h were chosen from a uniform distribution with the same average as in the simulations described above. This allowed us to investigate how polymorphism in the host population would change the average transmission of the parasite as well as its virulence as determined by the percentage mortality of the host. Fig. 3a shows the

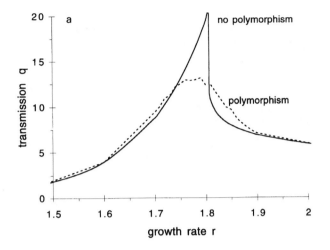

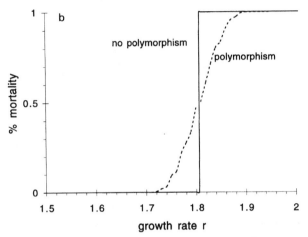

Fig. 3. Consequences of polymorphism in the host population. We plot the consequences of host polymorphism for (i) the (average) total transmission of parasite following infection for parasites with different growth rates (Fig. 3 *a*), and (ii) the percentage mortality following infection (Fig. 3 *b*). The solid line corresponds to a host population in the absence of polymorphism and the dashed line shows the consequence of introducing polymorphism into the host population. Polymorphism in the host population was introduced into our simulations by choosing the parameters h and k stochastically from a uniform distribution in the range $0.5 \ 10^{-3} < h < 1.5 \ 10^{-3}$ and $0.5 \ 10^{3} < k < 1.5 \ 10^{3}$ i.e. with the same average value as before. The mean transmission and percentage mortality was obtained for each growth rate by taking the average of 500 simulations. Other parameters same as in Fig. 1. We find that introducing polymorphism in the host population results in a decrease in the transmission of the parasite from infected hosts, a decrease in the growth rate of the parasite at which maximum transmission occurs.

consequences of incorporating polymorphism on the transmission of the parasites with different growth rates. We notice 2 effects. First, the maximum attainable transmission is lower in a polymorphic host population than in a monomorphic one. This occurs because the parasite cannot optimize to all

host genotypes at the same time. Second, the optimal growth rate of the parasite is lower in a polymorphic host population than in a genetically homogeneous host population. This occurs because in this model, the drop in transmission is more rapid for parasites growing slightly faster than the optimum rate than for parasites growing slightly slower than the optimum growth rate: parasites growing faster than the optimal rate kill the host and lose all subsequent transmission opportunities from that host. Fig. 3 *b* shows the consequences of incorporating polymorphism on the percentage mortality (measured in terms of the fraction of infections which end in the lethal within-host parasite density being reached) for parasites with different growth rates. In the absence of polymorphism there is a discontinuous variation in mortality with growth rate, with parasite-induced mortality undergoing a jump from 0 to 100% when the growth rate crosses an optimal value for transmission. The maximum transmission occurs at the highest growth rate for which the parasite does not kill the host. Introducing polymorphism into the host population gives a more realistic outcome in that the average percentage mortality increases gradually from 0 to 100% as the parasite growth rate increases.

A similar effect could be obtained by introducing variability into this model in other ways. This includes variability in the other parameters of the model (such as s or d) or the initial conditions such as the size of the inoculum or the precursor frequency of antigen-specific immune cells. We also note that the prediction that polymorphism lowers the optimal growth rate depends on the assumption that parasite transmission stops with the death of the host; if death does not completely curtail transmission, host polymorphism might increase the optimal growth rate. This will occur in the model if, for example, the transmission following host death is greater than the transmission which occurs as the parasite is being cleared by the immune system after the peak of parasitaemia. In this case the rate of growth at which transmission is maximized will be equal to the rate of growth at which the host is just killed, which is infinitesimally more than that for which it just survives. If we incorporate polymorphism under these circumstances the optimal growth rate will be slightly higher than that in the absence of polymorphism.

The model presented here only touches on some of the consequences of polymorphism in the host population for the parasite. We do not, for example, consider the factors responsible for the generation and maintenance of this polymorphism in the host population, or the interaction between polymorphism in the host and parasite populations (Frank, 1993). There is considerable scope for further theoretical and experimental research on these problems.

PARASITE EVOLUTION IN RESPONSE TO THE IMMUNE SYSTEM

Parasites are exposed to strong selection pressure that drives the evolution of mechanisms to counter the immune response. In this section we use the model described in the previous section to investigate the evolution of parasites which cause acute infections. Later we discuss the limitations of this model and consider the consequences of incorporating complexities in the immune system and parasite populations.

Evolution in parasites causing acute infections

Above we described a simple model for the within-host dynamics of parasites during acute infections. We now consider how parasite evolution in response to the immune system can be described in terms of changes in the parameters of this model. The model can incorporate 2 broad categories of parasite evolution. The first class consists of those responses which inhibit the generation of an immune response and the second class consists of defences which provide resistance to attack by an immune response following its generation.

Parasite responses that prevent the generation of immune responses can be described by changes in the parameters k, s and d, which determine the rate of generation of immune responses. The changes might involve a decrease in s, the maximum rate of growth of immune cells, or an increase in k, the parasite density at which the immune response grows at half its maximum rate, or an increase in d, the rate of death of immune cells. Some of the mechanisms by which the parasites could make these changes include: (i) by mimicking host proteins; (ii) by masking common antigens by binding with host proteins such as the serum proteins fibrinogen and fibrin (Whitnack, Dale & Beachey, 1984); (iii) the production of proteins which interfere with cytokine signalling pathways and the generation of the immune response (Hsu *et al.* 1990) or susceptibility of cells to killing (Wold & Gooding, 1991); (iv) interference in the processing and presentation of antigen on the MHC (Ehrlich, 1995).

Parasite responses which provide protection from immune attack can be described by a decrease in h, the rate constant for the clearance of the parasite by the immune response. Some of the mechanisms by which this could occur include: (i) the production of polysaccharide capsules (Finlay & Falkow, 1989); (ii) the generation of proteases that cleave immunoglobulins (Plaut, 1983); (iii) reducing the presentation of antigen on the MHC Class I molecules on the surface of infected cells (Ehrlich, 1995).

Some mechanisms may both reduce immunogenicity and provide resistance to immune attack. For example: (i) the masking of common antigens may both reduce the rate of generation of humoral immune responses and the rate at which antibody clears the pathogen, and (ii) intracellular parasites such as viruses may both slow down the generation of CTL responses as well as killing by CTL by down-regulating the expression of class I MHC molecules on the surface of infected cells. The action of natural killer (NK) cells may counter this evasion mechanism, as down-regulation of MHC expression renders the cell sensitive to killing by NK cells. One virus countermeasure is the generation of MHC class I homologues which may interfere with the presentation of antigen on the MHC and thus provide some resistance to attack by CTL, as well as providing resistance to attack by NK cells (Farrell *et al.* 1997; Reyburn *et al.* 1997).

What might we expect to be the consequences of a decrease in immunogenicity and resistance to immune attack for the evolution of the parasite? We now consider the relative transmission of the original and evolved strain from hosts infected with either strain alone. We consider the within-host competition between these 2 strains later. The relative transmission of the original and evolved strain from infected hosts depends on whether the parasite had reached its optimal growth rate prior to the appearance of a variant which provokes less immune-mediated killing, or whether co-evolution of the host had kept the growth rate of the original strain to a lower rate than optimal for parasite transmission. If the parasite's growth rate is less than the optimal rate, than parasite variants that avoid or resist the immune response will result in an increase in transmission compared to the original variant. If the parasite is at its optimum, then any change in parameters of the parasite will result in a reduction in fitness in terms of total transmission, even if it is selected at the within-host level. However, further evolution from this stage with reduced fitness can lead to a new higher optimum. For example, if the parasite is at its optimal growth rate, evolution of the parasite to resist immune attack (by a decrease in h) will result in lowering the total transmission of the parasite from the host because the lethal density is crossed before the immune system controls the evolved variant. If this resistance to immune attack is followed by a reduction in the growth rate of the parasite then a new optimum can be attained with a higher transmission than at the earlier optimum. These scenarios are illustrated in Fig. 4.

Further complexities

The model presented above is limited to describing the evolution of genetically invariant parasites which have relatively simple interactions with the immune response that are captured by eqns (1) and (2). In this section we begin by outlining the limitations of the model and describe some of the parasite

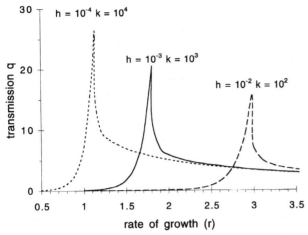

Fig. 4. Consequences of evolution of parasite and host. Evolution of the parasite to avoid stimulating or evading immune responses will result in an increase in the parameter k and a decrease in the parameter h, while evolution of the host immune response to be more rapid and efficient in clearing the parasite will result in a decrease in the parameter k and an increase in the parameter h. In Fig. 4 we plot the consequences of these changes for the transmission of parasites from an infected host. We note that only if the parasite growth rate is below the optimum value will parasite evolution to avoid the immune response be selected in the absence of a concomitant decrease in the growth rate of the parasite.

mechanisms for the evasion of immune responses which cannot be described by the model of an acute infection.

Implicit in much evolutionary theory about virulence is the assumption that the parasite has optimized its transmission from a host. Many parasite characters that enhance virulence may be incidental to the transmission and long-term fitness of the parasite (Levin & Bull, 1994; Lipsitch & Moxon, 1997). For example, the systemic sequelae of syphilis infection which are responsible for much of its pathology and virulence probably do not enhance transmission but reflect spread of the bacteria to new niches (tissues) within individuals. Other examples may include polio, HIV, meningitis caused by normally commensal bacteria and infections by organisms whose transmission chain does not normally include the given host (Levin & Bull, 1994; Lipsitch & Moxon, 1997). Careful consideration of the relationship between symptoms and transmission is necessary when applying any theoretical framework to understanding the evolution of individual parasites.

Another class of limitations of our simple model pertains to the nature of the immune response. There have been many mathematical models describing more complex immune responses. Particularly important is the nature of the proliferation function for T cells. Our model assumes that the *per capita* rate of T cell proliferation is a simple saturating

function of the antigen density. More complex immune response functions might include: (i) nonspecific immune responses (Antia & Koella, 1994); (ii) interaction between antigen presentation and the T cell response (Keverekidis *et al.* 1988; Fishman & Perelson, 1993; De Boer & Perelson, 1995); (iii) high dose suppression (McLean & Kirkwood, 1990; Schweitzer & Anderson, 1992); (iv) competition and cross regulation between different T cells (Fishman & Perelson, 1994); (v) the interactions between B and T cells required for the generation of antibody responses (Oprea & Perelson, 1996); (vi) adaptation of the immune response (Grossman & Paul, 1992; Grossman & Singer, 1996) and affinity maturation (Agur *et al.* 1991; Fishman & Perelson, 1995); (vii) age-structure of immune cells (Pilyugin *et al.* 1997).

A third class of limitations of the model pertains to the parasite growth within the host. Our model assumes that the parasite does not change antigenically and that it replicates at a constant rate r. In reality the growth of a parasite within the host may be considerably more complex, and involve mechanisms to avoid the immune response. Some of these mechanisms to avoid the immune response (which are not described in terms of changes in the parameters h and k of our model) include: (i) antigenic variation within the host; (ii) multiple life-stages within the host; (iii) dormant stages which are sequestered from the immune response; (iv) the ability to kill the immune response. We now briefly describe these parasite responses to the immune system.

Antigenic variation. Many parasites change their antigenic properties during the course of infection (e.g. *Trypanosoma brucei*, *Plasmodium falciparum*, HIV). The immune response of a host against 1 antigenic variant may have limited effect on other variants of the same parasite. Thus, following infection the rapid production of immunologically-distinct antigenic variants prior to immune clearance can potentially prolong an infection. Mathematical models have proved useful for our understanding of antigenic variation. Agur *et al.* (1989) used a simulation model to examine the dynamics of antigenic variation during trypanosome infections. Based on this model they proposed: (i) that the switch from the expression of 1 variable surface glycoprotein (VSG) to another involved an intermediate stage with the expression of both VSGs, and (ii) there may be some ordering to the appearance of antigenic variants. Antia *et al.* (1996) used a simple model to consider the consequences of variant-specific and cross-reactive immunity in controlling infection with antigenically varying infections. Based on this model they suggested that: (i) cross-reactive immunity may play an important role in control of the infection; (ii) there may be a maximum number of different variants which will be selected and this number would be approximately equal to the ratio of the immunogenicity of variant-specific antigens to the common or cross-reactive antigens; (iii) antigenically varying parasites can also cause a down-regulation of all immune responses by 'overwhelming' the immune system with hundreds of different antigens – if there is a carrying capacity to the immune response then immune homeostasis mechanisms could result in the downregulation of all immune responses when the parasite density is high. This would provide a simple explanation for the generalized immunosuppression reported during trypanosome infections of cattle (Vickerman & Barry, 1982; Sztein & Kierszenbaum, 1993).

Multiple life stages. Parasites can have complex life cycles which include different life-stages for growth within a host and for transmission between hosts. Malaria parasites, for example, reproduce asexually within a host, but only a distinct form, the gametocytes, allows transmission via the mosquito vector, where gamete fusion and meiosis take place. Similarly, microsporidia grow by fission within a host, but must develop into spores to allow transmission. The model described above suggested that, to maximize transmission, parasites will grow at a rate that is a compromise between the generation of high densities and prolonging survival within the host. If the growth and transmission forms of the parasite are immunologically distinct, as seems to be the case in the malaria parasite (Wakelin, 1997), maximization of density of the transmissible form can lead to a different form of compromise. Mathematical models have been used to consider the optimal pattern of resource allocation between growth to transmission stages in order to maximize transmission (Koella & Antia, 1995). These models have been used to consider the optimal pattern of replication and transmission stages where the switch from growth to transmission stages occurs at a constant rate or is time and density dependent. The models suggest that if a parasite can vary its investment into transmission during the course of infection it should delay investment into transmission until the growth stages reach a high density (just short of the lethal density) or till shortly before the growth stage is cleared by the immune response.

Dormancy/latency. Bacteria such as *Mycobacterium tuberculosis* and *M. leprae* can avoid the immune response by remaining dormant within cells and in the case of tuberculosis by remaining sequestered within calcified nodules in the lung (Grange, 1992; Britton *et al.* 1994). These dormant stages can be modelled by including them as a separate population (Antia *et al.* 1996). These models suggest that the dormant stage can prevent the immune response

from clearing the bacteria, and allowing the persistence of both bacteria and immunity in a steady state. It appears that only much later in disease is this balance between bacteria and immune system lost (see below).

Polarization between Th1 and Th2 responses. Microparasites can directly interfere with the generation and balance between the various components of the immune response. While some of these have been described in terms of changes in the parameters h and k of the acute infection model, others are more complex. One potentially important mechanism parasites may employ is to alter the balance between Th1 and Th2 responses (Abbas *et al.* 1996; Forsthuber *et al.* 1996; Mosmann & Sad, 1996). The polarization between Th1 and Th2 responses has been suggested to play a role in the pathogenesis of many infectious diseases including HIV/AIDS (Clerici & Shearer, 1993), leishmaniasis (Kemp *et al.* 1996) and infections with helminths (Maizels *et al.* 1993). A number of models has been proposed to describe the dynamics of Th1 and Th2 responses (Morel *et al.* 1992; Fishman & Perelson, 1994), and further development of these models may prove useful to understanding diseases characterized by polarized Th1/Th2 responses. The paradigm of T cell subsets which produce mutually antagonistic cytokines has also been extended to include CD8[+] T cell populations, and type 2 cytokine responses by CD8[+] cells have been suggested to play an important role in leprosy (Salgame *et al.* 1991; Yamamura *et al.* 1991). Particularly intriguing are the oscillations between lepromatous and tuberculoid forms of leprosy (Bloom *et al.* 1992) which are associated with type 2 and type 1 cytokine secretion patterns. However, models suggest that the mutual antagonism between type 1 and type 2 response patterns makes either type of response stable and reduces the likelihood of a spontaneous shift to the other type of response. At present, it is unclear what causes a shift between these response patterns (Antia *et al.* 1996).

Parasites can use more than one of the above mechanisms to counter the effect of the immune system. In some cases a single feature of the parasite can provide protection in more than one way. The capsules of bacteria, for example, may reduce the immune response by masking immunogenic proteins and may also provide protection to immune attack by antibody and complement. Other parasites may have evolved several features in combination to foil immune attack. The repertoire of responses of the HIV virus, for example (Dalgleish *et al.* 1990; Rosenberg & Fauci, 1990; McLean, 1993) includes several mechanisms; (1) HIV produces antigenic variants, escape mutants that are not recognized by the immune response to earlier strains; (2) HIV kills CD4-bearing T lymphocytes which regulate immune responses; (3) the HIV surface glycoprotein gp120, which is responsible for binding to the CD4 receptor on T cells, appears to resemble the major histocompatibility proteins that play a key role in the discrimination of self and parasite antigens, and in doing so may interfere with the normal functioning of the immune system; (4) following invasion of CD4 cells the virus may remain latent within these infected T cells, emerging at a later time, when the cell is activated; and (5) the HIV virus causes destruction of the architecture in the germinal centres of lymph nodes. The contribution of these various mechanisms in causing the onset of AIDS is at present unclear, nor is it clear whether these responses are evolved features that enhance transmission of the virus. Several exploratory mathematical models have been developed to understand better the dynamics of the virus following infection up to the development of AIDS (Reibnegger *et al.* 1989; Nowak *et al.* 1990; De Boer & Boerlijst, 1993; McLean, 1993; Frost & Mclean 1994*a*; Nowak *et al.* 1995; Mittler *et al.* 1996). The antigenic diversity models (Nowak *et al.* 1990; De Boer & Boerlijst, 1993) have proved particularly useful in understanding how the ability of HIV to generate antigenic variants together with its ability to kill immune cells may result in the overwhelming of the immune response. Models have also proved useful in understanding both the rates of turnover of virus and CD4$^+$ cells during the course of infection (Frost & McLean 1994*b*; Ho *et al.* 1995; Wei *et al.* 1995; Perelson *et al.* 1996) as well as the ascent of resistance subsequent to treatment (McLean & Nowak 1992; Frost & McLean 1994*b*; Ho *et al.* 1995; Wei *et al.* 1995; De Jong *et al.* 1996; Perelson *et al.* 1996).

WITHIN- AND BETWEEN-HOST DYNAMICS OF PARASITES

Models like those considered so far make the implicit assumption that parasites evolve to maximize their transmission over the course of the infection. This assumption is valid when the parasite population within a host is genetically homogenous. Under these conditions competition between different parasites leads to 'competitive exclusion' between different parasite genotypes, with the genotype which has the highest transmission dominating and all others being driven to extinction (Bremermann & Thieme, 1989). This assumption of genetic homogeneity of the parasite within a host may fail in 2 ways: the parasite may evolve to produce novel genotypes during the course of an infection within the host, or the host may be infected by more than one parasite genotype. We call the former within-host evolution and the latter multiple infection. Genetic variation within a host most frequently occurs during infections by rapidly mutating viruses such as the HIV virus. Multiple infection of a host

with more than one parasite genotype may occur in the initial inoculum or by sequential infection. Multiple infection of a host with more than one genotype is thought to occur in infections such as malaria and HIV. When a single host is infected with more than one parasite genotype either by within-host evolution or superinfection, there will be another level at which selection operates: that of the parasite genotypes within a host. The importance of these multiple levels of selection in infectious disease has been pointed out several times before (Lewontin, 1970; Levin & Pimentel, 1981).

Within-host competition

We have seen that, in the case of acute infections, increases in growth rate or decreases in the tendency to elicit or be killed by an immune response will be selected on the epidemiological (between-host) level up to the point at which the parasite just avoids killing the host and thereby maximizes total transmission. We now consider how the generation of novel genotypes within a host (by mutation, recombination, or any other means) affects the outcome of selection on these characters. Competition between parasite genotypes which vary in growth rate will result in the parasite with the most rapid growth rate being selected for within a host; if 2 parasite genotypes are present, the one with the higher growth rate (and all other parameters equal) will always produce more transmissible stages than one with a lower growth rate in a mixed infection. This is true despite the fact that the total transmission from the mixed infection might be less than the total transmission from a 'pure' infection with only the slower-growing variant. Bonhoeffer & Nowak (1994*b*) have modelled the consequences of this within-host evolution and have shown that the balance between selection for maximum transmission on the epidemiological level and selection of the parasite population within a host for a faster growth can result in the maintenance of polymorphism in the parasite population and an increase in the level of virulence of the parasite.

The competition between parasite genotypes with different abilities to avoid generating or evading immune responses may be more complex, depending on whether the change is 'selfish' or 'unselfish' (Bonhoeffer & Nowak, 1994*a*). A 'selfish' mechanism of immune evasion confers an advantage only on those parasites which carry the gene coding for the mechanism, while 'unselfish' mechanisms confer a general advantage on the entire parasite population within a host. For example, viral gene products that block the antiviral action of interferons or other cytokines within an infected cell or down-regulate viral peptide presentation on the surface of a cell are selfish, because they specifically protect their carriers (Wold & Gooding, 1991). By contrast, secretion of soluble cytokine receptors that modulate the fever response (Alcami & Smith, 1996) would be unselfish, because the effect should be the same on viruses carrying the gene and on those for which it is deleted. Selfish strategies will be selected within a host, but unselfish ones will be selectively neutral within a host: that is, selection will not increase the frequency of viruses carrying the gene for unselfish characters within a single infected host, because such viruses will not have any advantage over their counterparts which do not have the character. Although they focus on viruses, similar distinctions could be made for bacterial parasites: possession of a capsule that inhibits phagocytosis is clearly a selfish strategy, while secretion of a protease that cleaves antibodies might be unselfish. While Bonhoeffer & Nowak (1994*a*) described a model for a persistent infection, similar conclusions would be reached by using an acute infection model like the one suggested here, in which the parasite is eventually cleared unless it kills the host first. Selfish strategies of immune regulation would always be advantageous within a host in an acute infection, for the same reason described above for increases in growth rate. Unselfish strategies would confer no advantage in within-host competition since they would affect the net growth of both parasite types equally, but would be advantageous in homogeneous infections (as described above) as long as the parasite was not already at its optimum. Care should be taken in using models to determine which evasion mechanisms are selfish and which are unselfish. Based on models which assume that the within-host environment is homogeneous, one may expect that the secretion of factors which impair the ability of the immune system to kill parasites (e.g. by the inhibition of the complement pathway) will be unselfish as they will confer equal benefit to all parasites within the host. However, if these secreted components have a relatively short half-life then they may act over relatively short distances, and confer a greater advantage on the parasites carrying them than to other parasites (Bonhoeffer & Nowak, 1994*a*). Competition experiments, in which experimental animals receive a mix of viruses carrying a gene and viruses deleted for the gene, could distinguish whether the mechanism encoded by the gene is selfish or unselfish; similar experiments have been tried in at least one system (Alcami & Smith, 1996).

Gooding (1992) has pointed out that the mechanisms of immune evasion may either increase or decrease the virulence of an infection. This is because for certain infections, the damage to the host is caused principally by an effective immune response (Buchmeier *et al.* 1980; Gooding, 1992), rather than directly by the replication of the parasite. In such cases, the predictions of our model would not be applicable, since we have defined virulence as a direct consequence of parasite numbers.

Multiple infections of hosts

Several theoretical studies have considered the effect of within-host competition resulting from multiple infections on the evolution of virulence. All these studies have shown that when hosts can be multiply infected, selection will favour strains of higher virulence (faster within-host growth rate) than that which would maximize total propagule production (Levin & Pimentel, 1981; Nowak & May, 1994; May & Nowak, 1995; Van Baalen & Sabelis, 1995).

Two separate selective processes are responsible for this effect (Van Baalen & Sabelis, 1995). First, as described above, within-host competition between strains of a parasite infecting a host seems likely to favour faster-growing strains at the expense of slower-growing ones; as a result, growth at a relatively slow rate, which may be optimal when a particular strain is alone in a host, can become disadvantageous when a second strain is present (Bremermann & Pickering, 1983). Second, at an epidemiological level, a host harbouring a slower-growing strain will not transmit a full complement of that strain if it becomes infected with a rapid-growing strain that kills it more quickly. This second effect will hold even if no within-host competition between strains occurs; as a result of the risk of multiple infection, the transmission value of persisting for a long time within a host diminishes (Van Baalen & Sabelis, 1995). As a result, the possibility of superinfection can select for increased virulence even if the increased virulence is not correlated with increased transmissibility to uninfected hosts (Nowak & May, 1994; May & Nowak, 1995).

The various models constructed to account for multiple infections contain different assumptions about the within-host interactions of different parasite strains. However, all the models make 2 predictions: first, that the level of virulence selected will be higher than that which would prevail if multiple infections are impossible; and second, that parasite populations will be polymorphic, in contrast to the simple competitive exclusion principle that prevails when only single infections are possible.

CONCLUSIONS

We have illustrated how a simple model of an acute infection in a vertebrate host can provide a useful tool to aid our understanding of the evolution of parasites causing acute infections. By gradually introducing additional factors (such as host polymorphism and superinfection) to the simple model we are able to understand better the consequences of these complexities for the evolution of the parasite. As the complexities of the interaction of the immune system with individual microparasites become elucidated in more detail, it will be a challenge to keep the theoretical models simple enough to understand, yet complex enough to describe the host–parasite interaction in a way useful for understanding and for guiding further research.

ACKNOWLEDGEMENTS

Both R. A. and M. L. thank the EcLF for discussions. This work was supported by NIH grants R29-GM-54268 to R. A. and grant F32-GM-19182 to M. L.

REFERENCES

ABBAS, A. K., MURPHY, K. M. & SHER, A. (1996). Functional diversity of helper T lymphocytes. *Nature* **383**, 787–793.

AGUR, Z., ABIRI, D. & VAN DER PLOEG, L. H. T. (1989). Ordered appearance of antigenic variants of African trypanosomes in a mathematical model based on a stochastic process and immune-selection against putative switch intermediates. *Proceedings of the National Academy of Sciences, USA* **86**, 9626–9630.

AGUR, Z., MAZOR, G. & MEILIJSON, I. (1991). Maturation of the humoral immune response as an optimization problem. *Proceedings of the Royal Society (London)* B **245**, 147–150.

ALCAMI, A. & SMITH, G. L. (1996). A mechanism for the inhibition of fever by a virus. *Proceedings of the National Academy of Sciences, USA* **93**, 11029–11034.

ANDERSON, R. M. (1995). Evolutionary pressures in the spread and persistence of infectious agents in vertebrate populations. *Parasitology* **111**, S15–S31.

ANDERSON, R. M. & MAY, R. M. (1982). Coevolution of hosts and parasites. *Parasitology* **85**, 411–426.

ANTIA, R. & KOELLA, J. C. (1994). A model of non-specific immunity. *Journal of Theoretical Biology* **168**, 141–150.

ANTIA, R., KOELLA, J. C. & PERROT, V. (1996). Models of the within-host dynamics of persistent mycobacterial infections. *Proceedings of the Royal Society (London)* B **263**, 257–263.

ANTIA, R., LEVIN, B. R. & MAY, R. M. (1994). Within-host population dynamics and the evolution and maintenance of microparasite virulence. *American Naturalist* **144**, 457–472.

ANTIA, R., NOWAK, M. A. & ANDERSON, R. M. (1996). Antigenic variation and the within-host dynamics of microparasites. *Proceedings of the National Academy of Sciences, USA* **93**, 985–989.

BLOOM, B., MODLIN, R. L. & SALGAME, P. (1992). Stigma variations: observations on suppressor T cell and leprosy. *Annual Reviews of Immunology* **10**, 453–488.

BONHOEFFER, S. & NOWAK, M. A. (1994a). Intra-host versus inter-host selection: viral strategies of immune function impairment. *Proceedings of the National Academy of Sciences, USA* **91**, 8062–8066.

BONHOEFFER, S. E. & NOWAK, M. A. (1994b). Mutation and the evolution of virulence. *Proceedings of the Royal Society (London)* B **258**, 133–140.

BREMERMANN, H. J. & PICKERING, J. (1983). A game theoretical model of parasite virulence. *Journal of Theoretical Biology* **100**, 411–426.

BREMERMANN, H. J. & THIEME, H. R. (1989). A competitive-exclusion principle for pathogen virulence. *Journal of Mathematical Biology* **27**, 179–190.

BRITTON, W. J., ROCHE, P. W. & WINTER, N. (1994). Mechanisms of persistence of mycobacteria. *Trends in Microbiology* **2**, 284–288.

BUCHMEIER, M. J., WELSH, R. M., DUTKO, F. J. & OLDSTONE, M. B. A. (1980). The virology and immunobiology of LCMV infection. *Advances in Immunology* **30**, 275–331.

BURNET, F. M. (1970). *Immunological Surveillance.* Pergamon, New York.

CLERICI, M. & SHEARER, G. (1993). A Th1→Th2 switch is a critical step in the etiology of HIV infection. *Immunology Today* **14**, 107–111.

DALGLEISH, A. G., MANCA, F. & HABESHAW, J. A. (1990). The pathogenesis of HIV-induced disease. In *AIDS and the New Viruses* (ed. A. G. Dalgleish & R. A. Weiss), pp. 111–124. Academic Press, New York.

DE BOER, R. J. & BOERLIJST, M. C. (1993). Diversity and virulence thresholds in AIDS. *Proceedings of the National Academy of Sciences, USA* **94**, 544–548.

DE BOER, R. J. & PERELSON, A. S. (1995). Towards a general function describing T cell proliferation. *Journal of Theoretical Biology* **175**, 567–576.

DE JONG, M. D., VEENSTRA, J., STILIANAKIS, N. I., SCHUURMAN, R., LANGE, J. M., DE BOER, R. J. & BOUCHER, C. A. (1996). Host–parasite dynamics and outgrowth of virus containing a single K70R amino acid change in reverse transcriptase are responsible for the loss of human immunodeficiency virus type 1 RNA load suppression by zidovudine. *Proceedings of the National Academy of Sciences, USA* **93**, 5501–5506.

EBERT, D. & HAMILTON, W. D. (1996). Sex against virulence: the coevolution of parasitic diseases. *Trends in Ecology and Evolution* **11**, 79–82.

EHRLICH, R. (1995). Selective mechanisms utilized by persistent and oncogenic viruses to interfere with antigen processing and presentation. *Immunologic Research* **14**, 77–97.

EWALD, P. W. (1987). Transmission modes and evolution of the parasitism–mutualism continuum. *Annals of the New York Academy of Sciences* **503**, 295–306.

FARRELL, H. E., VALLY, H., LYNCH, D. M., FLEMING, P., SHELLAM, G. R., SCALZO, A. A. & DAVIS-POYNTER, N. J. (1997). Inhibition of natural killer cells by a cytomegalovirus MHC class 1 homologue *in vivo*. *Nature* **386**, 510–514.

FINLAY, B. B. & FALKOW, S. (1989). Common themes in microbial pathogenicity. *Microbiological Reviews* **52**, 210–230.

FISHMAN, M. A. & PERELSON, A. S. (1993). Modeling T cell-antigen resenting cell interactions. *Journal of Theoretical Biology* **160**, 311–342.

FISHMAN, M. A. & PERELSON, A. S. (1994). Th1/Th2 cross regulations. *Journal of Theoretical Biology* **170**, 25–56.

FISHMAN, M. A. & PERELSON, A. S. (1995). Lymphocyte memory and affinity selection. *Journal of Theoretical Biology* **173**, 241–262.

FORSTHUBER, T., YIP, H. C. & LEHMANN, P. V. (1996). Induction of Th1 and Th2 immunity in neonatal mice. *Science* **271**, 1728–1730.

FRANK, S. A. (1993). Evolution of host-parasite diversity. *Evolution* **47**, 1721–1732.

FROST, S. D. & MCLEAN, A. R. (1994 *a*). Germinal center destruction as a major pathway of HIV pathogenesis. *Journal of AIDS* **7**, 236–244.

FROST, D. D. & MCLEAN, A. R. (1994 *b*). Quasispecies dynamics and the emergence of drug resistance during zidovudine therapy of HIV infection. *AIDS* **8**, 323–332.

GOODING, L. R. (1992). Virus proteins that counteract host immune defenses. *Cell* **71**, 5–7.

GRANGE, J. M. (1992). The mystery of the mycobacterial persistor. *Tubercle and Lung Disease* **73**, 249–251.

GROSSMAN, Z. & PAUL, W. E. (1992). Adaptive cellular interactions in the immune system: the tunable activation threshold and the significance of subthreshold responses. *Proceedings of the National Academy of Sciences, USA* **89**, 10365–10369.

GROSSMAN, Z. & SINGER, A. (1996). Tuning of activation thresholds explains flexibility in the selection and development of T cells in the thymus. *Proceedings of the National Academy of Sciences, USA* **93**, 14747–14752.

HALDANE, J. B. S. (1949). Disease and evolution. *La Ricerca Scientifica* **19**, 68–76.

HILL, A. V. (1996). Genetic susceptibility to malaria and other infectious diseases: from the MHC to the whole genome. *Parasitology* **112**, S75–S84.

HO, D. D., NEUMANN, A. U., PERELSON, A. S., CHEN, W., LEONARD, J. M. & MARKOWITZ, M. (1995). Rapid turnover of plasma virions and CD4 lymphocytes in HIV-1 infection. *Nature* **373**, 123–126.

HSU, D. H., DE WAAL MALEFYT, R., FIORENTINO, D. F., DANG, M. N., VIEIRA, P., DE VRIES, J., SPITS, H., MOSMANN, T. R. & MOORE, K. W. (1990). Expression of interleukin-10 activity by Epstein–Barr virus protein BCRF1. *Science* **250**, 830–832.

KEMP, M., THEANDER, T. G. & KHARAZMI, A. (1996). The contrasting roles of CD4+ T cells in intracellular infections in humans: leishmaniasis as an example. *Immunology Today* **17**, 13–16.

KEVEREKIDIS, I. G., ZECHA, A. D. & PERELSON, A. S. (1988). Modeling dynamical aspects of the immune response. 1. T cell proliferation and the effect of IL-2. In *Theoretical Immunology* (ed. A. S. Perelson), pp. 167–197. Addison-Wesley, Reading, MA, USA.

KOELLA, J. C. & ANTIA, R. N. (1995). Optimal pattern of replication and transmission for parasites with two stages in their life cycle. *Theoretical Population Biology* **47**, 277–291.

LENSKI, R. E. & MAY, R. M. (1994). The evolution of virulence in parasites and pathogens: reconciliation between two competing hypotheses. *Journal of Theoretical Biology* **169**, 253–265.

LEVIN, B. R. & BULL, J. J. (1994). Short-sighted evolution and the virulence of pathogenic microorganisms. *Trends in Microbiology* **2**, 76–81.

LEVIN, S. & PIMENTEL, D. (1981). Selection of intermediate rates of increase in parasite–host systems. *American Naturalist* **117**, 308–315.

LEWONTIN, R. C. (1970). The units of selection. *Annual Review of Ecology and Systematics* **1**, 1–18.

LIPSITCH, M. & MOXON, R. E. (1997). Virulence and transmissibility of pathogens: what is the relationship? *Trends in Microbiology* **5**, 31–37.

LIPSITCH, M., SILLER, S. & NOWAK, M. A. (1996). The evolution of virulence in pathogens with vertical & horizontal transmission. *Evolution* **50**, 1729–1741.

MAIZELS, R. M., BUNDY, D. A. P., SELKIRK, M. E., SMITH, D. F. & ANDERSON, R. M. (1993). Immunological modulation and evasion by helminth parasites in human populations. *Nature* **365**, 797–805.

MAY, R. M. & ANDERSON, R. M. (1983). Parasite–host coevolution. In *Coevolution* (ed. D. J. Futuyama & M. Slatkin), pp. 186–206. Sinauer, Sunderland, Massachusetts.

MAY, R. M. & NOWAK, M. A. (1995). Coinfection and the evolution of parasite virulence. *Proceedings of the Royal Society (London)* B **261**, 209–215.

MCLEAN, A. R. (1993). The balance of power between HIV and the immune system. *Trends in Microbiology* **1**, 9–13.

MCLEAN, A. R. & KIRKWOOD, T. L. B. (1990). A model of human immunodeficiency virus (HIV) infection in T helper cell clones. *Journal of Theoretical Biology* **147**, 177–203.

MCLEAN, A. R. & NOWAK, M. A. (1992). Competition between zidovudine-sensitive and zidovudine-resistant strains of HIV. *AIDS* **6**, 71–79.

MITTLER, J., LEVIN, B. & ANTIA, R. (1996). T-cell homeostasis, competition and drift: AIDS, HIV-accelerated senescence of the immune repertoire. *Journal of AIDS* **12**, 233–248.

MOREL, B. F., KALAGNANAM, J. & MOREL, P. A. (1992). Mathematical modeling of Th1–Th2 dynamics. In *Theoretical and Experimental Insights into Immunology* (ed. A. S. Perelson & G. Weisbuch), pp. 171–190. Springer-Verlag, Berlin, Heidelberg.

MOSMANN, T. R. & SAD, S. (1996). The expanding universe of T-cell subsets: Th1, Th2 and more. *Immunology Today* **17**, 138–146.

NOWAK, M. A. & MAY, R. M. (1994). Superinfection and the evolution of parasite virulence. *Proceedings of the Royal Society (London)* B **255**, 81–89.

NOWAK, M. A., MAY, R. M. & ANDERSON, R. M. (1990). The evolutionary dynamics of HIV-1 quasispecies and the development of immunodeficiency disease. *AIDS* **4**, 1095–1103.

NOWAK, M. A., MAY, R. M., PHILLIPS, R. E., ROWLAND-JONES, S., LALLOO, D. G., MCADAM, S., KLENERMAN, P., KOPPE, B., SIGMUND, K., BANGHAM, C. R. M. & MCMICHAEL, A. J. (1995). Antigenic oscillations and shifting immunodominance in HIV-1 infections. *Nature* **375**, 606–611.

OPREA, M. & PERELSON, A. S. (1996). Exploring the mechanisms of primary antibody responses to T cell-dependent antigens. *Journal of Theoretical Biology* **181**, 215–236.

PARHAM, P. & OHTA, T. (1996). Population biology of antigen presentation by MHC class 1 molecules. *Science* **272**, 67–74.

PERELSON, A. S., NEUMANN, A. U., MARKOWITZ, M., LEONARD, J. M. & HO, D. D. (1996). HIV-1 dynamics *in vivo*: virion clearance rate, infected cell lifespan, and viral generation trends in microbiology. *Science* **271**, 1582–1586.

PILYUGIN, S., MITTLER, J. & ANTIA, R. (1997). Modelling T-cell proliferation: an investigation of the consequences of the Hayflick limit. *Journal of Theoretical Biology* **186**, 117–129.

PLAUT, A. G. (1983). The IgA proteases of pathogenic bacteria. *Annual Reviews of Microbiology* **37**, 603–622.

REIBNEGGER, G., FUCHS, D., HAUSEN, A., WERNER, E. R., WERNER-FELMAYER, G., DIERICH, M. P. & WACHTER, H. (1989). Stability analysis of simple models for immune cells interacting with normal pathogens and immune system retroviruses. *Proceedings of the National Academy of Sciences, USA* **86**, 2026–2030.

REYBURN, H. T., MANDELBOIM, O., VALES-GOMEZ, M., DAVIS, D. M., PAZMANY, L. & STROMINGER, J. L. (1997). The class I MHC homologue of human cytomegalovirus inhibits attack by natural killer cells. *Nature* **386**, 514–517.

ROSENBERG, Z. F. & FAUCI, A. S. (1990). Immunopathogenic mechanisms of HIV infection: cytokine induction of HIV expression. *Immunology Today* **11**, 176–180.

SALGAME, P., ABRAMS, J. S., CLAYBERGER, C., GOLDSTEIN, H., CONVIT, J., MODLIN, R. L. & BLOOM, B. R. (1991). Differing lymphokine profiles of functional subsets of human CD4 and CD8 cell clones. *Science* **254**, 279–282.

SASAKI, A. & IWASA, Y. (1991). Optimal growth schedule of pathogens within a host: switching between lytic and latent cycles. *Theoretical Population Biology* **39**, 201–239.

SCHWEITZER, A. N. & ANDERSON, R. M. (1992). The regulation of immunological responses to parasitic infections and the development of tolerance. *Proceedings of the Royal Society (London)* B **247**, 107–112.

SEGER, J. (1992). Evolution of exploiter–victim relationships. In *Natural Enemies: The Population Biology of Predators, Parasites and Diseases* (ed. M. J. Crawley), pp. 3–25. Blackwell, Oxford.

SZTEIN, M. B. & KIERSZENBAUM, F. (1993). Mechanisms of development of immunosuppression during trypanosome infections. *Parasitology Today* **9**, 425–428.

TOPLEY, W. W. C. (1919). The spread of bacterial infection. *Lancet* **ii**, 1–5.

VAN BAALEN, M. & SABELIS, M. W. (1995). The dynamics of multiple infection and the evolution of virulence. *American Naturalist* **146**, 881–910.

VICKERMAN, K. & BARRY, J. D. (1982). African trypanosomiasis. In *Immunology of Parasitic Infections* (ed. S. Kohen & K. S. Warren), pp. 204–260. Blackwell Scientific Publications, Oxford.

WAKELIN, D. (1997). *Immunity to Parasites*. Cambridge University Press, Cambridge.

WEI, X., GHOSH, S. K., TAYLOR, M. E., JOHNSON, V. A., EMINI, E. A., DEUTSCH, P., LIFSON, J. D., BONHOEFFER, S., NOWAK, M. A., HAHN, B. H., SAAG, M. S. & SHAW, G. M. (1995). Virus dynamics in human immunodeficiency virus type 1 infection. *Nature* **373**, 117–122.

WHITNACK, E., DALE, J. B. & BEACHEY, E. H. (1984). Common protective antigens of group A streptococcal M proteins masked by fibrinogen. *Journal of Experimental Medicine* **159**, 1201–1212.

WOLD, W. S. & GOODING, L. R. (1991). Region E3 of adenovirus: a cassette of genes involved in host immunosurveillance and virus–cell interactions. *Virology* **184**, 1–8.

YAMAMURA, M., UYEMURA, K., DEANS, R. J., WEINBERG, K., REA, T. H., BLOOM, B. R. & MODLIN, R. L. (1991). Defining protective responses to pathogens: cytokine profiles in leprosy lesions. *Science* **254**, 277–279.

Parasite immune evasion and exploitation: reflections and projections

R. T. DAMIAN

Department of Cellular Biology, University of Georgia, Athens, GA 30602, USA

SUMMARY

Recent developments in parasite immune evasion and exploitation are reviewed with special reference to the papers presented in this volume. Parasites, broadly defined, of animals with good immune responses have evolved many strategies that adapt them to survive and reproduce. These strategies may be passive, or may involve active intervention with host immune regulation, and can be categorized as immune evasion, immune exploitation and molecular piracy. The concept of immune evasion began with Paul Ehrlich's demonstration of antigenic variation in African trypanosomes and was reinforced by later ideas on molecular mimicry. Molecular mimicry is updated in the light of recent discoveries about degeneracy and plasticity of TCR/MHC-peptide recognition. Possible connections between two of its postulated consequences, evasion and autoimmunity, are discussed. Another putative consequence of molecular mimicry, host antigenic polymorphism, is also updated. The concept of exploitation of host immune responses by parasites has been reinforced by new data on its first known examples, especially the immune dependence of schistosome egg excretion. Newer examples include use of host cytokines as parasite growth factors, virokines, viroreceptors and helminth pseudocytokines. Finally, questions of host gene capture by viruses and possible horizontal gene transfer between host and parasite mediated by retroviruses are examined. The latter is compared with molecular conservation as a source of molecular mimicry and other aspects of host–parasite coevolution.

Key words: Parasite–host coevolution, immune evasion, immune exploitation, molecular piracy, molecular mimicry, autoimmunity.

INTRODUCTION

I was both pleased and honoured to have been asked by the editors of this volume to contribute an end-piece to this excellent and stimulating series of papers on the exploitation of immune and other defence mechanisms by parasites. The 'reflections' part of their charge was relatively benign and easy to do. 'Projections' always carry a certain risk, although the risk is minimal for one nearing the end of one's career!

The papers in the present volume cover nearly the gamut of parasites, as broadly defined, from relatively simple viruses to complex helminths. They also cover a wide range of possible mechanisms of host–parasite adaptation, with the adaptation to the host immune response the focus. As these articles abundantly demonstrate, parasites have a variety of strategies that may adapt them to evade and even exploit the immune response of their hosts. Some of these strategies are rather passive, while others may involve active manipulation, as it were, of the host's defensive responses. These strategies include immune evasion (Ogilvie & Wilson, 1976; Bloom, 1979; Dessaint & Capron, 1993), immune exploitation (Hayunga, 1979; Damian, 1987 a), molecular piracy (Ahuja & Murphy, 1993), and strategies that defy categorization. We employ pigeonholes, to be sure, but only for convenience. As many of the papers herein presented show, biological realities often resist our pigeonholing attempts. They all have in common, however, lessons to teach us about the

abilities of parasites to manipulate the immune response, and as one paper points out, they could well teach us lessons about the immune response itself (Barry and McFadden, this volume).

The concept of immune evasion by parasites has been around for a very long time. In fact, although it seems generally to have been forgotten, it was Paul Ehrlich (honoured as the 'Father of Immunology' at the First International Congress of Immunology in 1971), who discovered the first and still one of the best examples of immune evasion by a parasite. This is of course the phenomenon of antigenic variation in the African trypanosomes. In his Nobel Lecture, delivered in Stockholm in December of 1908 (subsequently published in 1909 and 1910), Ehrlich summarized experiments done by himself and his colleagues, Franke, Röhl and Gulbransen on the 'disappearance of receptors' clearly demonstrating the ability of these organisms to shift their antigenicity *in vivo*. It is perhaps unfortunate that Ross & Thomson (1910) are so often given the credit for the discovery of antigenic variation in trypanosomes, even though the term 'antigen' was never mentioned in their paper. Their contribution, however, was important enough: their now-classic oscillating parasitaemia curve in a case of sleeping sickness, which of course is best explained by variant antigen-driven immune responses.

Hommel (this volume) re-examines the question of antigenic variation in malaria parasites, and concludes that modulation of ligand expression on *Plasmodium* goes beyond simple immune escape to become rather 'an elegant process of host–parasite adaptation', or immune exploitation if you will. This process is co-dependent as well upon parasite-driven modulation of host cellular receptors for the parasite's changing ligands.

The idea of molecular mimicry (Damian, 1964), also known as adaptation tolerance (Sprent, 1962), immunoselection for eclipsed antigens (Damian, 1962), or selection for fitness antigens (Dineen, 1963) has historical significance in that it added to the then special case of antigenic variation in African trypanosomes 'to create a new conceptual environment... [that] encouraged... new questions... [and] approaches to uncover other candidate evasive mechanisms' (Damian, 1987b).

It may be that molecular mimicry has enjoyed somewhat greater success as a paradigm for autoimmunity than for parasite immune evasion. One of the clearest early enunciations of this aspect of molecular mimicry was that of Rowley & Jenkin (1962), with a more modern restatement by Oldstone (1987). Some of the more classic examples (Oldstone, 1989) of potential inciting agents for autoimmune diseases through a molecular mimicry mechanism include group A streptococci (rheumatic fever, glomerulonephritis, etc.), mycobacteria (rheumatoid arthritis), and *Klebsiella pneumoniae* (Reiter's Syndrome, ankylosing spondylitis). In multiple sclerosis (MS), T cell clones specific for an encephalitogenic epitope of myelin basic protein were found to strongly recognize certain viral peptides (Fujinami & Oldstone, 1985; Wucherpfennig & Strominger, 1995), and molecular mimicry by *Campylobacter jejuni* has been suggested to be a factor in the development of Guillain–Barré Syndrome (Yuki et al. 1993). Among traditional parasites, a recent report demonstrates that the presence of *Trypanosoma cruzi* amastigotes in the heart is both necessary and sufficient to induce cardiomyopathy in a murine model for human Chagas' disease (Tarleton, Zhang & Downs, 1997). The authors thus argue against molecular mimicry-based autoimmunity as playing a significant role in this disease model. On the other hand, the characteristic occurrence of neuropathies in Chagas' disease (Köberle, 1968), along with the existence of several reports that not only show molecular mimicry between *T. cruzi* and components of the nervous system, but also show some functional relevance of autoreactive antibodies or T cell lines (Hontebeyrie-Joskowicz et al. 1987; Ferrari et al. 1995; Elies et al. 1996) are sure to keep the controversy surrounding the significance of autoimmunity in Chagas' disease alive for some time to come.

Basic to the autoimmune concept of molecular mimicry is the premise that mimicked parasite epitopes would be sufficiently different from host epitopes to result in immune stimulation, which the host's self or 'model' epitopes would normally fail to do. It is now recognized that the essential difference could reside at the level of antigen processing and presentation route in the case of extracellular parasites (Davies, 1997). But the idea of subtler differences between model and mimic at the peptide level has recently been strengthened by the discovery that T cell receptor (TCR) recognition of MHC-bound peptides is surprisingly quite degenerate. Cogent examples for autoimmunity include the study of Wucherpfennig & Strominger (1995), who analysed myelin basic protein (MBP) peptide structural motifs for both major histocompatibility complex (MHC) class II binding and TCR recognition. They found that a T cell clone (TCC) from an MS patient that was specific for MBP recognized eight distinct but structurally related viral and bacterial peptides. Later, using the powerful combinatorial peptide library approach, Hemmer et al. (1997) showed that an even higher level of degeneracy was allowed in epitope recognition by autoreactive T cells. Some of their artificial ligands were even more potent agonists for the human TCC they used than its nominal self epitope, MBP(86–96). In another report (Ausubel et al. 1996), it was shown that the TCR–MHC complex was highly plastic due to the existence of two functional pockets in the TCR. The knowledge that peripheral autoreactive T cells exist in health (Fredrikson et al. 1994; Wekerle et al. 1996), and the degeneracy and plasticity of TCR recognition disclosed by the elegant studies mentioned above, raise important new questions about how autoimmunity is normally avoided. They also suggest new ways in which immune evasion through molecular mimicry could take place. Such structural approaches to molecular mimicry could lead to new insights into the relationship between two of its postulated consequences, autoimmunity and immune evasion (Damian, 1964). For example, Barnaba & Sinigaglia (1997) have suggested, based upon the findings and quantitative arguments of Hemmer et al. (1997), that microbially cross-activated peripheral T cells could be anergized by a pool of cross-reactive self peptides, acting as altered peptide ligands (ALPs). In this case, autoimmunity would be switched off while at the same time immune clearance of pathogens would be inhibited. Thus two formerly disparate aspects of molecular mimicry are united in this concept, which suggests that subtle control mechanisms must exist to allow a compromise between these two consequences of immune recognition. How some of these ALP-based control mechanisms, ALPs in this case of the circumsporozoite protein of *Plasmodium*, could lead to immune evasion is the subject of the report by Plebanski, Lee & Hill in this volume.

Baum, Davies & Peakman (1996) have recently presented a provocative proposition that molecular mimicry may actually involve 'three-way mimicry' between autoantigen, foreign infectious peptide, and self MHC peptide. This idea has important implications for the question of the role of infectious diseases in maintaining host antigenic, including MHC, polymorphisms. The paper by Plebanski *et al.* (this volume) is important because it stresses the possible significance of parasite and host polymorphisms in host–parasite adaptation. Long ago it was suggested that biochemical (Haldane, 1949) and antigenic (Damian, 1962, 1964) polymorphisms of pathogens could favour the development of host biochemical and antigenic polymorphisms. These ideas have stimulated much discussion (Snell, 1968; Wills, 1991; Apanius *et al.* 1997). They remain controversial and occasionally resurface in more modern formulations (Murphy, 1993, 1994). The report of Hill *et al.* (1991) on the association of certain HLA antigens with protection against malaria in west Africa still provides some of the best evidence for the role of infectious diseases driving human MHC polymorphisms. Antia and Lipsitch (this volume) present a mathematical model for an acute infection in a host with a good immune system, and expand their model to include complicating considerations such as host polymorphisms.

The original conceptions of molecular mimicry were of a passive evasive tactic to escape from, or reduce, immune detection of parasites through the possession of shared host–parasite epitopes ('eclipsed antigens' [Damian, 1964] or 'eclipsed epitopes' [Damian, 1989]). This original concept was subsequently expanded to include escape through a more aggressive form of mimicry (Damian, 1979, 1984). This was necessitated by the finding of parasite-mimicked epitopes of host immuno-regulatory molecules or their receptors; in the early days, molecules such as α_2-macroglobulin (Damian, Greene & Hubbard, 1973), murine MHC (Sher, Hall & Vadas, 1978; Gitter & Damian, 1982; Simpson *et al.* 1983), and immunoglobulin and complement receptors (Kemp *et al.* 1977; Torpier, Capron & Ouaissi, 1979; Tarleton & Kemp, 1981). Grencis & Entwistle (this volume) present the interesting finding of an IFN-γ-like molecule being produced by the intestinal nematode, *Trichuris muris*. Whether this is an example of homology (conservation) or convergence (mimicry), as discussed below, remains to be determined, as does the molecule's functional relevance. These authors speculate that an inappropriate Th1 immune response may be stimulated if the *T. muris* molecule proves to be a functional mimic. Pritchard, Hewitt & Moqbel (this volume) refer to such parasite products as 'pseudocytokines', and they discuss other pseudocytokines that could influence the balance between Th1 and Th2 T cell subsets, particularly within the framework of helminth-induced allergy.

Still later, the concept of molecular mimicry was further broadened to include receptor mimicry to gain immunologically privileged intracellular sites

(Damian, 1988, 1989). Numerous examples of receptor mimicry were reviewed by Hall (1994).

As was pointed out a decade ago (Damian, 1987*a*), the concept of the exploitation of host immune responses by parasites had been 'in the air' for a while. This concept may have gotten a fillip by the evidence gathered together for it, by its naming, and a formal definition given it in that paper. As was stated, 'to qualify as an example of immune exploitation, the parasite's adaptation must be specifically directed to take over some facet of host immune responsiveness in such a way that the parasite's life needs are met, its reproduction is facilitated, or its propagation to new hosts is made possible. In other words, it must be an obligatory interaction between parasite and host, mediated through both the parasite's antigenicity and the host's immune recognition system.' Discoveries made since 1987 have made it clear that the above definition should be broadened to include non-antigen-specific mechanisms of exploitation that involve important players in the host immune system. In particular, examples of cytokines serving as parasite growth factors (Mazingue *et al.* 1989; Amiri *et al.* 1992; McKerrow, this volume) should be included.

The exploitation of the host immune response as a means for the propagation of parasites to new hosts is the subject of the article by Doenhoff (this volume) in which he examines relationships between schistosome egg-induced granulomas and the rate of egg excretion in mice. This expands upon his earlier elegant demonstrations of the immune dependence of egg excretion in experimental schistosomiasis (Doenhoff *et al.* 1978, 1986; Doenhoff, Hassounah & Lucas, 1985). As mentioned by Doenhoff (this volume), a recent study done in Kenya in which concurrent HIV and schistosome infections were studied (Karanja *et al.* 1997), strongly suggests that the excretion of *Schistosoma mansoni* eggs in humans is also dependent upon an intact immune response. Doenhoff (this volume) also reviews evidence that coinfection with HIV and *Mycobacterium tuberculosis* compromises transmission of tubercle bacilli, presumably through a similar mechanism of granuloma inhibition in immunocompromised patients.

Two of the most suggestive, possible earlier examples of immune exploitation were immune granuloma-mediated egg excretion in schistosomiasis and nutritional use of specific antibodies by taeniid tapeworm cystercerci (Damian, 1987*a*). The former is brought up to date by Doenhoff (this volume). The evidence is becoming compelling that the immune schistosome egg granuloma does subserve this function. The latter possibility has received some subsequent experimental support (Hayunga, Sumner & Letonja, 1989) although specificity still remains an issue. It seems a problem deserving of further investigation.

Possible cases of immune exploitation in other

host–parasite systems are revealed in the articles by
Launois, Louis & Milon, by Mosser & Brittingham
and by Maillard, Luthi, Acha-Orbea & Diggelmann
(this volume). Maillard *et al.* review microbial
superantigens (SAg) and discuss their possible roles
in transmission. The clearest example of immune-
dependent viral transmission is provided by the
mouse mammary tumour virus, whereas bacteria
may be favoured by superantigen production
through immune deviation, although this is con-
jectural at this time. These authors also refer to the
work of Lafon *et al.* (1994) who showed that in the
murine rabies model, SAg-reactive cells are responsible for limb paralysis. One may wonder if this
phenomenon could play a role in rabies transmission
from mice to carnivores in nature. Wild rodents are
found infected with the rabies virus (Winkler, 1991)
and could possibly infect foxes through the oral
route, as well as by biting (Blancou, Aubert & Artois,
1991).

The papers by Launois *et al.* and by Mosser &
Brittingham present two very different strategies
used by *Leishmania* parasites to exploit and evade the
immune response to their advantage. Launois *et al.*
discuss how CD4 T cell subset polarization through
early parasite manipulation of host cytokines leads to
dermal parasite expansion in a Th2 environment,
which should favour infection of vector sandflies. In
a Th1 environment, the parasites are driven to
deeper tissues, where they can quietly persist until
possible reactivation by incompletely known mechanisms. The relationship of these options to the
epidemiology of leishmaniasis should be a fascinating
subject for future investigation. In relation to the
persistence of pathogens in macrophages, the article
by Hilbi, Zychlinsky & Sansonetti (this volume)
argues that macrophage apoptosis in shigellosis is
proinflammatory and that it ultimately leads to
clearance of the infection. They review macrophage
apoptosis in other infections including those by
protozoan parasites: the evidence suggests that the
phenomenon could play an important role in immune evasion and parasite persistence.

The contribution by Mosser & Brittingham (this
volume) focuses on how *Leishmania* promastigotes
gain entry into mononuclear phagocytes by sub-
verting and exploiting the hosts' complement opso-
nizing system. Related to this is the article by Füst
(this volume) which discusses the mechanisms and
implications of the enhancement of HIV infections
by antibodies or antibody/complement mediated
infection of target cells. One of the values of a
collection such as is found in the present volume is
the juxtaposition of topics usually separated by
artificial boundaries. Surely the time has come to
study more thoroughly concurrent infections in both
human populations and experimental animals.

The paper by Stewart & Heppner (this volume)
stands somewhat apart from the others in that it

examines the role of the immune response in the
progression of a cancer so far unlinked to a known
infectious agent, namely, cancer of the breast. An
impressive amount of evidence is brought together
to support the hypothesis that the progression of this
type of cancer is actually facilitated by the immune
response. Interestingly, the breast itself is identified
as an immunological organ, and this emphasis should
benefit infectious disease specialists and parasitologists, who often must consider the trans-
mammary passage of pathogens.

In his opening chapter in this volume, Locksley
refers to the possibility of convergent evolution of
host and parasite genomes through long association
as a factor in the development of parasite evasive
mechanisms and chronicity, a concept discussed at
length by Klein & O'Huigin (1994). An alternative
explanation is provided by molecular conservation,
which has been discussed within the context of
molecular mimicry on several occasions (Damian,
1988, 1989, 1991). The problem, like many other
problems in evolutionary biology, including molecular evolution (Doolittle, 1994), is to distinguish
between convergence (mimicry) and conservation
(homology) as responsible for shared host–parasite
epitopes, or other molecules. Recent advances in
evolutionary developmental biology have emphasized and affirmed the stem position of the Platy-
helminthes in the evolution of other metazoan
animals, including both other parasitic helminths
and the vertebrate hosts that include ourselves (Field
et al. 1988; Raff, 1996). Most of the basic biochemistry and organization of metazoan life was
already well-established in the primitive flatworms
and passed on to their wormy, parasitic descendants.
In a sense, we are parasitized by our 'ancestors'! A
case in point is the mediation of apoptosis by
conserved cysteine protease genes including that for
IL-1β-converting enzyme and the *Caenorhabditis
elegans* cell death gene *ced-3* (mentioned by Hilbi *et
al.* this volume). Therefore, like Locksley (this
volume), we should be less than surprised about
reports of cytokine-like or cytokine-receptor-like
molecules in helminths.

However, I am more a follower of Howell (1985)
than of Locksley (this volume) on the possible
importance of gene capture in the development of
evasive or exploitative mechanisms by parasites.
Howell suggested that horizontal transfer of DNA
between hosts and parasites could result in immune
evasion and other adaptations for successful parasitism, but also in certain cases giving advantages to
hosts. He implicated retroviruses as likely agents of
this genetic exchange. The recent discovery of the
first invertebrate retrovirus, the gypsy element of
Drosophila (Kim *et al.* 1994; Song *et al.* 1994, 1997),
should make us slow to dismiss horizontal transfer of
genes between hosts and parasites as being of
potential importance in their mutual adaptation.

When considering viruses, the evidence for gene capture or 'molecular piracy' is overwhelming (Barry & McFadden, this volume). Captured molecules include both 'virokines' (Kotwal & Moss, 1988) and 'viroreceptors' (Upton *et al.* 1991), and many strategies for immune evasion and exploitation are possible, as discussed by Barry & McFadden.

In considering horizontal gene transfer, the results of a group of Japanese workers may be very significant (Tanaka *et al.* 1989; Nara *et al.* 1990; Iwamura *et al.* 1991, 1995; Irie & Iwamura, 1993). They have reported on the occurrence of host DNA sequences, including sequences shared with murine type A and C retrovirus, in schistosome genomes. These reports have engendered considerable debate (Simpson & Pena, 1991; Iwamura & Irie, 1992; Clough, Drew & Brindley, 1996), centering around whether or not the host-type sequences are actually host-derived contaminants. Nevertheless, these reports have been valuable in re-focusing attention on Howell's proposition. I can only agree with Iwamura & Irie (1992) that further 'progress in this field must be expected'.

Controversies like the one mentioned above, the question of autoimmunity in Chagas' disease, and many others spice up parasitology and infectious disease research, and make the study of parasite–host interactions ever more exciting. This spice further enhances the olio that contains many challenging scientific problems for the intensely curious and the intensely humane: problems of intrinsic and emerging importance for global public health in an uncertain, exponentially changing world (Garrett, 1994). The integration of diverse areas of research specialization and approaches as evidenced in the present volume will play a huge part in taking parasitological research to the next level, and one can only wonder at what wonders the next decadal collection focusing on parasite immune evasion and exploitation will contain.

ACKNOWLEDGEMENTS

I thank Michael J. Doenhoff and Leslie H. Chappell for their kind invitation to contribute to this volume. I am indebted to Haini Cai, Jorge Morales-Montor and Steve Toenjes for bringing certain references to my attention.

REFERENCES

AHUJA, S. K. & MURPHY, P. M. (1993). Molecular piracy of mammalian interleukin-8 receptor type B by herpesvirus saimiri. *Journal of Biological Chemistry* **268**, 20691–20694.

AMIRI, P., LOCKSLEY, R. M., PARSLOW, T. G., SADICK, M., RECTOR, E., RITTER, D. & McKERROW, J. H. (1992). Tumour necrosis factor α restores granulomas and induces parasite egg-laying in schistosome-infected SCID mice. *Nature* **356**, 604–607.

APANIUS, V., PENN, D., SLEV, P. R., RUFF, L. R. & POTTS, W. K. (1997). The nature of selection on the major histocompatibility complex. *Critical Reviews in Immunology* **17**, 179–224.

AUSUBEL, L. J., KWAN, C. K., SETTE, A., KUCHROO, V. & HAFLER, D. A. (1996). Complementary mutations in an antigenic peptide allow for crossreactivity of autoreactive T-cell clones. *Proceedings of the National Academy of Sciences, USA* **93**, 15317–15322.

BARNABA, V. & SINIGAGLIA, F. (1997). Molecular mimicry and T cell-mediated autoimmune disease. *Journal of Experimental Medicine* **185**, 1529–1531.

BAUM, H., DAVIES, H. & PEAKMAN, M. (1996). Molecular mimicry in the MHC: hidden clues to autoimmunity? *Immunology Today* **17**, 64–70.

BLANCOU, J., AUBERT, M. F. A. & ARTOIS, M. (1991). Fox rabies. In *The Natural History of Rabies*, 2nd Edition (ed. Baer, G. M.), pp. 257–290. Boca Raton, Florida: CRC Press.

BLOOM, B. R. (1979). Games parasites play: how parasites evade immune surveillance. *Nature* **279**, 21–26.

CLOUGH, K. A., DREW, A. C. & BRINDLEY, P. J. (1996). Host-like sequences in the schistosome genome. *Parasitology Today* **12**, 283–286.

DAMIAN, R. T. (1962). A theory of immunoselection for eclipsed antigens of parasites and its implications for the problem of antigenic polymorphism in man. *The Journal of Parasitology* **48** (2, section 2), 16 (Abstract).

DAMIAN, R. T. (1964). Molecular mimicry: antigen sharing by parasite and host and its consequences. *The American Naturalist* **98**, 129–149.

DAMIAN, R. T. (1979). Molecular mimicry in biological adaptation. In *Host–Parasite Interfaces: At Population, Individual and Molecular Levels* (ed. Nickol, B. B.), pp. 103–126. New York: Academic Press.

DAMIAN, R. T. (1984). Immunity in schistosomiasis: a holistic view. *Contemporary Topics in Immunobiology* **10**, 359–420.

DAMIAN, R. T., (1987a). The exploitation of host immune responses by parasites. *The Journal of Parasitology* **73**, 3–13.

DAMIAN, R. T. (1987b). Molecular mimicry revisited. *Parasitology Today* **3**, 263–266.

DAMIAN, R. T. (1988). Parasites and molecular mimicry. In *Molecular Mimicry in Health and Disease* (ed. Lernmark, Å., Dyrberg, T., Terenius, L. & Hokfelt, B.), pp. 211–218. Amsterdam: Excerpta Medica.

DAMIAN, R. T. (1989). Molecular mimicry: parasite evasion and host defense. In *Molecular Mimicry: Cross-Reactivity between Microbes and Host Proteins as a Cause of Autoimmunity* (ed. Oldstone, M. B. A.), pp. 101–115. Berlin: Springer-Verlag.

DAMIAN, R. T. (1991). Tropomyosin and molecular mimicry. *Parasitology Today* **7**, 96.

DAMIAN, R. T., GREENE, N. D. & HUBBARD, W. J. (1973). Occurrence of mouse α₂-macroglobulin antigenic determinants of *Schistosoma mansoni* adults, with evidence on their nature. *The Journal of Parasitology* **59**, 64–73.

DAVIES, J. M. (1997). Molecular mimicry: Can epitope mimicry induce autoimmune disease? *Immunology and Cell Biology* **75**, 113–126.

DESSAINT, J.-P. & CAPRON, A. (1993). Survival strategies of parasites in their immunocompetent hosts. In *Immunology and Molecular Biology of Parasitic Infections* (ed. Warren, K. S.), pp. 87–99. Oxford: Blackwell Scientific Publications.

DINEEN, J. K. (1963). Antigenic relationship between host and parasite. *Nature* **197**, 471–472.

DOENHOFF, M. J., HASSOUNAH, O. A. & LUCAS, S. B. (1985). Does the immunopathology induced by schistosome eggs potentiate parasite survival? *Immunology Today* **6**, 203–206.

DOENHOFF, M. J., HASSOUNAH, O., MURARE, H., BAIN, J. & LUCAS, S. (1986). The schistosome egg granuloma: immunopathology in the case of host protection or parasite survival? *Transactions of the Royal Society of Tropical Medicine and Hygiene* **80**, 503–514.

DOENHOFF, M., MUSALLAM, R., BAIN, J. & McGREGOR, A. (1978). Studies on the host–parasite relationship in *Schistosoma mansoni*-infected mice: the immunological dependence of parasite egg excretion. *Immunology* **35**, 771–778.

DOOLITTLE, R. F. (1994). Convergent evolution: the need to be explicit. *Trends in Biochemical Sciences* **19**, 15–18.

EHRLICH, P. (1909). Ueber Partial-funktionen der Zelle. *Münchener medizinische Wochenschrift* **5**, 217–222.

EHRLICH, P. (1910). The partial-function of cells. In *Studies in Immunity*, 2nd Edition (ed. Ehrlich, P.), pp. 676–693. New York: Wiley.

ELIES, R., FERRARI, I., WALLAKUT, G., LEBESGUE, D., CHIALE, P., ELIZARI, M., ROSENBAUM, M., HOEBEKE, J. & LEVIN, M. J. (1996). Structural and functional analysis of the B cell epitopes recognized by anti-receptor autoantibodies in patients with Chagas' disease. *The Journal of Immunology* **157**, 4203–4211.

FERRARI, I., LEVIN, M., WALLAKUT, G., ELIES, R., LEBESGUE, D., CHIALE, P., ELIZARI, M., ROSENBAUM, M. & HOEBEKE, J. (1995). Molecular mimicry between the immunodominant ribosomal protein P0 of *Trypanosoma cruzi* and a functional epitope of the human β_1-adrenergic receptor. *Journal of Experimental Medicine* **182**, 59–65.

FIELD, K. G., OLSEN, G. J., LANE, D. J., GIOVANNONI, S. J., GHISELIN, M. T., RAFF, E. C., PACE, N. R. & RAFF, R. A. (1988). Molecular phylogeny of the animal kingdom. *Science* **239**, 748–753.

FREDRICKSON, S., SODERSTROM, M., HILLERT, J., SUN, J.-B., KALL, T.-B. & LINK, H. (1994). Multiple sclerosis: occurrence of myelin basic protein peptide-reactive T cells in healthy family members. *Acta Neurologica Scandinavica* **89**, 184–189.

FUJINAMI, R. S. & OLDSTONE, M. B. A. (1985). Amino acid homology between the encephalitogenic site of myelin basic protein and virus: Mechanism for autoimmunity. *Science* **230**, 1043–1045.

GARRETT, L. (1994). *The Coming Plague: Newly Emerging Diseases in a World Out of Balance*. New York: Farrar, Straus and Giroux.

GITTER, B. D. & DAMIAN, R. T. (1982). Murine alloantigen acquisition by schistosomula of *Schistosoma mansoni*: Further evidence for the presence of K, D and I region gene products on the tegumental surface. *Parasite Immunology* **4**, 383–393.

HALDANE, J. B. S. (1949). Disease and evolution. In *Symposium sui Fattori Ecologici e Genetici della Speciazione negli Animali, Pallanza, Italy, 1948. La Ricerca Scientifica* **19** (supplement), 68–76.

HALL, R. (1994). Molecular mimicry. *Advances in Parasitology* **34**, 81–132.

HAYUNGA, E. G. (1979). Observations on the intestinal pathology caused by three caryophyllid tapeworms of the white sucker *Catostomus commersoni* Laecpede (sic). *Journal of Fish Diseases* **2**, 239–248.

HAYUNGA, E. G., SUMNER, M. P. & LETONJA, T. (1989). Evidence for selective incorporation of host immunoglobulin by strobilocerci of *Taenia taeniaeformis*. *The Journal of Parasitology* **75**, 638–642.

HEMMER, B., FLECKENSTEIN, B. T., VERGELLI, M., JUNG, G., McFARLAND, H., MARTIN, R. & WEISMULLER, K.-H. (1997). Identification of high potency microbial and self ligands for a human autoreactive class II-restricted T cell clone. *Journal of Experimental Medicine* **185**, 1651–1659.

HILL, A. V. S., ALLSOP, C. E. M., KWIATKOWSKI, D., ANSTEY, N. M., TWUMASI, P., ROWE, P. A., BENNETT, S., BREWSTER, D., McMICHAEL, A. J. & GREENWOOD, B. M. (1991). Common West African HLA antigens are associated with protection from severe malaria. *Nature* **352**, 595–600.

HONTEBEYRIE-JOSKOWICZ, M., SAID, G., MILON, G., MARCHAL, G. & EISEN, H. (1987). L3T4$^+$ T cells able to mediate parasite-specific delayed-type hypersensitivity play a role in the pathology of experimental Chagas' disease. *European Journal of Immunology* **17**, 1027–1033.

HOWELL, M. J. (1985). Gene exchange between hosts and parasites. *International Journal for Parasitology* **15**, 597–600.

IRIE, Y. & IWAMURA, Y. (1993). Host-related DNA sequences are localized in the body of schistosome adults. *Parasitology* **107**, 519–528.

IWAMURA, Y. & IRIE, Y. (1992). Heterogeneity of host-related DNA sequences in schistosomes. *Parasitology Today* **8**, 90.

IWAMURA, Y., IRIE, Y., KOMINAMI, R., NARA, T. & YASURAOKA, K. (1991). Existence of host-related DNA sequences in the schistosome genome. *Parasitology* **102**, 397–403.

IWAMURA, Y., YONEKAWA, H. & IRIE, Y. (1995). Detection of host DNA sequences including the H–2 locus of the major histocompatibility complex in schistosomes. *Parasitology* **110**, 163–170.

KARANJA, D. M. S., COLLEY, D. G., NAHLEN, B. L., OUMA, J. H. & SECOR, W. E. (1997). Studies on schistosomiasis in western Kenya. I. Evidence for immune-facilitated excretion of schistosome eggs from patients with *Schistosoma mansoni* and human immunodeficiency virus coinfections. *American Journal of Tropical Medicine and Hygiene* **56**, 515–521.

KEMP, W. M., MERRITT, S. C., BOGUCKI, S. C., ROSIER, M. S. & SEED, J. R. (1977). Evidence for adsorption of heterospecific host immunoglobulin on the tegument of *Schistosoma mansoni*. *The Journal of Immunology* **119**, 1849–1854.

KIM, A., TERZIAN, C., SANTAMARIA, P., PELISSON, A., PRUD'HOMME, N. & BUCHETON, A. (1994). Retroviruses in invertebrates: the gypsy retrotransposon is apparently an infectious retrovirus of *Drosophila melanogaster*. *Proceedings of the National Academy of Sciences, USA* **91**, 1285–1289.

KLEIN, J. & O'HUIGIN, C. (1994). MHC polymorphism and parasites. *Philosophical Transactions of the Royal Society of London, Series B* **346**, 351–358.

KOBERLE, F. (1968). Chagas' Disease and Chagas' Syndromes: the pathology of American trypanosomiasis. *Advances in Parasitology* **6**, 63–116.

KOTWAL, G. J. & MOSS, B. (1988). Vaccinia virus encodes a secretory polypeptide structurally related to complement. *Nature* **335**, 176–178.

LAFON, M., SCOTT-ALGARDA, D., MARCHE, P. N., CAZENAVE, P. A. & JOUVIN-MARCHE, E. (1994). Neonatal deletion and selective expansion of mouse T cells by exposure to rabies virus nucleocapsid superantigen. *Journal of Experimental Medicine* **180**, 1207–1215.

MAZINGUE, C., COTTREZ-DETOEUF, F., LOUIS, J., KWEIDER, M., AURIAULT, C. & CAPRON, A. (1989). *In vitro* and *in vivo* effects of interleukin 2 on the protozoan parasite leishmania. *European Journal of Immunology* **19**, 487–491.

MURPHY, P. M. (1993). Molecular mimicry and the generation of host protein diversity. *Cell* **72**, 823–826.

MURPHY, P. M. (1994). Viral imitations of host defense proteins. Flattery that turns to battery. *Journal of the American Medical Association* **271**, 1948–1952.

NARA, T., IWAMURA, Y., TANAKA, M., IRIE, Y. & YASURAOKA, K. (1990). Dynamic changes of DNA sequences in *Schistosoma mansoni* in the course of development. *Parasitology* **100**, 241–245.

OGILVIE, B. M. & WILSON, R. J. M. (1976). Evasion of the immune response by parasites. *British Medical Bulletin* **32**, 177–181.

OLDSTONE, M. B. A. (1987). Molecular mimicry and autoimmune disease. *Cell* **50**, 819–820.

OLDSTONE, M. B. A. (ed.) (1989). Molecular mimicry. Cross-reactivity between microbes and host proteins as a cause of autoimmunity. *Current Topics in Microbiology and Immunology* **145**. Berlin: Springer-Verlag.

RAFF, R. A. (1996). *The Shape of Life*. Chicago: University of Chicago Press.

ROSS, R. & THOMSON, D. (1910). A case of sleeping sickness studied by precise enumerative methods: regular periodical increase of the parasites disclosed. *Proceedings of the Royal Society of London (Biology)* **82**, 411–415.

ROWLEY, D. & JENKIN, C. R. (1962). Antigenic cross-reaction between host and parasite as a possible cause of pathogenicity. *Nature* **193**, 151–154.

SHER, A., HALL, B. F. & VADAS, M. A. (1978). Acquisition of murine major histocompatibility complex gene products by schistosomula of *Schistosoma mansoni*. *Journal of Experimental Medicine* **148**, 46–57.

SIMPSON, A. J. G. & PENA, S. D. J. (1991). Host-related DNA sequences in the schistosome genome? *Parasitology Today* **7**, 266.

SIMPSON, A. J. G., SINGER, D., MCCUTCHAN, T. F., SACKS, D. L. & SHER, A. (1983). Evidence that schistosome MHC antigens are not synthesized by the parasite but

are acquired from the host as intact glycoproteins. *The Journal of Immunology* **131**, 962–965.

SONG, S. U., GERASIMOVA, T., KURKELOS, M., BOEKE, J. D. & CORCES, V. G. (1994). An env-like protein encoded by a *Drosophila* retroelement: evidence that gypsy is an infectious retrovirus. *Genes and Development* **8**, 2046–2057.

SONG, S. U., KURKULOS, M., BOEKE, J. D. & CORCES, V. G. (1997). Infection in the germ line by retroviral particles produced in the follicle cells: a possible mechanism for the mobilization of the gypsy retroelement of *Drosophila*. *Development* **124**, 2789–2798.

SNELL, G. D. (1968). The H–2 locus of the mouse: Observations and speculations concerning its comparative genetics and its polymorphism. *Folia Biologica (Prague)* **14**, 335–338.

SPRENT, J. F. A. (1962). Parasitism, immunity and evolution. In *The Evolution of Living Organisms, Symposium of the Royal Society, Melbourne* (ed. Leeper, G. W.), pp. 149–165. Melbourne: Melbourne University Press.

TANAKA, M., IWAMURA, Y., AMANUMA, H., IRIE, Y., WATANABE, M., WATANABE, T., UCHIYAMA, Y. & YASURAOKA, K. (1989). Integration and expression of murine retrovirus-related sequences in schistosomes. *Parasitology* **99**, 31–38.

TARLETON, R. L. & KEMP, W. M. (1981). Demonstration of IgG-Fc and C3 receptors on adult *Schistosoma mansoni*. *The Journal of Immunology* **126**, 379–384.

TARLETON, R. L., ZHANG, L. & DOWNS, M. O. (1997). 'Autoimmune rejection' of neonatal heart transplants in experimental Chagas' disease is a parasite-specific response to infected host tissue. *Proceedings of the National Academy of Sciences, USA* **94**, 3932–3937.

TORPIER, G., CAPRON, A. & OUAISSI, M. A. (1979). Receptor of IgG (Fc) and human β_2-microglobulin on *S. mansoni* schistosomula. *Nature* **278**, 447–449.

UPTON, C., MACEN, J. L., SCHREIBER, M. & MCFADDEN, G. (1991). Myxoma virus expresses a secreted protein with homology to the tumor necrosis factor receptor gene family that contributes to viral virulence. *Virology* **184**, 370–382.

WEKERLE, H., BRADL, M., LININGTON, C., KAAB, G. & KOJIMA, K. (1996). The shaping of the brain-specific T lymphocyte repertoire in the thymus. *Immunological Reviews* **11**, 231–243.

WILLS, C. (1991). Maintenance of multiallelic polymorphism at the MHC region. *Immunological Reviews* **124**, 165–220.

WINKLER, W. G. (1991). Rodent rabies. In *The Natural History of Rabies*, 2nd Edition (ed. Baer, G. M.), pp. 405–410. Boca Raton, Florida: CRC Press.

WUCHERPFENNIG, K. W. & STROMINGER, J. L. (1995). Molecular mimicry in T cell-mediated autoimmunity: viral peptides activate human T cell clones specific for myelin basic protein. *Cell* **80**, 695–705.

YUKI, N., TAKI, T., INAGAKI, F., KASAMA, T., TAKAHASHI, M., SAITO, K., HANDA, S. & MIYATAKI, T. (1993). A bacterium lipopolysaccharide that elicits Guillain–Barré syndrome has a GM1 ganglioside-like structure. *Journal of Experimental Medicine* **178**, 1771–1775.

Printed in the United States
By Bookmasters